Migrant Ecologies

 **Perspectives on the Global Past**

Anand A. Yang and Kieko Matteson
SERIES EDITORS

Migrant Ecologies

Environmental Histories of the Pacific World

Edited by James Beattie, Ryan Tucker Jones,
and Edward Dallam Melillo

University of Hawaiʻi Press
Honolulu

Paperback edition 2023

Printed in the United States of America

First printed, 2023

Library of Congress Cataloging-in-Publication Data

Names: Beattie, James, editor. | Jones, Ryan Tucker, editor. | Melillo, Edward D., editor.
Title: Migrant ecologies : environmental histories of the Pacific world / edited by James Beattie, Ryan Tucker Jones, Edward Dallam Melillo.
Other titles: Perspectives on the global past.
Description: Honolulu : University of Hawai‘i Press, 2022. | Series: Perspectives on the global past | Papers from a symposium held at Amherst College in 2015. | Includes bibliographical references and index.
Identifiers: LCCN 2022026302 (print) | LCCN 2022026303 (ebook) | ISBN 9780824891060 (hardback) | ISBN 9780824892258 (pdf) | ISBN 9780824892906 (epub) | ISBN 9780824894849 (kindle edition)
Subjects: LCSH: Human ecology—Pacific Area—History—Congresses. | Ecology—Pacific Area—History—Congresses. | Pacific Ocean—Environmental conditions—Congresses. | LCGFT: Conference papers and proceedings.
Classification: LCC GF798 .M53 2022 (print) | LCC GF798 (ebook) | DDC 304.209164—dc23/eng20220922
LC record available at https://lccn.loc.gov/2022026302
LC ebook record available at https://lccn.loc.gov/2022026303

ISBN 9780824894207 (paperback)

Cover art: *Te Ha o te Marama* series, ink and oil on board with waka, by Simon Kaan (2021). Courtesy of artist.

University of Hawai‘i Press books are printed on acid-free paper and meet the guidelines for permanence and durability of the Council on Library Resources.

To all of the students who have helped us come to know the Pacific World.

Contents

Map of the Pacific World

Geoffrey Wallace

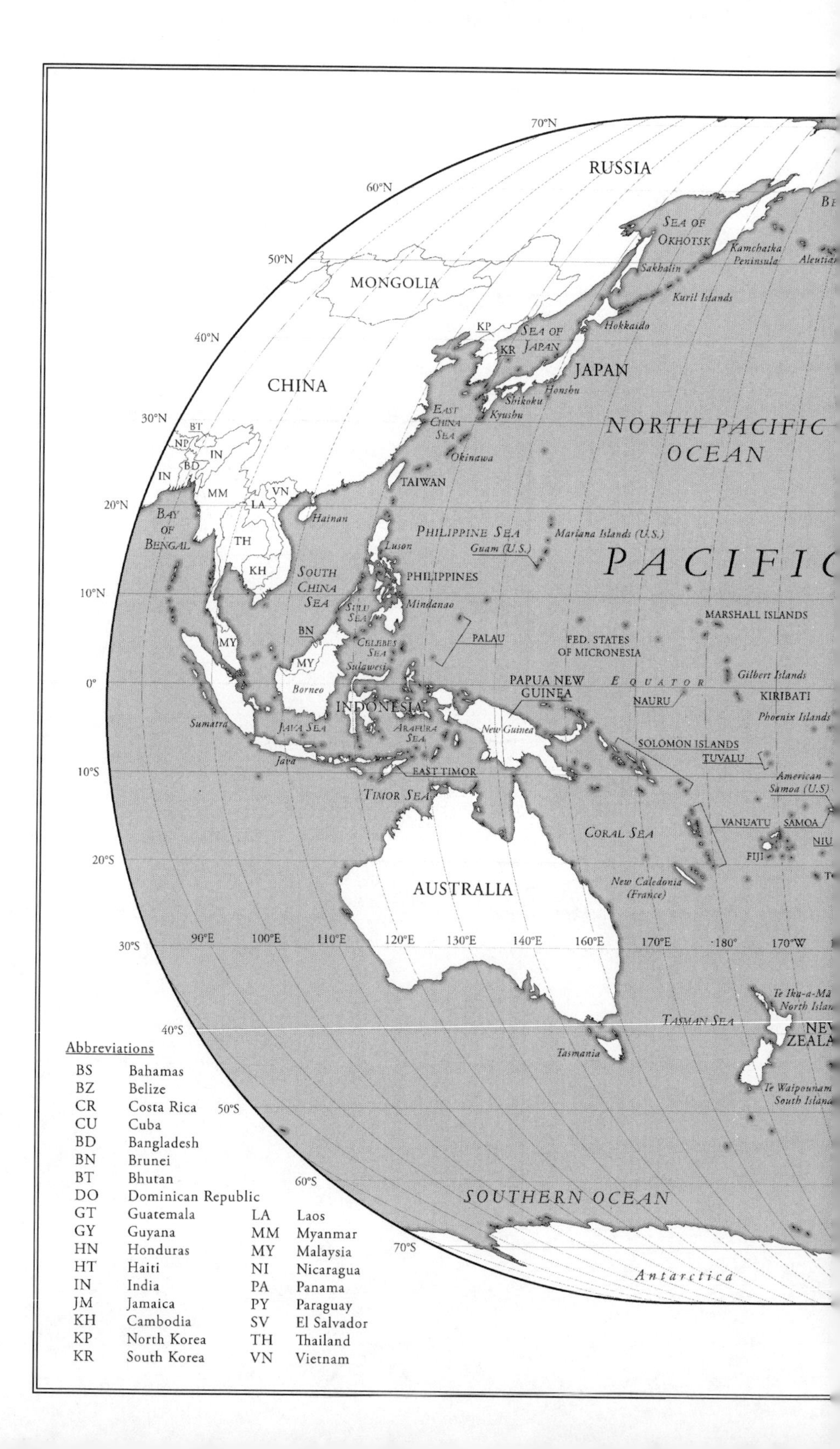

70°N
RUSSIA
60°N
SEA OF OKHOTSK
Kamchatka Peninsula
Sakhalin
50°N
MONGOLIA
Kuril Islands
40°N
KP
SEA OF JAPAN
KR
Hokkaido
JAPAN
CHINA
Honshu
Shikoku
Kyushu
30°N
BT
NP
IN
BD
IN
EAST CHINA SEA
NORTH PACIFIC
OCEAN
Okinawa
TAIWAN
MM
VN
LA
20°N
BAY OF BENGAL
TH
Hainan
PHILIPPINE SEA
Mariana Islands (U.S.)
Guam (U.S.)
Luzon
PACIFIC
KH
SOUTH CHINA SEA
PHILIPPINES
10°N
SULU SEA
Mindanao
MARSHALL ISLANDS
BN
PALAU
MY
CELEBES SEA
FED. STATES OF MICRONESIA
MY
Sulawesi
0°
PAPUA NEW GUINEA
EQUATOR
Gilbert Islands
Borneo
KIRIBATI
NAURU
INDONESIA
Phoenix Islands
Sumatra
JAVA SEA
ARAFURA SEA
New Guinea
SOLOMON ISLANDS
TUVALU
Java
EAST TIMOR
10°S
American Samoa (U.S)
TIMOR SEA
VANUATU
SAMOA
CORAL SEA
FIJI
20°S
New Caledonia (France)
AUSTRALIA
30°S
90°E
100°E
110°E
120°E
130°E
140°E
160°E
170°E
180°
170°W
Te Ika-a-Māui North Island
40°S
TASMAN SEA
Tasmania
Te Waipounamu South Island
50°S
60°S
SOUTHERN OCEAN
70°S
Antarctica
Abbreviations
BS Bahamas
BZ Belize
CR Costa Rica
CU Cuba
BD Bangladesh
BN Brunei
BT Bhutan
DO Dominican Republic
GT Guatemala
GY Guyana
HN Honduras
HT Haiti
IN India
JM Jamaica
KH Cambodia
KP North Korea
KR South Korea
LA Laos
MM Myanmar
MY Malaysia
NI Nicaragua
PA Panama
PY Paraguay
SV El Salvador
TH Thailand
VN Vietnam

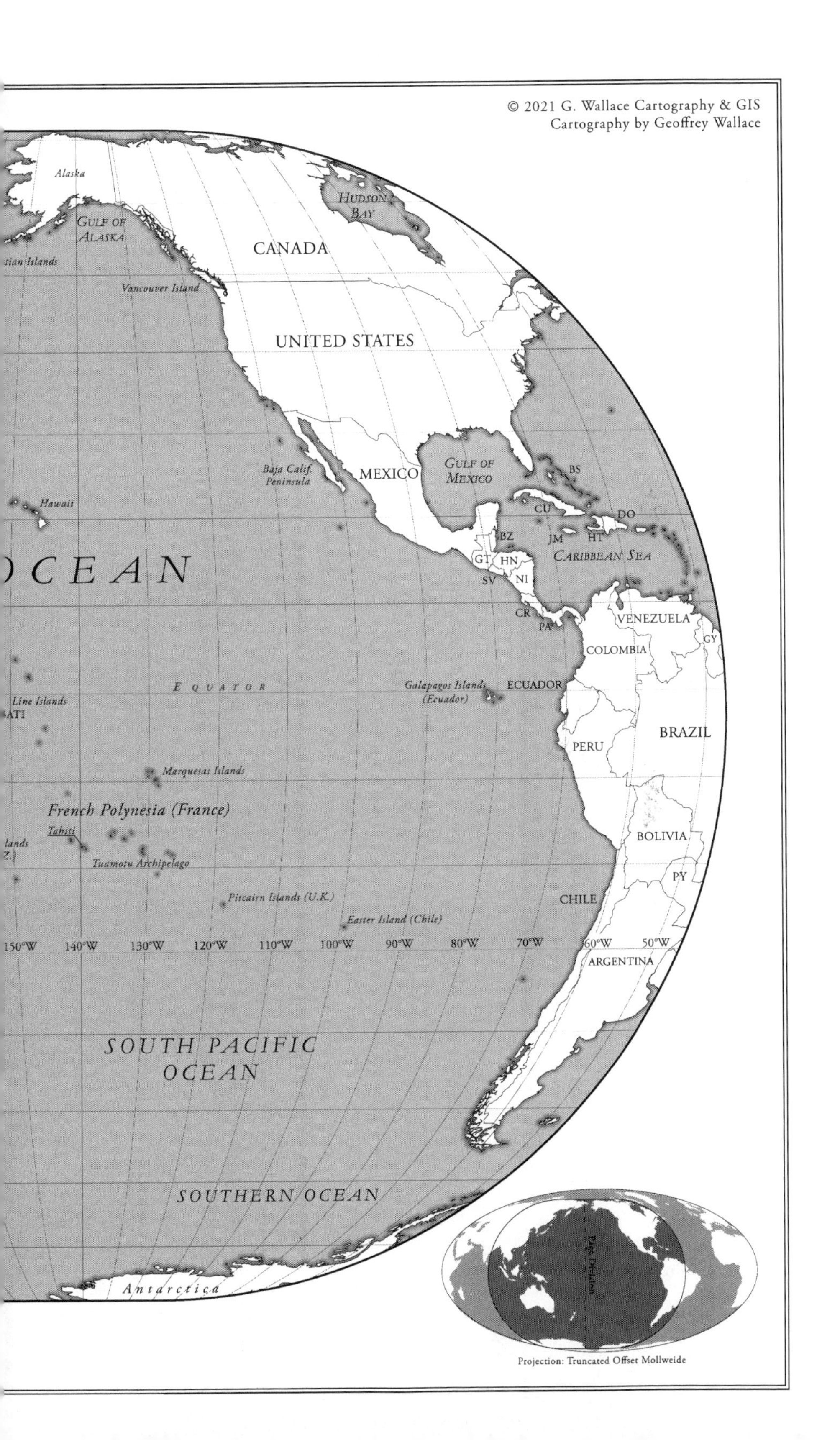

© 2021 G. Wallace Cartography & GIS
Cartography by Geoffrey Wallace
Alaska
Hudson Bay
Gulf of Alaska
CANADA
Vancouver Island
UNITED STATES
Baja Calif. Peninsula
MEXICO
Gulf of Mexico
BS
Hawaii
CU
DO
BZ
JM
HT
GT
HN
Caribbean Sea
SV
NI
CR
PA
VENEZUELA
GY
COLOMBIA
Equator
Galapagos Islands (Ecuador)
ECUADOR
Line Islands
BRAZIL
PERU
Marquesas Islands
French Polynesia (France)
Tahiti
BOLIVIA
Tuamotu Archipelago
PY
Pitcairn Islands (U.K.)
CHILE
Easter Island (Chile)
150°W
140°W
130°W
120°W
110°W
100°W
90°W
80°W
70°W
60°W
50°W
ARGENTINA
SOUTH PACIFIC OCEAN
SOUTHERN OCEAN
Antarctica
Page Division
Projection: Truncated Offset Mollweide

Preface

J. R. McNeill

Compared to the Indian Ocean over the last two millennia or the Atlantic over the past five hundred years, the Pacific lacks a coherent history. Its size and the variety of peoples living in and around it make its history seem disjointed in comparison to other ocean basins. Yet Pacific history acquires a measure of unity and coherence when it becomes Pacific environmental history.[1] Even a vast ocean that covers one-third of the globe features some commonalities. One is the ring of fire, the active volcanic zone that encircles the ocean, from New Zealand to Kamchatka to Alaska to Chile. Another is the El Niño/Southern Oscillation (ENSO) pattern that scrawls its signature across nearly half of the Pacific, most powerfully at equatorial latitudes but at times as much as 35 degrees latitude North and South of the equator. A third is the mere fact of distance. For many people living in the Pacific, most voyages were long voyages, requiring—until recently—careful assembly of resources and detailed knowledge of the sea. Different peoples in and around the Pacific adjusted to these geographical realities in different ways, influenced by their religions, cultures, technologies, and much else. But almost all denizens of the region, from Rongelap and Rapa Nui to Auckland and Anchorage, have long reckoned with a distinctive set of environmental challenges and opportunities presented by the particularities of the Pacific.

This book, arising from a symposium held at Amherst College in 2015, displays some of the best new work in Pacific environmental history. It brings the islands and the rim together in a diverse array of chapters that do justice to the uniqueness of specific times and places yet also—both collectively and within some of the chapters—to the coherence of the region.

The chapters operate at every spatial scale, from the intensely local to the pan-Pacific with global reach. That variety enables readers to see how certain patterns and forces reveal themselves more clearly on smaller or larger scales. Some chapters uncover obscure connections across the wide Pacific. As readers will discover, Americans advised Japanese authorities on the colonization of Hokkaido in the late nineteenth century, and Māori hunting of sooty shearwaters in New Zealand probably reduced their populations in North America in the fifteenth century. Other chapters illuminate the ways in which the Pacific became an experimental playground, for the introduction of alien species or for testing of novel weapons, and how people responded to these experiments. Still others reveal new aspects of extractive economies in the Pacific, whether of sandalwood, tuna, pearls, or whales. Several of the chapters point to the power of unintended consequences, often a strong theme in the genre of environmental history but especially so when societies, microbes, and economies collide and so many of the protagonists are so ignorant of the peoples, cultures, and ecosystems they are encountering.

While some of the chapters reach back into time, most of them focus on the last two hundred fifty years, an era of heightened instability in the Pacific both ecologically and politically. The main reason for these twin forms of instability is the systematic intrusion into the Pacific of peoples and power from elsewhere, or, put differently, the globalizing currents of modern Pacific history.

While few parts of the Pacific were truly isolated from the wider world before the late eighteenth century, the connections between Pacific communities and one another, and between Pacific communities and the rest of the planet, intensified dramatically and disruptively in the last two hundred fifty years. In general terms, dramatic disruptions brought by tighter links to the wider world are a routine experience in world history, at least since the dawn of agriculture some eleven thousand years ago. People have long found their lives recast almost overnight by new ideas, diseases, crops, technologies, weapons, products, and market links. All these themes appear in one or more of the chapters that follow. But every instance played out differently, as the chapters also show.

Pacific environmental history originated not with historians but in the work of historical anthropologists and archaeologists working exclusively on the islands and usually on periods before contact with the wider world.[2] However, as the concept of a Pacific Rim took hold in the wider public arena in the 1980s and 1990s, environmental historians—like historians generally—came to frame work on places such as California, Peru, and Japan as Pacific, rather than exclusively American, Latin American, or East Asian. Additionally, attention to the nineteenth and twentieth centuries, rarely the concern of archaeologists or historical anthropologists working on the Pacific, helped environmental historians to see the connections that united the Rim and the Islands, in forms such as fishing and whaling, mining and agriculture, biological invasions and anti-nuclear protests, and much else. This book advances that process, by focusing mainly on recent centuries and collectively, and in some cases individually, linking the islands and the Rim into an increasingly coherent and connected environmental Pacific—just as ecological, social, economic, and political processes have long done.

Notes

1. Recent contrasting perspectives appear in Gregory Cushman, *Guano and the Opening of the Pacific World: A Global Ecological History* (New York: Cambridge University Press, 2013); David Igler, *The Great Ocean: Pacific Worlds from Captain Cook to the Gold Rush* (New York: Oxford University Press, 2013); and Paul D'Arcy, *The People of the Sea: Environment, Identity, and History in Oceania* (Honolulu: University of Hawai'i Press, 2008).

2. This tradition dates back at least to the 1940s. The most influential practitioner of late has been Patrick V. Kirch. See, for example, his *Historical Ecology in the Pacific Islands: Prehistoric Environmental and Landscape Change* (New Haven, CT: Yale University Press, 1997).

INTRODUCTION

Environmental Histories of the Pacific World

James Beattie, Ryan Tucker Jones, and Edward Dallam Melillo

In 1994, a young environmental historian, John McNeill, wrote the first large-scale environmental history of the Pacific. In a crisp fifty pages, McNeill's "Of Rats and Men: A Synoptic Environmental History of the Island Pacific" gave structure and coherence to eighty million years of geological, natural, and human history in Oceania. McNeill identified speed of transport as the most important factor in bringing change to Pacific environments, with the fast-moving steamship effecting the most radical transformations. Together, the various waves of migrants that have alighted on Pacific Islands brought the region's "ecological harmonization," though certainly not its harmony. Many of the changes were wrenching and—because of the Pacific's long history of isolation from humans—for such species as New Zealand's moa, they were terminal. The very scope and velocity of these changes, McNeill thought, posed "seductive charms for those interested in environmental history."[1]

What is remarkable is that McNeill—whose expertise is global, but hardly centered on the Pacific—should be the first to conceive of the Pacific's environmental history in such grand terms. Of course, "Of Rats and Men" relied on deeply researched examinations of individual islands, produced mostly by anthropologists and ecologists, such as Atholl Anderson, Patrick Kirch, Patrick Nunn, and F. R. Fosberg. Excellent work in historical geography, in particular R. G. Ward's (underappreciated) 1972 *Man in the Pacific*, set the stage for today's environmental historians.[2] But it took someone with McNeill's eye for large-scale patterns to see the common currents in this wide variety of literature. His article has since played a key role in the development of environmental history. It is usually the first piece consulted by those curious about this vast part of the world, and it makes the Pacific readily amenable to analyses of global environmental change.[3]

McNeill was right, too, that the Pacific would seduce environmental historians. In the three decades since "Of Rats and Men," research has proliferated on the roles of introduced species in shaping Pacific societies, the crucial importance of Pacific institutions of landholding, and the more hidden history of the Pacific's underwater environments.[4] Historians, including the editors of this volume, have built on McNeill's expansive Pacific purview, bringing environmental histories of the Humboldt Current, California, Alaska, maritime East Asia, Chile, Peru, littoral Australasia, and even New England and the Indian Ocean world into conversation with Oceania.[5] Far fewer, though, have followed McNeill's lead by attempting to organize this even larger and more historically inclined scholarship into a larger story. Paul D'Arcy's 2012 book chapter, "Oceania: The Environmental History of One-Third of the Globe," is a stellar exception, one that stressed the longer histories of interconnectedness among islands, the ecological diversity of these islands, and their more endogenous environmental instability, while re-affirming the necessity of approaching this history from an interdisciplinary perspective.[6] In 2014, Ryan Tucker Jones attempted to synthesize the modern history of the entire Pacific Ocean, from Alaska to New Zealand, through a focus on energy flows, finding broad concordance in species extinction and maritime pollution.[7] The contributors to *Migrant Ecologies* build on this developing scholarship, while also offering far more space for discordance and local specificity.

When "Of Rats and Men" was published, few scholars remarked upon the fact that influential and important strands of Pacific history were founded on a resistance to precisely the kind of synthetic approaches that McNeill and others had developed. D'Arcy hinted at some of these counterclaims when he urged greater attention to Indigenous perspectives in Pacific environmental histories. Others, such as Basil Davidson and Greg Dening, devoted their scholarly careers to narrating, without ever claiming to fully encapsulate, radically altern Indigenous worlds. Those antagonisms between approaches have become clearer in recent decades. Perhaps the most outspoken protagonist is historian David Hanlon, whose 2017 article "Losing Oceania to the Pacific and the World" argued passionately for the practitioners of his discipline to resist large-scale conceptualizations. In doing so, Hanlon called out Armitage and Bashford's *Pacific Histories* (which included Jones' synoptic view of the environment) as well as Matt Matsuda's integrative *Pacific Worlds*.[8] While not specifically addressing environmental historians, Hanlon's arguments imply opposition to history as conceived by McNeill. Hanlon foregrounded Indigenous Pacific conceptions of the natural world, such as "the possibility of other realms and dimensions of being," that fit uneasily in accounts of Pacific environments informed primarily by Western science. This "more radical Indigenous historiography" in the Pacific is, for Hanlon, fundamentally incommensurable with global history.[9] He likely would add that they are at odds with larger visions of Pacific environmental history, as well.

Hanlon's protests have merit, and they are repeated, implicitly, in several of this volume's chapters. N. Ha'alilo Solomon analyzes the difficult history of Maunalua Bay, not far from Hanlon's own Hawaii Kai, with reference to highly localized understandings of environmental change. But, as many other chapters here demonstrate, Pacific peoples' regional and global linkages have played crucial roles in their long-term histories. Those transregional connections could be tragic, as Benjamin Madley shows in his history of epidemics. Conversely, they could activate the resilience and creativity of Pacific societies. Melillo's account of the marketing genius displayed by O'ahu's nineteenth-century Governor Boki attests to this. Nor have global epistemologies, such as Western science, been entirely deaf to local and Indigenous difference. The Pacific tracings of the Anthropocene, as detailed by Ruth Morgan, confirm as much.

We believe the best way for Pacific environmental history to grapple with this tension between global forces and Indigenous ways of acting, valuing, and knowing is to let a multitude of these perspectives coexist. A collected volume of essays can be awkward and disjointed, but it can also acknowledge the unresolved contradictions that exist within a field of knowledge. Likewise, it can celebrate the diversity of experience that characterizes the astonishingly heterogenous Pacific World. While acknowledging this breadth, we also feel that the essays here make a collective argument, namely, that ecological processes operated over the last two hundred years in such a way that the Pacific Ocean provides a necessary framework for understanding local environmental histories.

Here we take inspiration from one of the most important figures in Pacific history, Epeli Hau'ofa. If Hau'ofa is rarely cited by environmental historians, that is perhaps because attention has focused on his seminal essay "Our Sea of Islands," instead of his equally powerful "We Are the Ocean." In the latter essay, Hau'ofa argues for a regional Pacific identity, "anchored in our common inheritance of a very considerable portion of Earth's largest body of water: the Pacific Ocean." Among the sources of his inspiration for this assertion were scientific conceptions of the ocean's centrality to human life.[10] "All our cultures," Hau'ofa claimed, "have been shaped in fundamental ways by the adaptive interactions between our people and the sea that surrounds our island communities." An identity based on these interactions would facilitate a more powerful collective Pacific voice in global affairs, as well as advance the protection of the most important resource in most Pacific peoples' lives. This identity would be in "addition to the other identities" of

Pacific peoples and could unite people north of the equator as well. Such alliances also have the advantage of not deriving from explicitly colonial frameworks.[11] Characteristically, Hau'ofa left the details of an oceanic identity, including its historical component, rather vague. As he admitted, "What we lack is the conscious awareness of it, its implications, and what we could do with it."[12] This volume hopes to provide some measure of historical detail to facilitate this awareness.

If the ocean is not central to every chapter in this book, it is the medium that connects them all. And, as Hau'ofa noted, "The water that washes and crashes on our shores is the water that washes and crashes on the coastlines of the whole Pacific Rim from Antarctica to New Zealand, Australia, Southeast and East Asia, and right around to the Americas." Indeed, Hau'ofa saw the ocean as "our route to the rest of the world."[13] As environmental historians, we have understood the Pacific Ocean less metaphorically than Hau'ofa did and conceived of it more tangibly as a force that connects many peoples and stories, even if it does not do so in equal measure in all times and places.

It is intriguing, in retrospect, to note that McNeill's "Of Rats and Men" (1994) and Hau'ofa's "Our Sea of Islands" (1993) appeared within a year of each other. Hau'ofa's gifts, like McNeill's, lay, above all, in his appetite for the big story. In the early 1990s, as Cold War paradigms faded in favor of a quicker, and in some ways more ruthless, era of hyper-globalization, the Pacific region had already begun to emerge as an appealing counterpoint. Thirty years later, the global scale of environmental catastrophe has become clearer. The first impacts of climate change disproportionately center on Pacific Islands and the waters that wash their shores. In this context, these earlier works appear prophetic. We hope *Migrant Ecologies* may, in its simultaneous wide scope and attention to difference, keep this conversation alive by offering a dialogue with the past, present, and future of the islands and seas that animate the great Pacific.

In the pages that follow, not all of the protagonists are human. In chapter 1, Ryan Tucker Jones traces the migrations of sooty shearwaters, whales, and tuna, showing their tremendous range. For example, shearwaters can undertake a staggering forty-thousand-mile migration to the Pacific's four corners. "To look at a map of Pacific animals' migrations," he argues, "is to see the Pacific's limits traced and its spaces filled." The astounding migration of Pacific fauna connected human communities, often with unexpected ecological, social, and political consequences. Reports of teeming pods of whales enticed New England's whaling fleets around Cape Horn and into the Pacific, marking growing European and North American interest in the resources of this vast ocean. Jones also makes the telling point that

"Pacific humans, while possessing venerable migratory histories of their own, have not been nearly as mobile as their maritime co-predators."

Pacific explorers and subsequent hunters who pursued whales, seals, sea otters, sandalwood, sea cucumbers, guano, and other oceanic resources created intentional and unintentional migrant ecologies. We know this thanks to a rich body of scholarship that integrates disciplines such as ecology, linguistics, ethnography, history, and archaeology.

In chapter 2, Gregory Rosenthal contends that human, animal, plant, and microbial diasporas mobilized "globetrotting natures." For well over one thousand years, he notes, "Pacific peoples have been at the forefront of ecological change and biological transformations in the Pacific Ocean." As well as deliberate introductions of animals, such as chickens and dogs, Pacific Islanders transported a host of so-called canoe plants around the Pacific. Sweet potatoes, breadfruit, and taro are just some of the dozens of canoe plants set in motion by these travelers.

In James Beattie's explorations of how Chinese immigrant communities transformed societies and ecosystems in the South Pacific from the 1790s through the 1920s (chapter 3), he illuminates the environmental histories of other Pacific peoples on the move. Focusing on the boom-and-bust cycles that drove resource frontiers, especially gold mining, Beattie frames these developments as ecocultural networks that connected the flows of labor, immigration, and capital in the process of commodifying nature. Chinese demand, not European finance, was the main driver of these circuits, and Beattie illustrates this claim with a case study of Chew Chong, a Chinese entrepreneur who sold New Zealand's edible wood ear fungus in his home country.

Another, much less perceptible, menagerie of organisms was also circulating in the Pacific during the nineteenth century. In chapter 4, Benjamin Madley turns his historian's microscope on the pathogens that entered California between 1828 and 1844, wreaking havoc on Native communities. Madley frames these lethal encounters within the context of colonial disease ecologies—the human regimes that magnified four devastating epidemics—examining how microbes were potent allies to the outsiders who invaded North America's west coast.

The entwined ecological and human dimensions of nineteenth-century violence tie Madley's investigations to the topics at the heart of chapter 5. Here, Lissa Wadewitz unravels the paradox of wonder and sentimentality that accompanied the blood sport of animal hunting in the Pacific. As she asserts, the relationships that whalers developed with their prey also have much to tell us about human interactions and social stratification aboard the seafaring vessels that pursued these intrepid creatures for years on end.

Transpacific hierarchies of power were on display in other nineteenth- and early twentieth-century contexts, as well. In Samoa, Holger Droessler (chapter 6) finds that retention of land ownership enabled Samoans to weather some of the cascade of ecological and social impacts that followed development of a European-led cash-crop economy. This commodity frontier was first largely based on cotton, then copra. Migrant ecologies, in the form of invasive plants, insects, and disease, as well as coercive colonial powers (American, German, New Zealand), undermined Samoan social and agricultural systems. Notably, copra replaced local coconut plantations, and a cash-crop economy transferred power to males rather than females, while populations reeled from the devastating impact of the 1918 influenza epidemic.

In chapter 7, Hannah Cutting-Jones examines attempts to create a citrus industry on the Cook Islands between 1900 and 1970. This South Pacific Island group became a British possession in 1900 and was administered by New Zealand from 1901 until its independence in 1965. As Cutting-Jones notes, "The migrant ecology envisioned and imported by missionaries, Europeans, and New Zealanders—one that attempted to implement industrial export agriculture and transform traditional land tenure—failed, due to a large extent on Cook Islanders' resistance." The key, as she observes, was that Cook Islanders retained land title. In addition, they deliberately sabotaged groves of citrus, and undermined attempts to develop monocultures, by continuing to practice traditional forms of horticulture, which involved growing different types of plants on the same plot.

Traditionally, histories have examined nineteenth-century European colonial expansion in the Pacific.[14] Much less well studied until recently have been the ecological dimensions of non-Western expansion in the Pacific prior to the twentieth-century Japanese Empire.[15] In chapter 8, Katsuya Hirano presents a re-reading of Pacific history, by examining Japanese conquest in northern Japan on the island of Ezo. Hirano focuses on the significance of Meiji government use of "the concept of *terra nullius* to displace Ainu and expropriate their land." A term originating in European law, *Terra nullius*—or right of discovery—deemed that occupation and agriculture signaled ownership. Infamously, the concept was used to justify European annexation of Australia, on the basis that since Indigenous Australians did not cultivate land, they therefore could not own it. Hirano's study adds to a lively body of global work examining what John C. Weaver has characterized as the "great land rush" of European powers for overseas territory beginning in the seventeenth century and dramatically accelerating into the nineteenth.[16]

So far, chapters have examined Pacific migrant ecologies in the formal and informal empire building of Britain, the United States, and New Zealand. In chapter 9, William Cavert turns his gaze to French colonial enterprise in the Tuamotus, "an archipelago as big as Europe and governed largely from distant Pape'ete by one naval officer." Cavert examines the tragic story of mismanagement of pearl-bearing oyster fisheries in this island group and presents a fascinating case study of metropole-colonial tensions. Initially, he shows, colonial officials welcomed the services of metropolitan overseas naturalists, but when they presented management suggestions at odds with what locals wanted, authorities instead turned to local (European) experts. Tragically, at this same time, the introduction of diving suits in the late 1880s created an ecological crisis. These new-fangled contraptions gave fishers access to new oyster beds. Increased production of pearls masked overfishing by giving "the impression that the lagoons were stable and not on the verge of exhaustion."

Distant locales are also connected by oceans in Edward Melillo's history of Kona coffee, which "complicates our basic assumptions about environmental change in the Pacific." He tells a story that walks the ephemeral line between local and global in the Pacific, charting Kona coffee's introduction through global linkages and its subsequent associations with Hawai'i's hyper-local volcanic microclimates. Using Hawaiian-language newspapers, among other sources, Melillo demonstrates how it was an *ali'i* (royal), namely O'ahu's Governor Boki, who did the crucial work of translating between those scales, organizing the beans' journey from Brazil to Hawai'i, and Hawaiian entrepreneurs—alongside Japanese and others—who tended and promoted the plants. The migrant ecology of coffee plants intersected with both the voyages of humans and their rootedness in specific soils of the Pacific.

If Kona coffee was one delicious, stimulating result of global and local intersections, migratory ecologies could also bring devastation to Hawai'i. As N. Ha'alilio Solomon reveals in his examination of Maunalua, O'ahu—now called Hawaii Kai—introduced species, residential development, and stream channelization have had profound, often detrimental effects on Hawai'i's nearshore ecosystems. These changes were particularly intense from the 1960s on, a function of the islands' urbanization and integration into a growing tourist economy. Solomon uses endonyms, traditional place names, and Indigenous narratives, alongside careful archival research, to re-create Maunalua and recover the holistic Hawaiian landscapes and concepts that integrated land and sea. As he notes, such attendance to local place, termed "kuleana" by Hawaiians, helps re-establish an ethics of responsibility toward the land and sea that colonial schemes of private property and profit seem to reject.

The pelagic Pacific inscribed somewhat different environmental stories in the twentieth century. There, extraction, rather than development, continued to be humanity's most consequential role as an ecological agent.

Kristen Wintersteen explores the antagonism between American fishermen and South American Pacific states that grew from their divergent interests in the Southeastern Pacific. The "tuna wars" of the 1940s–1970s saw Ecuador, Peru, and Chile articulate a new relationship to oceanic space—territorial claims 200 miles offshore—that infuriated North Americans who considered the ocean's biomass as a resource available to anyone who invested the capital and labor to capture it. This history, too, was shaped by the migratory patterns of tuna, as well as the pronounced oceanographic fluctuations that El Niños cause in the Humboldt Current. Here, migrating North American fishermen pursued migrating tuna straight into webs of conflict around how tightly oceanic space would be attached to Pacific shores.

Pacific species not subjected to human predation could also entangle national interests. Emily O'Gorman examines scientific practice, racial ideology, and national exceptionalism in the twentieth-century history of the Latham's snipe, a bird that migrates every year between Australia and Japan. Australians interpreted this seasonal visitor alternately as an avatar of their own migrations, a dangerous disease vector, and an object of conservationist concern. The last development, which occurred in the 1970s, led to the country's first migratory bird treaty with Japan. This product of Australia's integration into a larger East Asian bioregion demonstrates how international migrations could lead to domestic innovations, as the treaty introduced new ways of dealing with wetlands. As O'Gorman writes, a look at migratory birds in the Pacific "challenges simpler stories of exceptionalism or cosmopolitanism and points to the need to think both together."

Franko Zelko's chapter on the fallout from the heyday of Pacific nuclear testing combines the macro with the micro environmental scale. Here, the distance between those scales is even greater, and their integration is even more insidiously tight. Invisible radioactive particles produced in the Pacific quite literally entered every human body around the globe after the 1950s. Zelko points to another poignant irony, that the science used to diagnose the global spread of strontium-90, and later to oppose nuclear testing, came from seminal ecological studies done on the very local Bikini Atoll lagoon. Its intense contamination resulting from the 1954 Castle Bravo test provided an ideal laboratory for such studies. The humans, Marshall Islanders in this case, who had inhabited the island, endured their own migrations, first forced to vacate Bikini and later sent to Washington, DC, to serve as test subjects for studies on the effects of radiation on the human body.

Finally, Ruth Morgan addresses one of the largest, and most controversial, frameworks developed for human history. This is the concept of the Anthropocene, the notion that human impact on the globe has become so pervasive as to constitute a new geological epoch. Morgan cuts the concept down to size, arguing that historians need to "place" the Anthropocene and emphasize the "material differences in its manifestation around the globe." Through the 1980s, scientists and novelists alike considered the South Pacific a refuge from Northern Hemisphere environmental problems. Even as those claims have receded in the recognition of the planetary scale on which climate change operates, differences remain. One has been Australia's outsized contribution to global carbon dioxide emissions, another that Pacific islands have experienced some of the earliest, and worst, impacts of sea level rise. With the Anthropocene, as with so many other environmental histories examined in this volume, the tension between the local and the global is irresolvable. The dual scales exist inside each other, while it is the Pacific Ocean that encircles, interpenetrates, and connects them and their histories. Morgan refers to Fijian prime minister Frank Bainamarama, who told the assembled United Nations in 2017 that we are "all in the same canoe." But Morgan and the other authors in this volume offer a more precise analysis. We are all buffeted—sometimes swamped and sometimes propelled—by an ocean rich with inscrutable intention, but we travel the waves of past, present, and future in boats of our own design.

Notes

1. J. R. McNeill, "Of Rats and Men: A Synoptic Environmental History of the Island Pacific," *Journal of World History* 5, no. 2 (Fall 1994): 299–349.

2. R. G. Ward, ed., *Man in the Pacific: Essays on Geographical Change in the Pacific Islands* (Oxford: Clarendon Press, 1972).

3. Crucial accounts of global history that cite McNeill include: Kenneth Pomeranz, *The Great Divergence: China, Europe, and the Making of the Modern World Economy* (Princeton, NJ: Princeton University Press, 2000); Patrick Manning, *Navigating World History: Historians Create a Global Past* (New York: Palgrave Macmillan, 2003); Jerry Bentley, "Sea and Ocean Basins as Frameworks of Historical Analysis," *Geographical Review* 89, no. 2 (April 1999): 215–224; Paul Sutter, "What Can Environmental Historians Learn from Non-US Environmental Historiography?" *Environmental History* 8, no. 1 (January 2003): 109–129; David Christian, "World History in Context," *Journal of World History* 14, no. 4 (December 2003): 437–458; Tony Ballantyne, ed., *Science, Empire and the European Exploration of the Pacific* (Aldershot: Ashgate Publishing, 2004); among others.

4. Jennifer Newell, *Trading Nature: Tahitians, Europeans, and Ecological Exchange* (Honolulu: University of Hawai'i Press, 2010), John Ryan Fischer, *Cattle Colonialism: An Environmental History of the Conquest of California and Hawai'i* (Chapel Hill: University of North Carolina Press, 2015); Lissa Wadewitz, *The Nature of Borders: Salmon, Boundaries, and Bandits on the Salish Sea* (Seattle: University of Washington Press, 2012); Hannah Cutting-Jones, "Feasts of Change: Food and History in the Cook Islands, 1825–1975," unpublished dissertation (University of Auckland, 2018); Tamatoa Bambridge, ed., *The Rahui: Legal Pluralism in Polynesian Traditional Management of Resources and Territories* (Acton, A.C.T.: ANU Press, 2016).

5. Edward Melillo, *Strangers on Familiar Soil: Rediscovering the Chile-California Connection* (New Haven, CT: Yale University Press, 2015); Melillo, "Making Sea Cucumbers out of Whales' Teeth: Nantucket Castaways and Encounters of Value in Nineteenth-Century Fiji," *Environmental History* 20, no. 30 (2015): 449–474; Ryan Tucker Jones, *Empires of Extinction: Russians and the North Pacific's Strange Beasts of the Sea* (Oxford: Oxford University Press, 2014), Ryan Tucker Jones, "A Whale of a Difference: Southern Right Whale Culture and the Tasman World's Living Terrain of Encounter," *Environment and History* 25 (2019): 185–218; James Beattie, *Empire and Environmental Anxiety: Health, Science, Art and Conservation in South Asia and Australasia, 1800–1920* (Houndmills, UK: Palgrave-Macmillan, 2011); Beattie, "Biota Barons, 'Neo-Eurasias' and Indian-New Zealand Informal Eco-cultural Networks, 1830s–1870s," *Global Environment* 13 (2020): 134–165; Greg Cushman, *Guano and the Opening of the Pacific World: A Global Ecological History* (Cambridge: Cambridge University Press, 2013); David Igler, *The Great Ocean: Pacific Worlds from Captain Cook to the Gold Rush* (Oxford: Oxford University Press, 2013).

6. Paul D'Arcy, "Oceania: The Environmental History of One-Third of the Globe," in *A Companion to Global Environmental History*, ed. J. R. McNeill and Erin Stewart Mauldin (Chichester, UK: Wiley, 2012).

7. Ryan Tucker Jones, "The Environment," in *Pacific Histories: Ocean, Land, People*, ed. David Armitage and Alison Bashford (New York: Palgrave Macmillan, 2014).

8. David Hanlon, "Losing Oceania to the Pacific World," *The Contemporary Pacific* 29, no. 2 (2017): 286–318.

9. Hanlon, 299.

10. Epeli Hau'ofa, "The Ocean in Us," in *We Are the Ocean*, Epeli Hau'ofa (Honolulu: University of Hawai'i Press, 2008), 53.

11. Hau'ofa, "The Ocean in Us," 42–44; for arguments of a Pacific environmental history and identity connecting both the North and South Pacific, see Ryan Tucker Jones, "Kelp Highways, Siberian Girls in Maui, and Nuclear Walruses: The North Pacific in a Sea of Islands," *Journal of Pacific History* 49, no. 4 (2014): 373–395.

12. Hau'ofa, "The Ocean in Us," 54.

13. Hau'ofa, "The Ocean in Us," 55.

14. Note, for example, *Pacific Empires: Essays in Honour of Glyndwr Williams*, ed. Alan Frost and Jane Samson (Melbourne: Melbourne University Press, 1999); and David Mackay, *In the Wake of Cook: Exploration, Science, and Empire, 1780–1801* (London: Croom Helm, 1985).

15. For an exception, see the chapters in *Japan at Nature's Edge: The Environmental Context of a Global Power*, ed. Ian Jared Miller, Julia Adeney Thomas, and Brett L. Walker (Honolulu: University of Hawai'i Press, 2013).

16. John C. Weaver, *The Great Land Rush and the Making of the Modern World* (Montréal: McGill-Queen's University Press, 2003).

ONE

Long-Distance Animal Migration and the Creation of a Pacific World

A History in Three Species

Ryan Tucker Jones

When in 2012 scientists caught tuna off the coast of Oregon and found elevated levels of radioactivity derived from Japan's Fukushima Daiichi meltdown the year before, they had not only brought home some dodgy sashimi.[1] In demonstrating conclusively that these large predatory fish migrate frequently from one side of the vast Pacific to the next, the scientists had also pulled up a significant clue about the past and present of the Pacific World. Animal migrations like these, which are becoming increasingly well known and whose scope continues to surprise researchers, have the potential to reshape historians' conceptions of Pacific history (see figure 1.1).

The creation of—or even the existence of—a "Pacific World" is a question that has preoccupied scholars to a much greater degree than existential doubts have bothered historians of other oceanic basins. Economic historian Eric Jones and colleagues have written that "there can be no meaningful history of the whole Rim or Basin [of the Pacific] since there has never been such an integrated unit," while environmental historian David Igler worries that "numerous issues urge caution against embracing a concept like *the* Pacific World."[2] Matt Matsuda carefully delineates the Pacific as a space of "multiple translocalisms" and stresses the radically different experience of people around the ocean.[3] Finally, criticizing both Igler and Matsuda for too hearty an embrace of the Pacific, David Hanlon has pointed to the "methodological shortcomings of both a Pacific history and a Pacific Worlds approach."[4] Part of the angst around the Pacific World concept comes from the size of the ocean itself, which, if considered a coherent whole, constitutes the largest geographical feature on Earth. As many scholars have rightly pointed out, geographical concepts have no inherent meaning, but

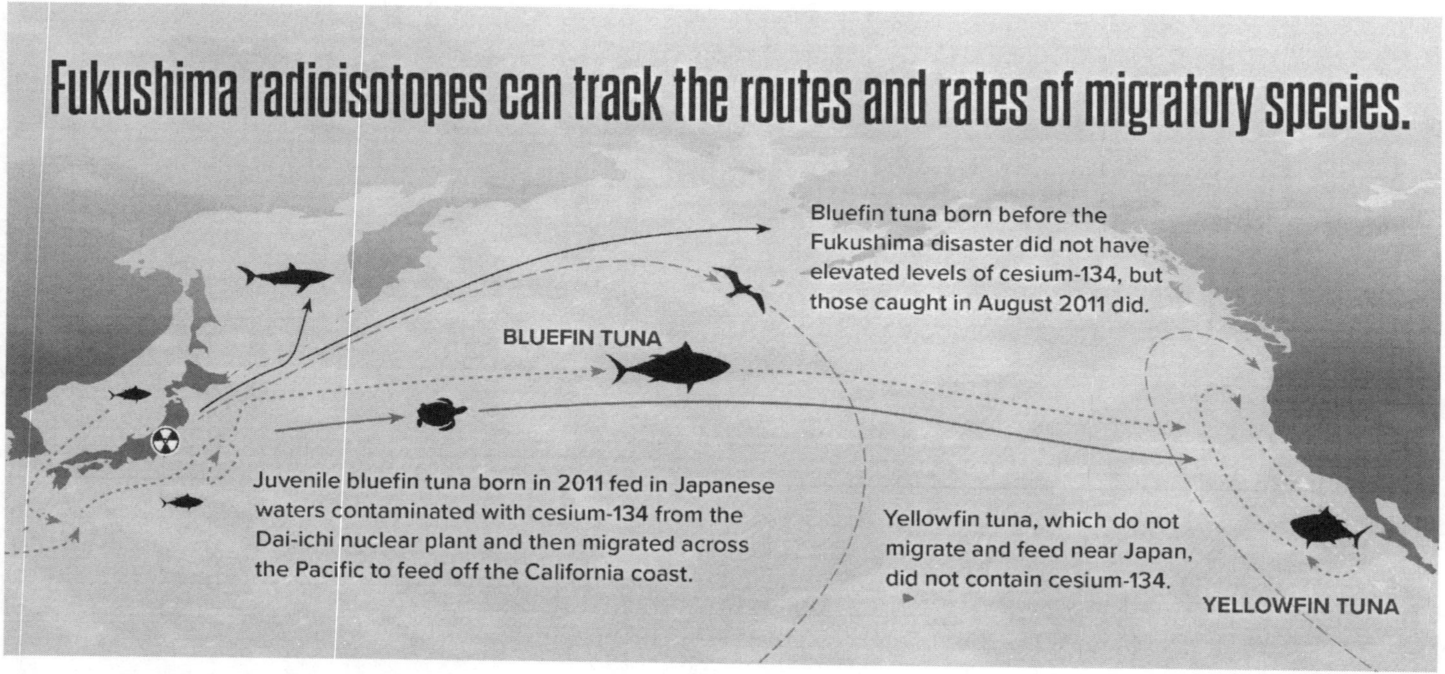

Figure 1.1. Migration of a radioactive tuna.

are rather created and sustained or rejected through the practices of human culture.[5] The concept of oceanic basins is itself relatively recent, replacing ancient Western notions of one gigantic encircling sea and more recent conceptions of much smaller navigational basins united by prevailing winds and routes of trade.[6] There is much, then, to recommend skepticism about the coherence of histories laid out over such large and contingent spaces.

From the perspective of environmental history, however, there are some strong, but still mostly unexplored, arguments for considering as a coherent historical unit the Pacific Ocean as it is depicted on maps humans make. Many of these arguments are coming from marine science. On the broadest scale, maps of tsunami energy propagation in the Pacific give powerful testimony to the way local events can radiate their effects to every corner of the ocean, effects that either do not trickle into any other oceans or rapidly lose their power once they leave the cartographic Pacific. Because of the coherence of the Pacific, humans in Japan, for example, have to care about the effects of earthquakes in Alaska, Chile, or New Zealand, but not those in Western Indonesia (see figure 1.2). In important ways, the geographical Pacific periodically gathers all the humans living on its shores into communities experiencing similar, sometimes transformative, events.[7]

Perhaps the most promising new avenue of scientific research for Pacific historians involves long-distance animal migrations. Thanks to concerted efforts to tag and track large Pacific predatory marine animals, along with other research programs, we now have a picture of the exceptional migratory patterns of animals around the entire ocean.[8] From great white sharks to stormy petrels to gray whales to tuna, many of the consumers of the Pacific's biomass visit distant corners of the ocean in pursuit of prey or breeding and birthing opportunities. The scope of these migrations is startling (and has surprised many of the researchers), topped by the forty-thousand-mile journeys sooty shearwaters execute every other year to the four corners of the ocean. At the same time, very few—if any—ever leave the Pacific, for reasons that are not always clear, but which have to do with the geographical constraints imposed by the continents and islands bounding the ocean, as well as the lack of food available in the polar exits from the ocean. To look at a map of Pacific animals' migrations is to see the Pacific's limits traced and its spaces filled. These animals inhabit—and have long inhabited—a Pacific World, one that also has drawn human histories into its web.

The important insights these migrations can offer to Pacific historians are twofold. First, these migratory species significantly reduce the amount of ecological discontinuity between the vast spaces of the Pacific Ocean. While local, relatively small species in different parts of the ocean may vary

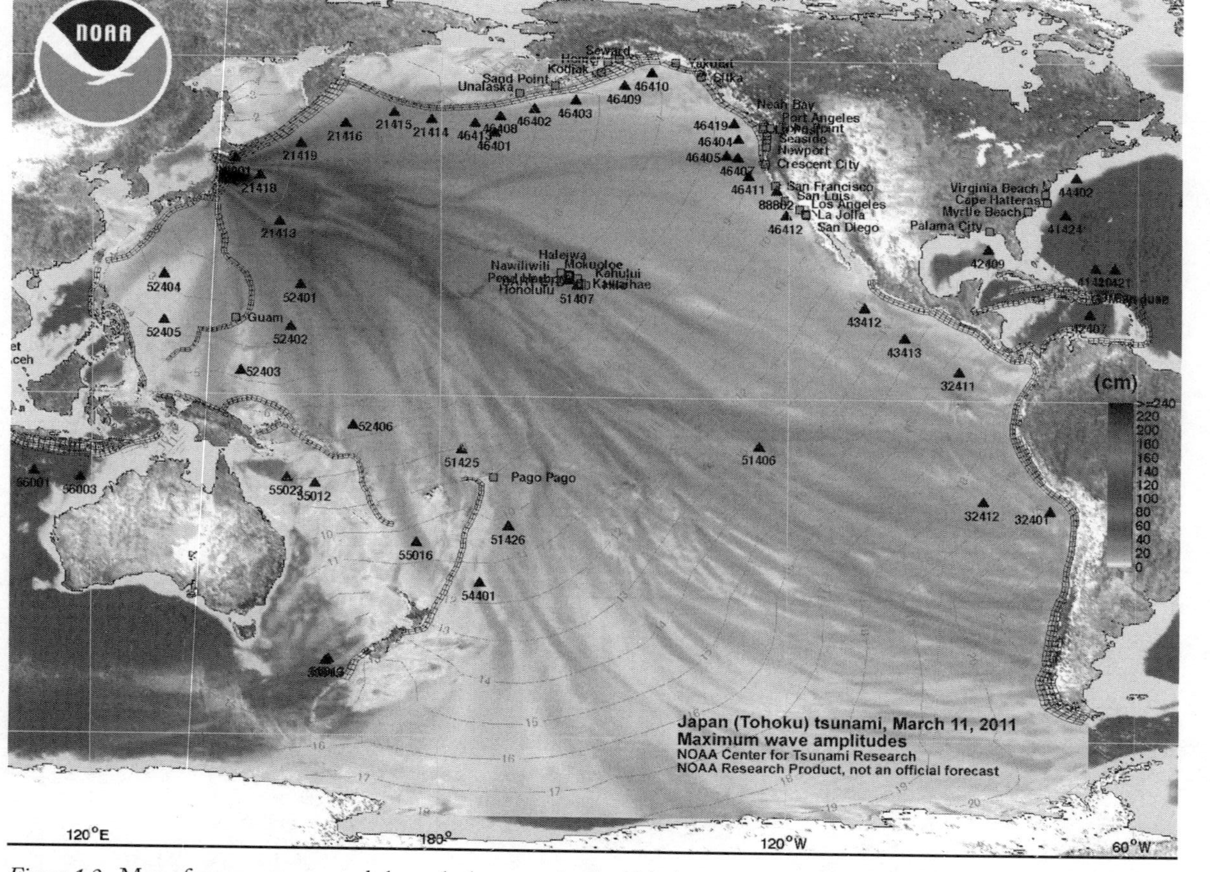

Figure 1.2. Map of energy propagated through the ocean by the Tohoku tsunami of March 11, 2014. The linear boundary at the Strait of Magellan may be considered imaginary, but in large part the tsunami's energy did not impact other oceans, while reaching nearly every Pacific coastline.

substantially across latitudinal and longitudinal gradients, they often face a common suite of large predators. As predators are usually the most important structuring factors in ocean ecosystems, this situation encourages convergent species evolution and convergent ecosystem assemblages in response to similar threats. Large predators also travel in response to changing oceanic conditions, and thus level out energy disparities around the Pacific. Secondly, large predators—which are frequently in turn preyed upon by humans—have produced common human experiences around the ocean and in some cases directly linked the activities of distant human societies. Pacific humans, while possessing venerable migratory histories of their own, have not been nearly as mobile as their maritime co-predators. A history of humans' place in larger ecosystems linked by animal migrations, then, reveals a Pacific Ocean much more deeply connected, and much earlier, than usually described.

Writing about the *yolla*, or shearwater, and the ways its travels connected vast spaces around the Pacific, ethnohistorian Greg Dening once opined, "There are many other tracks, too, of whales in seasonal migration, of tuna, of birds . . . mysteriously . . . directed by systems of knowledge."[9] This chapter attempts to take some of the mystery out of Dening's speculations, using new insights about these same three animals to describe a much more materially connected Pacific than even Dening imagined. It examines the long-distance migrations of sooty shearwaters, several species of whales, and tuna and traces their interaction with humans to argue for new chronologies and geographies of the Pacific World. These histories (meant to be broadly suggestive of wider trends rather than exhaustive accounts of Pacific integration) reveal that wide expanses of the Pacific have been integrated since at least c. 1200 CE, far earlier than claimed for most human-centric Pacific worlds. It is also apparent that the North Pacific has been better connected and more influential in the creation of these worlds than is commonly assumed. Taken together, these stories of sea-spanning animals and the ways humans dealt with them suggests that long-distance migration is one of the most important historical themes in the Pacific. Thus, only a frame as expansive as the Pacific itself can make sense of the thousands of local histories that almost never played out in isolation.

Seabirds: Sooty Shearwaters, Yolo, Muttonbirds, Tītī

No other bird has so consistently and for so long integrated Pacific ecologies as the sooty shearwater, a species that takes on central importance in Pacific history as well. These modestly sized birds (15–20 inches in length) have, from an unknown antiquity, migrated annually throughout nearly the entire width and length of the Pacific Ocean. Their travels trace a figure eight from the austral summer in New Zealand and the Bass Strait in Australia, northward to California and Alaska, westward to Siberia and Japan, then sometimes south to the Chilean coast before returning to New Zealand around September.[10] Shearwaters expend enormous amounts of energy and time to take advantage of seasonal abundances of fish and squid in transitional latitudinal zones in either hemisphere. In the North, they concentrate their feeding along the upwelling systems of the California, Japanese, and Kamchatka coasts and along the Emperor Seamounts in between the continents. In the South, they feed in the subantarctic Southern Ocean as well as the southern Polar Front.[11] This migrational-feeding strategy, despite its risks, has paid off handsomely for the species, which numbers among the most abundant in the world.

With such great numbers and such vast travels, shearwaters have played a key role in Pacific ecosystems from pole to pole and from hemisphere to hemisphere. In the words of one biologist, the birds "integrate peak oceanic resources on a global scale throughout the year."[12] The meaning of this phrase works two ways—first, the Pacific was an integrated space for shearwaters, a place that existed as a whole in whatever conception they may have formed of it. Secondly, shearwaters, through their wide-ranging generalist predation, have exerted a similar force throughout Pacific ecosystems on fish and squid populations. While shearwater numbers may be small in comparison with fish populations in the Pacific, predators exert an outsized influence on the makeup of ecosystems, especially when they feed at high trophic levels (the measurement of a number of species below them in a food chain), as sooty shearwaters do.[13] The numbers can be shocking. "Each year," according to recent research, sooty shearwater off the Oregon coast "consume . . . as much as 22% of the annual production of pelagic fish." While migrating in the boreal autumn, the birds "consume twice as much energy per day than the peak of any of the other species."[14] This impact—and the way it integrates the Pacific—can be shown in reverse as well, as shearwater numbers depend to a large degree on oceanic conditions far from their breeding grounds. Contemporary analysis of the bird's abundance in New Zealand, for example, has demonstrated that it is correlated with weather regimes in the North Pacific in the year prior.[15]

Thus, to understand species assemblage and abundance at nearly every point in the Pacific, it is not enough to know that place's primary productivity (the amount of energy available to living creatures), but also the presence of shearwaters. When humans began exploiting the birds in large numbers, they were unknowingly taking hold of one of the ocean's most powerful levers, with the potential to reshape ecosystems nearly everywhere from New

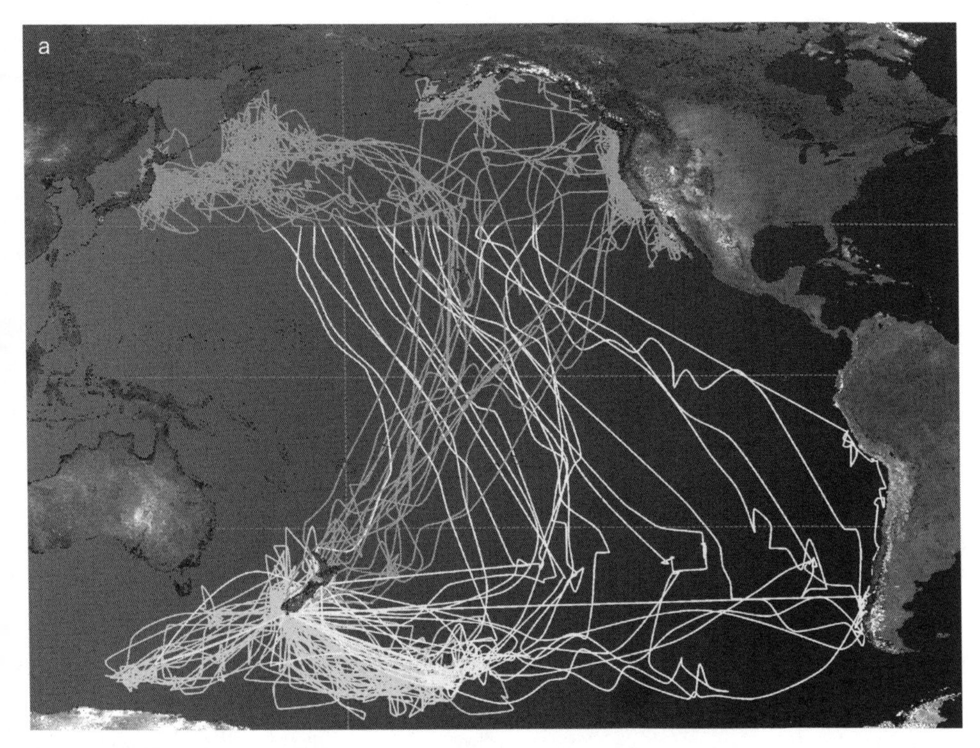

Figure 1.3. The migration of three twenty-first-century sooty shearwaters.

Zealand to Japan to Oregon to Chile. Shearwaters were, metaphorically, the local fault line that could send tsunami waves to unseen and still unimagined shores (see figure 1.3).

There is some disagreement about when the birds came under sustained exploitation, but all agree it resulted from the first great human migration in the Pacific—Polynesian voyaging. In particular, the tail end of the eruption of humans out of Melanesia and into the open Pacific, beginning around 1200 BCE, brought humans and shearwater histories together in Aotearoa New Zealand. Polynesian expansion everywhere decimated bird populations, as much through the introduction and depredations of the kiore—the Polynesian rat—as through human predation. If anything, New Zealand saw greater impacts on the aviafauna than elsewhere in the Pacific, as human immigrants found no alternative terrestrial sources of protein there.[16] Aotearoa's South Island (Te Wai Pounamou), and especially the far south's Murihuku, are the most important breeding places in Polynesia for shearwaters—called *tītī* by the Māori colonists and muttonbirds by later British colonists.[17]

Archaeologist Atholl Anderson deems the eighteenth century the most probable start for sustained Māori hunting of the birds, but archaeological evidence suggests that a date as early as 1470 might have seen at least intermittent harvesting.[18] Historian Michael Stevens believes that the harvest of shearwaters increased substantially with the arrival of Europeans to Murihuku in the early nineteenth century, as they brought with them dinghies that proved more effective than the Māori canoe, the *waka*.[19] On Rakiura/Stewart Island and the nearby "muttonbird" islands, Māori would raid shearwater burrows just as the fledglings (called muttonbirds at this stage) were getting ready to fly, thus at their fattest but still easy to catch. They were then eaten fresh or preserved in kelp, practices still vital to some *iwi* (tribal) identities. Estimates of the catch vary widely, but some early twentieth-century reports claim 200,000–250,000 birds taken annually.[20] This is a small dent indeed in an estimated global population of over twenty million, but combined with the introduction of rats, shearwater populations have likely suffered several serious depletions in their history since human colonization.[21] Given shearwaters' importance for Pacific ecosystems everywhere, much of the ocean must have changed as well.

While we can only surmise the Māori's prehistoric impact on the rest of the Pacific, at least one piece of concrete evidence has appeared. The onset of Murihuku muttonbirding, especially if assigned an early date, likely played a significant role in Northeastern Pacific human economies. In one of the most important midden sites on the Northwest Coast, near Grey's Harbor in present-day Washington State, shearwater remains appear as a significant prey species for inhabitants of the early tenth century—likely Coast Salish—but then nearly entirely disappear by the 1500s. Archaeologists rule out local overexploitation as a cause of shearwater decline, and though some of the disappearance could be due to a warming ocean, other similar birds that also depend on coastal upwelling do not drop out of the record. The most likely explanation for shearwater disappearance at Grey's Harbor, then, was human exploitation elsewhere. The only known location around the shearwater's range that was then experiencing increased hunting pressure was colonial Aotearoa. It seems likely, then, that Māori muttonbirding in Murihiku directly impacted Northwest Coast culture at least five hundred years ago. Whatever the cause of decline, the Salish people had to refocus hunting attention on smaller murres, scoters, and ducks.[22] This remarkable story presents an early lesson in the power of the ocean and the air to transmit change quickly and over vast spaces. As early as 1500 CE, before Magellan first sailed across the entire Pacific, these two distant corners of the ocean had been brought together, even if these two societies had no knowledge of each other.

Whales: Greys and Bowheads

As with human and shearwater migrations coming into contact, the movements of whales and humans in the North Pacific have deep linkages. Siberian peoples likely crossed the Bering Strait and migrated southward down the American coast just as gray whales were beginning to colonize the seas north of Mexico (a similar synchronization can be traced between human and salmon migration).[23] Scientists have recently determined the extent of these whales' epic peregrinations from Russia's Sakhalin Island all the way to the sheltered lagoons of Baja California, a journey of over fourteen thousand miles.[24] Meanwhile, from the Southern Hemisphere, some blue whales move from summer feeding grounds in the Antarctic to their presumptive breeding areas off Costa Rica. Whales normally do not feed in their warm-water haunts, so their impact on tropical oceans is slight, but they are the most important species in high-latitude Pacific ecosystems, where their eating and swimming rearranges life and churns up rich nutrients.[25]

Pacific peoples hunted whales in Japan, Chukotka, Alaska, British Columbia, and Washington State (and, speculatively, in Oregon) by at least eight thousand years ago, while Europeans first entered the Pacific for whales in the late eighteenth century.[26] Takes in lower latitudes were relatively small. Chukchi and Inupiat in the far north were more active, but as they mostly hunted bowheads, whose migrations are relatively short, their heavy impact on the whales' numbers barely affected others around the Pacific. But commercial whaling began reshaping the Pacific from the start; the hunt for migrating sperm whales off the coasts of Chile, Japan, and along the equator was the most important factor in bringing Euro-Americans in contact with Pacific Islanders. Hawaiians, Māori, Evenki (from Siberia), Fijians, and other Pacific people joined the scouring of the seas for moving wealth and energy, a large-scale integration of people and place that probably peaked in the mid-nineteenth century.[27]

The darkest side of these migrant ecologies stripped of whales became particularly apparent in the North Pacific after 1848. That year, Yankee whalers first began killing bowheads in the Bering Strait. Though initially the hunting ground was far from the Inupiat and Chukchi farther north who depended upon the whales, the killing choked off this narrow line of energy into the cold north, and those living in the north began to weaken and starve. In 1872, an American whaler wrote, "Twenty years ago whales were plenty and easily caught, but the whales have been destroyed and driven north, so that now the natives seldom get a whale."[28] Foreigners killed some 7,000–9,000 of the creatures, and by 1856 whalers abandoned the Bering Strait grounds.[29] In both Siberia and Alaska, people began to die in the thousands;

in the 1870s on St. Lawrence Island bodies were reportedly "everywhere in the village as well as scattered along in a line toward the graveyard for half a mile inland."[30] Fast-moving diseases buttressed this starvation and shared its dark, distant origination. Other species, such as fur seals and smaller whales, may have even then been taking advantage of this hole in the ecosystem to experience population growth, but few northern Pacific humans remained to profit from this possible boon.[31]

Elsewhere, Indigenous people both participated in pelagic Pacific whaling and felt its effects most keenly. The Makah people, with a whaling tradition stretching back at least two thousand years, began selling whale oil to European hunters in the 1840s. They continued killing seven to ten whales, mostly gray whales, per year through the early 1900s, equal to—or more than—European shore stations at the same time. However, the Euro-American pelagic hunt exploded at the same time, and—crucially—expanded to whales' calving grounds in the subtropical Pacific. In 1845, American whalers discovered the gray whale lagoons in Baja California, and in several short seasons devastated the population. Biologists estimate that gray whale numbers were reduced from around 24,000 to 2,000 by the 1880s, work done mostly far from the Makah homeland, but whose effects were felt perhaps most keenly there.[32] As historian Josh Reid has written, Makah whalers, unnerved by their declining success, engaged in ever more elaborate ritual preparation for the hunt to try to coax profits from the dwindling number of whales. Some blamed the intense brightness of a new lighthouse nearby for scaring away the whales, while one hunter claimed—perhaps not completely in jest—that the new smell of coffee on their breaths was keeping the whales away. But the real cause was thousands of miles distant. The Makah engaged in one last ceremonial hunt in 1928 and then ended whaling for the next seventy years.[33] The whales' disappearance also imperiled the Makahs' financial and political independence, a threat resulting in significant part because of the long-distance integration of the Pacific Ocean.

Bereft of whales, Makah and other Pacific people turned increasingly to fur seals, another long-distance migratory species, which ranges from California to Siberia. Now it was the Indians' turn to cut the thread of migration as they purchased mechanized ships capable of intercepting the seals while they migrated through the open waters of the North Pacific. "Sealing from ships allowed this tribal nation to expand Makah marine space," as Reid puts it, from local waters "to the south off the coast of California, to the north in the Bering Sea, and eastward [*sic*] to Japan."[34] Canadian First Nations people joined this hunt, which also brought in Japanese sealers before it was ended by international treaty in 1911. Thus, the Pacific's migratory connections not only had the potential to cut Indigenous people off from the ocean but also opened new opportunities for their own migrations.

These stories repeated themselves in altered form nearly a century later, as massive fleets of industrial whaleships colonized the Pacific, especially from the 1950s.[35] While the greatest beneficiaries of sail whaling might have been non-target species such as fins and seis who now had much more food to themselves, the industrial whaling of the twentieth century cut huge holes in every species. The Soviet Union made attempts to shield the Chukchi from a second starvation by guaranteeing them access to the hunt, and the United States tried to wean the Inupiat onto processed foods instead of whale, but neither could stop the changes that crested through ecosystems far from whaling's focal points in the Antarctic and the North Pacific. While some nations worried about pelagic whaling's effect on their own shore-based industries, it was whales' ceaseless movements that nations like Russia referenced to resist effective conservation of whales. If they restrained themselves from catching the creatures, the Soviets argued, then the whales would simply fall prey to the Japanese (with equivalent bogeys postulated elsewhere).[36]

As whales disappeared, fish took their place as the dominant factor in ecosystems from New Zealand to Alaska, boosting the commercial industries that were developing there from the mid-twentieth century.[37] At the same time, the mysterious declines of sea lions, fur seals, and sea otters dating from the 1970s have been traced to increased predation by killer whales that used to prey on their larger brethren and have now been forced to expand their diets down the food chain.[38] Whales thus integrate the Pacific not only over space but also over time, as the effects of a now-discontinued whaling industry continue to show themselves on the ocean forty years later. The eating away of these medium-sized mammals came just as Indigenous Pacific people, like Aleuts, Haida, and others, had returned to these creatures as part of reclaiming their pre-colonial cultural traditions. Thus, unlike muttonbirds, whales' contribution to Pacific integration has come largely in the form of tragedy—as the heralding angels of the Euro-American plagues of starvation, disease, and dispossession; and then in a second wave as the ghosts of ecosystems past. If Indigenous Pacific peoples often jumped at the chance to follow whales around the ocean, they then returned to local seas radically impoverished by their prey's absence.

Fish: Tuna

The removal of whales gave opportunities for other species, such as Pacific tuna, to boom. Tuna vie with whales and shearwaters for the extent of their distribution around the Pacific; found almost everywhere from Oregon to

Australia, they are especially abundant in one of the prime whaling grounds of the nineteenth century, "the Line," or the seas around many-islanded Micronesia—especially around the Federated States of Micronesia (FSM), Kiribati, the Marshall Islands, Palau—and parts of Melanesia, including Papua New Guinea and the Solomon Islands.[39] Tuna migration is also among the most complex in the Pacific. Notably, some tuna species migrate mainly east-west, rather than north-south as do whales and shearwaters. Yellowfin tuna (*Thunnus albacares*), commercially the most important species, consists of at least three sub-populations, with adults migrating and juveniles staying close to shore. Skipjacks (*Katsuwonus pelamis*) move east then back west to spawn in the tropical Pacific.[40] The largest and most valuable of the Pacific species—the bluefin (*Thunnus thynnus*)—spends its youth in the Western mid-latitude Pacific before traveling east to feed in the California Current (as the irradiated fish mentioned in the introduction helped demonstrate), with some adults then traveling to the South Pacific.[41] Like the other species discussed here, tuna feed at a high trophic level and thus exert powerful downward pressure on ecosystems. Similarly, by virtue of their long lives and long migrations, they "reduce temporal and spatial variations in ecosystem structure"—in other words, providing stability and similarity over large parts of the equatorial and temperate Pacific.[42]

While tuna have been important food for tropical Pacific Islanders for a long time, they were not staple foods for any, with the possible exception of those living in Kiribati.[43] The fish were found relatively far offshore and were in constant movement—Japanese, Micronesians, and Polynesians all caught them, but in small numbers. Even such small-scale work required years of study of complex tuna migration routes in order to become productive. Local tuna fisheries, though, would be quickly overwhelmed by the introduction of industrial trawling. For example, the traditional tuna fishing grounds of one man in Tokelau (a group of atolls north of Samoa) was big enough to occupy a modern purse seiner for less than an hour.[44] Those seiners, which deploy a wall of netting that scoops up huge numbers of fish, followed the initial movement of Japanese long-liners, catching fish with individual poles, into the North Pacific in the 1920s.[45] Soon thereafter, Japanese fishermen entered Southeast Asia and Micronesia, especially after inheriting the region's German colonies after World War I.

Meanwhile, scattered American ventures were beginning the first surveys of open-ocean tuna fishing as early as 1899, looking at various parts of the central Pacific from Guam to Tahiti to the Galapagos.[46] World War II's Pacific theater afforded more chances—and motivation—to follow up these lone ventures, and in 1944, a team of scientists led by University of

Washington biologist W. M. Chapman explored much of the central and western Pacific looking for commercial possibilities. As Chapman reported, "The region from the Gilberts east to the Marquesas and south through the Fiji, Cook, Tonga, Samoa, Society and Tuamoto [*sic*] islands have never been exploited commercially for tuna. . . . It is in this great area that the expansion of the American tuna fishery will first enter."[47] It did, especially in American Samoa, where Americans built a large cannery in 1949.[48] Nor did defeat in war long stem the tide of Japanese expansion, as their fishermen returned to these same waters from the 1950s, now establishing canneries and fishing bases as well. Soviet, South Korean, Taiwanese, and Chinese seiners followed from the 1980s. Filipino and Indonesian fishermen often did (and do) the work on board the seiners, making the tuna fishery another great motor of Pacific human migration.

A major question facing the early developers of the tuna fishery was the interrelationship between various fish populations. To their dismay, the fishermen learned there was little possibility of drawing clear lines between them.[49] What is more, large oceanographic changes, such as the El Niño phenomenon, could move tuna from one end of the Pacific to another for seasons at a time.[50] As the United States and Japan were increasingly engaged in catching tuna near other countries' shores, they dusted off arguments made in the whaling industry about the absurdity and impossibility of ceding control of fish to any one nation. However, determined to protect their own waters from Japanese salmon fishermen, who were then expanding into Siberia and Alaska, Americans also insisted on the right to protect fisheries imperiled by overharvest.[51]

But in an era of decolonization, the story of whaling would not repeat itself with tuna. Pacific Island countries and Latin American nations fronting the Pacific argued instead that they maintained exclusive rights to catch the fish when they were in territorial waters, as they constituted part of a country's biomass.[52] Each side, it should be noted, made arguments that were most likely to benefit their own fishermen or potential fishermen. In 1952, Ecuador and Peru, whose waters American tuna seiners were mercilessly exploiting for bait fish, cooperated with Chile, which was concerned about Soviet whaling ships, to insist on a 200-mile exclusive economic zone (EEZ) around their coastlines.[53] The United Nations ratified this concept in the 1982 Law of the Sea. Though the United States never signed the convention, it respects it in practice.

In this new regulatory environment, tuna fishing in the central Pacific exploded further. However, with a highly migratory fish like tuna, national jurisdiction proved only of minor assistance in dealing with the problem of management. The two-hundred-mile EEZ is equal to only a day-and-a-half's journey for a bluefin.[54] Thus, while EEZs have been profitable for Pacific Island nations, which can now lease out their territorial waters to better-capitalized fishing fleets, migrating fish have also encouraged something equally consequential for Pacific history—large-scale, regional cooperation. The South Pacific Forum, where regional leaders met, had existed since 1971, but it gained significant coherence and power in response to foreign tuna fleets. In 1976, Papua New Guinea and Fiji decided they needed to present a common front toward outside fishing nations, and discussion in the Forum led to the creation of the Pacific Islands Forum Fisheries Agency. The Agency has proved to be the strongest form of cooperation among Pacific nations, a "shining example" of regional cooperation.[55] Particularly notable was the so-called Palau Arrangement, instituted in 1992, which has been termed "the largest and most complex fishery management ever to be put in place."[56] The Arrangement enjoins cooperation among all member states to limit the overall fishing effort in place across a large portion of the tropical Pacific; as such, it benefits both Pacific Island states and the state of tuna stocks.

It is worth noting that none of this cooperation—which has rippled out from fisheries management to encompass many other matters—would have made much sense if tuna migrated from north to south instead of primarily from east to west. A north-south migrational pattern would instead have delivered fish to Japanese, Korean, and other trawlers whenever they left South Pacific waters and merely encouraged as quick and thorough a destruction of tuna stocks as possible.

In the early twenty-first century, the Forum Fisheries Agency has not yet solved the problem of overfishing tuna—researchers believe yellowfin, bluefin, bigeye, and even skipjack numbers are in decline.[57] Distant fishing by other outside fleets has seriously eroded tuna stocks that migrate past the Cook Islands, hampering plans to create a domestic industry there.[58] Still, in 2008, Kiribati created the world's largest Marine Protected Area, banning fishing in a tuna-rich 157,626 square mile portion of its EEZ (the Phoenix Island Protected Area), and in 2011, the Forum Fisheries Agency set aside a huge amount of the Eastern Pacific for hook-and-line fishing only.[59] These measures are interesting not only for their ambitious attempts at preserving an important part of the Pacific ecosystem but for the way they came out of regional agencies and concepts of marine tenure developed in large part in response to the migratory habits of tuna and—earlier—whales. If Pacific tuna do indeed survive into another century, it will be in large part thanks to the integration of Pacific politics and identities that responded to the integration of Pacific ecosystems.[60]

Conclusion

Of course, stories of the Pacific's expanse and expansive identities—from Polynesian migrations to post-colonial diasporas—have been the lifeblood of Pacific history since at least the 1990s.[61] Historians have described Pacific Worlds in constant motion, as people have created social connections around the ocean and formed various conceptions of large oceanic communities that sometimes qualify as an entire Pacific World.[62] A deeper focus on the migrations of non-humans reinforces these narratives. But it does more. It also helps us recognize that the Pacific World is not just a human creation, but a world that created connections between humans. These connections have not always been immediately apparent—Northwest coasters had no notion their shearwaters disappeared because of Māori migration—but they played important roles in the way Indigenous histories, Euro-American colonialism, and post-colonialism developed in the Pacific. The extraordinary integrating force possessed by migrating predators demonstrates that a closely connected Pacific World has long been in existence.

This Pacific World, though, does not stick to canonical geographies or chronologies. Muttonbird connections reveal that the North Pacific was long connected to the South; though the human migratory streams across the Bering Strait and through Melanesia were quite different, all alighted on a shared ocean. In fact, humans from Siberia to the Northwest Coast shared connections with each other and with humans in the South Pacific that were sometimes stronger than connections with those living behind the inland barrier of high mountains springing from the Pacific's tectonic hyperactivity.[63] North Pacific people again found themselves tightly bound to distant ecosystems as an era of sealing and whaling washed them into a colonial Pacific unified by a squalid search for living commodities at any cost.[64] Only in the late twentieth century, when industrial tuna fishing rose so quickly to such a great extent, do we see a South Pacific so interwoven. While Alaskans and Siberians, in particular, are still often thought of as particularly isolated, an awareness of the intense relationship they have often had with migratory animals makes their worlds begin to look much, much larger—in short, like a Pacific World.[65]

This Pacific World also has a distinct chronology. The histories of the three species discussed suggest the Pacific was first integrated in about 1500 at a few important birding sites around its four corners. As whaling exploded after c. 1815, important connections between the temperate and tropical Pacific were created, connections that did not always cross the equator, but looked very similar in both halves of the ocean, and that paralleled the increasing voyaging of Pacific peoples. Finally, an era of industrial fishing that began in earnest after 1945 left nearly no portion of the ocean detached from another, especially along the tuna migration lines across the equatorial and subtropical Pacific (and the salmon commons of the temperate North not discussed here). Perhaps this does not constitute one Pacific World, but it does hint at much larger and much deeper Pacific Worlds than historians have noticed.

Notes

1. In fact, in small quantities, the tuna was deemed safe for human consumption. See "Study Finds Only Trace Levels of Radiation from Fukushima in Albacore," http://oregonstate.edu/ua/ncs/archives/2014/apr/study-finds-only-trace-levels-radiation-fukushima-albacore (accessed January 14, 2016); see also Daniel J. Madigan, Zofia Baumann, and Nicholas S. Fisher, "Pacific Bluefin Tuna Transport Fukushima-Derived Radionuclides from Japan to California," *Proceedings of the National Academy of Sciences of the United States of America* 109, no. 24 (June 12, 2012): 9483–9486.

2. Eric Jones, Lionel Frost, and Colin White, *Coming Full Circle: An Economic History of the Pacific Rim* (Boulder, CO: Westview Press, 1993), 6; David Igler, *The Great Ocean: Pacific Worlds from Captain Cook to the Gold Rush* (Oxford: Oxford University Press, 2013), 11.

3. Matt Matsuda, *Pacific Worlds: A History of Seas, Peoples, and Cultures* (Cambridge: Cambridge University Press, 2012), 5.

4. David Hanlon, "Losing Oceania to the Pacific and the World," *The Contemporary Pacific* 29, no. 2 (2017): 290.

5. See especially Martin W. Lewis and Karen Wigen, *The Myth of Continents: A Critique of Metageography* (Berkeley: University of California Press, 1997).

6. Martin W. Lewis, "Dividing the Ocean Sea," *Geographical Review* 89, no. 2 (April 1999): 188–214.

7. For a history of tsunamis in the Pacific and transnational efforts to predict and react to them, see Walter C. Dudley and Min Lee, *Tsunami!* (Honolulu: University of Hawai'i Press, 1998); Laura Kong, ed., *Pacific Tsunami Warning System: A Half-Century of Protecting the Pacific, 1965–2015* (Honolulu: International Tsunami Information Center, 2015).

8. Tagging of Pelagic Predators website, www.gtopp.org (accessed January 7, 2016).

9. Greg Dening, *Beach Crossings: Voyaging across Times, Cultures, and Self* (Melbourne: Miegunyah Press, 2004), 15.

10. Scott A. Shaffer et al., "Migratory Shearwaters Integrate Oceanic Resources across the Pacific Ocean in an Endless Summer," *Proceedings of the National Academy of Sciences of the United States of America* 103, no. 34 (2006): 12799–12802, 12799.

11. H. Weimerskirch, "How Can a Pelagic Seabird Provision Its Chick When Relying on a Distant Food Resource? Cyclic Attendance at the Colony, Foraging Decision and Body Condition in Sooty Shearwaters," *Journal of Animal Ecology* 67 (1998): 99–109.

12. Shaffer et al., 12799.

13. Trophic level 4.3 according to one analysis carried out in the North Pacific. See Patrick Gould, Peggy Ostrom, and William Walker, "Food, Trophic Relationships, and Migration of Sooty and Short-Tailed Shearwaters Associated with Squid and Large-Mesh Driftnet Fisheries in the North Pacific Ocean," *Waterbirds* 23, no. 2 (2000): 165–186.

14. Stewart T. Schultz, *The Northwest Coast: A Natural History* (Seal Rock, OR: Timber Press, 2011), 94.

15. Rosemary Clucas, "Long-Term Population Trends of Sooty Shearwater (*Puffinus griseus*) Revealed by Hunt Success," *Ecological Adaptations* 21, no. 4 (June 2011): 1308–1326, 1323.

16. Especially since Polynesian colonists failed to bring pigs or chickens with them, or, as some speculate, simply let these species die when they encountered the massive, flightless, protein-rich moa birds.

17. There is some evidence shearwaters may have once been present in large numbers on New Zealand's North Island as well.

18. Atholl Anderson, "Historical and Archaeological Aspects of Muttonbirding in New Zealand," *New Zealand Journal of Archaeology and Natural History* 17 (1995): 35–55; Anderson, "The Origin of Muttonbirding in New Zealand," *New Zealand Journal of Archaeology* 22 (2000): 5–14; Michael Stevens, "Muttonbirds and Modernity in Murihuku: Continuity and Change in Kai Tahu Knowledge" (PhD diss., University of Otago, 2009), 57.

19. Stevens, 57.

20. Anderson, "Aspects of Muttonbirding," 38.

21. Anderson, "Origins of Muttonbirding in New Zealand," 9.

22. K. M. Bovy, "Global Human Impacts or Climate Change? Explaining the Sooty Shearwater Decline at the Minard Site, Washington State, USA," *Journal of Archaeological Science* 34 (2007): 1087–1097.

23. Dick Russell, *Eye of the Whale: Epic Passage from Baja to Siberia* (New York: Simon & Schuster, 2001), 23.

24. Jane J. Lee, "A Grey Whale Breaks the Record for Longest Mammal Migration," *National Geographic*, April 14, 2015; Bruce Mate et al., "Critically Endangered Western Gray Whales Migrate to the Eastern North Pacific," *Biology Letters* 11, no. 4 (April 2015), https://doi.org/10.1098/rsbl.2015.0071.

25. Though whale carcasses that sink to the bottom of the ocean have recently been found to be key components of ecosystems wherever they fall. See Craig R. Smith, "Bigger Is Better: The Role of Whales as Detritus in Marine Ecosystems," in *Whales, Whaling, and Oceanic Ecosystems*, ed. James A. Estes, Douglas P. Emaster, Daniel F. Doak, Terrie M. Williams, and Robert L. Brownell, Jr. (Berkeley: University of California Press, 2006), 286–302. For whales' active role in reshaping their environment, see Joe Roman et al., "Whales as Marine Ecosystem Engineers," *Frontiers in Ecology and the Environment* 12, no. 7 (September 2014): 377–385. Whales occupy a slightly lower trophic level than do shearwaters or tuna, at 3.2–4.4, though killer whales are even higher at 4.6; see D. Pauly, A. W. Trites, E. Capuli, and V. Christensen, "Diet Composition and Trophic Levels of Marine Mammals," *ICES Journal of Marine Science* 55 (1988): 467–481.

26. Victor Krupnik, *Arctic Adaptations: Native Whalers and Reindeer Herders of Northern Eurasia* (Hanover, NH: University Press of New England, 1993), 185; Margaret Lantis, "The Alaskan Whale Cult and Its Affinities," *American Anthropologist* 40, no. 3 (July–September 1938): 438–464; on Oregon, see Hannah P. Wellman, Torben C. Rick, Antonia T. Rodrigues, and Dongya Y. Yang, "Evaluating Ancient Whale Exploitation on the Northern Oregon Coast through Ancient DNA and Zooarchaeological Analysis," *Journal of Island and Coastal Archaeology* 12, no. 2 (2017): 255–275.

27. Arrell Morgan Gibson, *Yankees in Paradise: The Pacific Basin Frontier* (Albuquerque: University of New Mexico Press, 1993); Ryan Jones, "The Environment," in David Armitage

and Alison Bashford, *Pacific Histories: Oceans, Lands, People* (New York: Palgrave-Macmillan, 2014).

28. "Shipmaster," *Friend*, Honolulu (March 1, 1872).

29. John Bockstoce, *Whales, Ice, and Men: The History of Whaling in the Western Arctic* (Seattle: University of Washington Press, 1986), 346, 347.

30. Edward William Nelson, *The Eskimo about Bering Strait* (Washington, DC: Government Printing Office, 1900), 269.

31. The ecological dimensions of whaling and walrus-hunting are described in Bathsheba Demuth, *Floating Coast: An Environmental History of the Bering Strait* (New York: W. W. Norton, 2019), Chapters 2, 3.

32. Springer et al., "Whales and Whaling in the North Pacific Ocean and Bering Sea," in *Whales, Whaling, and Oceanic Ecosystems*, 246.

33. Josh Reid, *The Sea Is Our Country: The Maritime World of the Makahs* (New Haven, CT: Yale University Press, 2015), 172, 175.

34. Reid, 177.

35. For an overview of twentieth-century whaling, see John Nicolay Tonnessen and Arne Odd Johnsen, *The History of Modern Whaling* (Berkeley: University of California Press, 1982).

36. See, among many examples, V. K. Arsen'ev, "Doklad," April 7, 1921, State Archive of Primor'ye Krai, Vladivostok (GAPK), F. 633, Op. 4, No. 100, p. 42. For a comprehensive treatment of the commission that attempted to regulate whaling, the International Whaling Commission, see Kurkpatrick Dorsey, *Whales and Nations: Environmental Diplomacy on the High Seas* (Seattle: University of Washington Press, 2013).

37. Alan M. Springer, Gus B. Van Vliet, John F. Piatt, and Eric M. Danner, "Whales and Whaling in the North Pacific Ocean and Bering Sea," in *Whales, Whaling, and Oceanic Ecosystems*, 254.

38. Trevor A. Branch and Terrie M. Williams, "Legacy of Industrial Whaling. Could Killer Whales Be Responsible for Declines of Sea Lions, Elephant Seals, and Minke Whales in the Southern Hemisphere?" in *Whales, Whaling, and Oceanic Ecosystems*, 263.

39. Parzival Copes, "Tuna Fisheries Management in the Pacific Islands Region," in *Tuna Issues and Perspectives in the Pacific Islands Region*, ed. David J. Doulman (Honolulu: East-West Center, 1987), 17.

40. J. E. Bardach and Y. Matsuda, "Fish, Fishing, and Sea Boundaries: Tuna Stocks and Fishing Policies in Southeast Asia and the South Pacific," *GeoJournal* 4, no. 5 (1980): 467–478.

41. Daniel J. Madigan et al., "Reconstructing Transoceanic Migration Patterns of Pacific Bluefin Tuna Using a Chemical Tracer Toolbox," *Ecology* 95, no. 6 (June 2014): 1674–1683; Madigan, Baumann, and Fisher, "Pacific Bluefin Tuna Transport Fukushima-Derived Radionuclides from Japan to California."

42. The trophic level of most deep-sea tuna caught is 4.2, much higher than most near-shore fish and nearly the same as the 4.3 of sooty shearwaters. See Timothy E. Essington, "Pelagic Ecosystem Response to a Century of Commercial Fishing and Whaling," in *Whales, Whaling, and Oceanic Ecosystems*, 42.

43. John Sibert, "Biological Perspectives on Future Development of Industrial Tuna Fishing," in *Tuna Issues and Perspectives*, 39, 40; Roniti Teiwaki, *Management of Marine Resources in Kiribati* (Suva, Fiji: University of South Pacific Press, 1988), 19.

44. Robert Gillett, "Traditional Fishing in Tokelau," *South Pacific Regional Environment Programme*, Topic Review No. 27 (1985): 49.

45. Norio Fujinami, "Development of Japan's Tuna Fisheries," in *Tuna Issues and Perspectives*, 54.

46. Kate Barclay, "History of Industrial Tuna Fishing in the Pacific Islands: A HMAP Asia Project Paper," *Asia Research Centre Working Paper* 169 (December 2010): 4; August Felando, "U.S. Tuna Fleet Ventures in the Pacific Islands," in *Tuna Issues and Perspectives*, 94.

47. Quoted in Felando, 95.

48. Alfonso P. Galea'i, "American Samoa: The Tuna Industry and the Economy," in *Tuna Issues*, 191.

49. Felando, 104, 105.

50. Kate Barclay with Ian Cartwright, *Capturing Wealth from Tuna: Case Studies from the Pacific* (Canberra: ANU Press, 2007), 4.

51. This history, crucial for the development of modern fisheries around the world, is expertly narrated in Carmel Finley, *All the Fish in the Sea: Maximum Sustainable Yield and the Failure of Fisheries Management* (Chicago: University of Chicago Press, 2011).

52. Jon M. Van Dyke and Carolyn Nicol, "U.S. Tuna Policy: A Reluctant Acceptance of the International Norm," in *Tuna Issues and Perspectives in the Pacific Islands Regions*, ed. David J. Doulman (Honolulu: East-West Center, 1989), 105; Finley, *All the Fish in the Sea*, 127.

53. Finley, 127.

54. C. S. Wardle, J. J. Videler, T. Arimoto, J. M. Franco, and P. He, "The Muscle Twitch and the Maximum Swimming Speed of Giant Bluefin Tuna, *Thunnus thynnus* L." *Journal of Fish Biology* 35, no. 1 (July 1989): 129–137.

55. Quoted in Barclay and Cartwright, 18.

56. Transform Aqorau, "Recent Developments in Pacific Tuna Fisheries: The Palau Arrangement and the Vessel Day Scheme," *International Journal of Marine and Coastal Law* 24, no. 3 (2009): 557–581, 558.

57. Phil Crawford, "Pacific Island Countries Strive to Save Their Tuna Fisheries," *Pacific Ecologist* 20 (2011): 42.

58. Barclay and Cartwright, 63.

59. Paul Christopher, "Islands Champion Tuna Ban: Pacific Nations to Restrict Fishing across a Vast Swathe of International Waters," *Nature* 468 (2010): 7325; Phoenix Islands Protected Area, www.phoenixislands.org (accessed January 10, 2016).

60. This kind of correspondence between political and ecological systems is described as the best practice in conservation; see Eric A. Treml et al., "Analyzing the (Mis)Fit between the Institutional and Ecological Networks of the Indo-West Pacific," *Global Environmental Change* 31 (2015): 263–271.

61. See, among others, Epeli Hau'ofa, "Our Sea of Islands," *The Contemporary Pacific* 6, no. 1 (Spring 1994): 147–161; Dening, *Beach Crossings*; Matt Matsuda, "The Pacific," *American Historical Review* 111, no. 3 (June 2006): 758–780.

62. Katrina Gulliver, "Finding the Pacific World," *Journal of World History* 22, no. 1 (2011): 83–100.

63. This is an idea I developed at more length in Ryan Tucker Jones, "Kelp Highways, Siberian Girls in Maui, and Nuclear Walruses: The North Pacific in a Sea of Islands," *Journal of Pacific History* 49, no. 4 (2014): 373–395.

64. Jones, "The Environment," in Bashford and Armitage.

65. As two anthropologists recently put it, "The Aleutians are among the most isolated islands in the world. Only the Middle and South Pacific islands are more remote locales for human habitation." Allen P. McCartney and Douglas W. Veltre, "Aleutian Island Prehistory: Living in Insular Extremes," *World Archaeology* 30, no. 3 (1999): 503–515, 513.

Two

Many Diasporas
People, Nature, and Movement in Pacific History

Gregory Samantha Rosenthal

> The resources of Samoans, Cook Islanders, Niueans, Tokelauans, Tuvaluans, I-Kiribatis, Fijians, Indo-Fijians, and Tongans, are no longer confined to their national boundaries; they are located wherever these people are living permanently or otherwise . . . One can see this any day at seaports and airports throughout the central Pacific where consignments of goods from homes-abroad are unloaded, as those of the homelands are loaded. Construction materials, agricultural machinery, motor vehicles, other heavy goods, and myriad other things are sent from relatives abroad, while handcrafts, tropical fruits and rootcrops, dried marine creatures, kava and other delectables are despatched from the homelands.
>
> —*Epeli Hau'ofa (1993)*

> The land is the people, is the money, is the phosphate, is the farm, is the grain, is the cattle, is the development and pollution of a nation.
>
> —*Katerina Martina Teaiwa (2005)*

The above statements are not really lists. They are maps. Epeli Hau'ofa, in his essay "Our Sea of Islands," wrote that Oceania—the lived-in and peopled Pacific Ocean—was interconnected, not disconnected, by the vast amount of water lying between islands and continents. Not only did Pacific Islanders move through oceanic space, but nature moved, too: "construction materials, agricultural machinery, motor vehicles, other heavy goods," as well as "handcrafts, tropical fruits and rootcrops, dried marine creatures, kava and other delectables." Oceania was animated and integrated by people and nature in motion. Movement, circulation, travel, production, and consumption mapped out worlds much larger and less isolated than the "tiny worlds" usually ascribed to the histories of Island peoples and Island nature. One decade later, Banaban scholar Katerina Martina Teaiwa retold Hau'ofa's tale, this time adding another dimension: the land itself. In "Our Sea of Phosphate," Teaiwa meditates on the simultaneous, interdependent histories of human and non-human dispossession and dislocation as twentieth-century phosphate extraction pushed both Banaban people and Banaban land out to sea. People scattered and became migrant workers; capitalists turned the land into fertilizer and scattered it upon foreign lands. "The writing on diaspora," Teaiwa notes, "focuses on the movement of bodies, labour, peoples, ideas and cultural productions, usually as a consequence of colonialism and imperialism, but rarely of land in its physical sense." However, "the experiences of Banaban and Nauruan land mined, shipped, transformed into fertilizer and then literally scattered across the fields of Australia and New Zealand come close to the original meanings of *diaspeirein*." Diaspora: the scattering and sowing of people, like seeds; the scattering and sowing of nature to seed someone else's environment. Teaiwa writes, "The land is the people," and Hau'ofa claims, "We are the ocean." Movement, of both people and nature, animates the histories of Pacific Islands and Islanders, the ocean, the rim, and the world beyond.[1]

This chapter argues that movement has been—and remains—a key variable in Pacific Islander lives. In the Pacific, people and nature moved as in no other world-historical region.[2] More specifically, diaspora, as Teaiwa suggests, presents a useful lens for examining and analyzing the ways in which human migration is related to plant and animal migrations. In this chapter, I focus on Oceanian migrations and diasporas, with an emphasis on Polynesian historical experiences. The term "Polynesia"—meaning "many islands"—originated in early colonial definitions of Oceanic space; other terms, such as "Micronesia" (little islands) and "Melanesia" (black/dark islands), represent similarly colonial geographies. While some Pacific peoples have embraced these terms and others resist them, I want to suggest that these terms, even in their most literal sense, are simply inaccurate. Rather than thinking of Polynesia as "many islands," we might consider the entire Pacific to be a world of many diasporas.[3] There are Pacific Island people diasporas, plant diasporas, animal diasporas, even disease diasporas.[4] Below, I examine several of these globetrotting natures. I focus on Pacific peoples whose histories reflect unique connections and active integrations of distant and disparate peoples, places, and processes through human and non-human movement, just as Hau'ofa and Teaiwa have suggested. Not only were Pacific peoples agents of epic transoceanic migrations in millennia past, but these migrations continue apace in the twenty-first century. Pacific Islanders have carried nature with them and rearranged Pacific environments all across the ocean, from the past to the present, just as the environment has also guided and delimited the contours of human movement and migration upon the waves.[5]

The diasporic nature of Pacific environmental history is in stark contrast to two canonical ideas within both Pacific and environmental historiographies. These are, one, that outsiders rather than insiders effect ecological change; and, two, that Indigenous peoples' histories largely unfold in situ, or in place.[6] But the truth is that for the past one thousand years, Pacific peoples have been at the forefront of ecological change and biological transformations in the Pacific Ocean, and, although native to many islands, they have yet made the whole world their home, a world of many diasporas, a world made by people and nature in motion.

People in Diaspora

At a time when "Western civilization" was confined simply to the Mediterranean Basin, the seeds of a diaspora that would stretch thousands of miles across the Pacific Ocean in all directions took root in Fiji, Samoa, and Tonga. This triangle is known as the Polynesian homeland. Prior to the Common Era, Polynesian ancestors settled in these islands. They carried pigs, chickens, dogs, and other faunal stowaways, as well as over a dozen "canoe plants": taro, sugarcane, breadfruit, coconut, and others.[7] The first fifteen centuries of the Common Era were a time of both "rootedness" and "routedness" for Polynesian peoples. Complex societies developed in the homeland as well as in the Society and Marquesas Islands, while others continued to travel, discovering and colonizing islands as far away as Rapa Nui (Easter Island) in the southeast, Hawai'i in the north, and Aotearoa (New Zealand) in the southwest. These three points form the edges of what is known as the Polynesian Triangle, a cartographic representation of the world's most expansive human diaspora on the cusp of global early modernity. Long before 1492 and the famous Columbian exchange, Oceania was a theater of its own transoceanic exchanges. Not only did Polynesian migrants settle upon so many islands across the Pacific Ocean, bringing their own "portmanteau biota" with them—canoe plants and all—but they also created "neo-tropics" in far-flung places such as Hawai'i in the north and Aotearoa in the south (see figure 2.1).[8]

Pacific peoples also encountered the American continent and American nature. It should not be too shocking to imagine Indigenous Pacific Islanders sailing up to American shores before 1492. Polynesians were not the only maritime people in the Eastern Pacific Basin. For example, Chumash moved between the Channel Islands and the North American mainland for

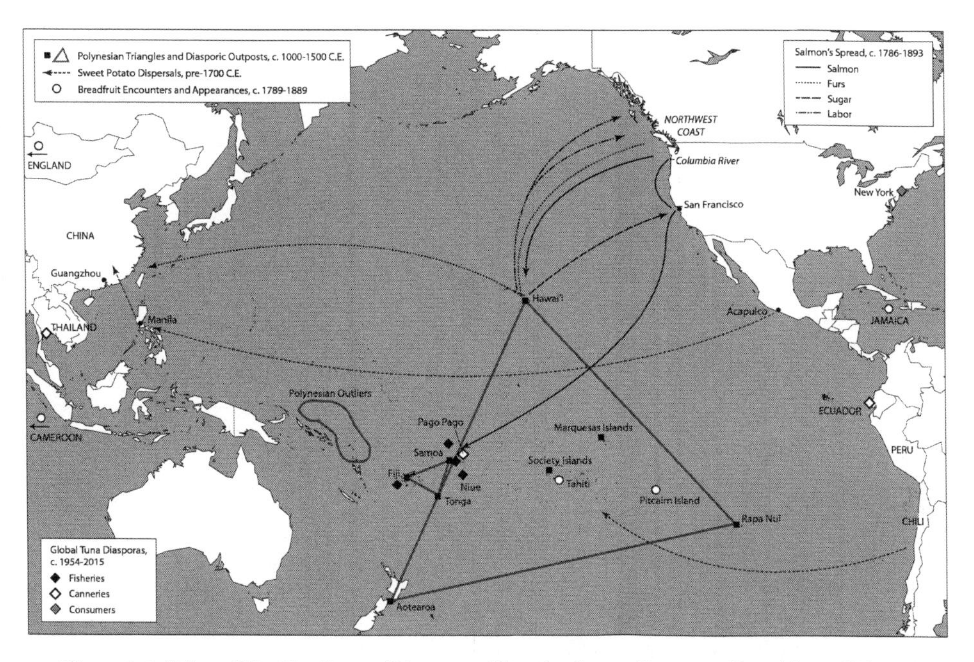

Figure 2.1. Map of Pacific Ocean Diasporas (People, Sweet Potatoes, Breadfruit, Salmon, Tuna). Map by Bill Nelson.

thousands of years, and there were many other Indigenous maritime migrants up and down the American Pacific coast from Alaska to Chile. Europeans, too, sailed to the northern reaches of North America in the centuries before Columbus' voyage. Rather than thinking of the Americas as a sealed vessel prior to 1492, historians are now recognizing the fluidity and mobility of peoples and natures in and out of the Americas in the centuries before Columbus sailed the ocean blue. In the case of the Polynesian Diaspora, researchers have found little convincing evidence that Polynesian peoples *lived* in the Americas before 1492—or that their DNA mixed with Indigenous Americans—but there is strong evidence that Polynesian animals came to South America before Spaniards did, and that a South American plant most certainly left the continent and traveled across the Pacific Ocean. The most cutting-edge research on these exchanges looks at what are called "commensal animals" and "commensal plants." These are plants and animals theorized to have moved with, and because of, humans, but, where solid evidence of human migrations is irretrievable, researchers look instead for the lingering presence of certain DNA in local plant and animal populations that point to genetic intermixing of native and non-native species. For example, researchers have studied whether Polynesian rats interbred with local South American rat populations in the period prior to Spanish conquest. Early evidence suggests they did. Whereas the "commensal" approach relies on genetic testing, archaeologists have also recently discovered strong evidence of Polynesian–South American contact in Chile in an even more tangible source: the existence of prehistoric Polynesian chicken bones. These bones have been dated to at least the early fifteenth century, if not earlier. The findings from these combined genetic and archaeological studies strongly suggest that Pacific peoples reached American shores prior to 1492. The Americas were just one node in the world's greatest maritime diaspora.[9]

A Sweet Potato Diaspora

Evidence left behind by Polynesians in South America is somewhat limited, yet there is abundant evidence of South American natures in Oceania. Look no further than the humble sweet potato. It was likely no later than the twelfth or thirteenth centuries that the South American sweet potato began to appear on Pacific Islands all across the Polynesian Diaspora. The best explanation for this comes from evidence that Polynesian peoples were regularly moving throughout this human web, connecting home islands with distant islands and even continents. Within centuries a veritable sweet potato diaspora fanned out across the Pacific. This was an Indigenous-people-powered environmental transformation. Pacific Islanders adopted the sweet potato into Indigenous agricultural practices and foodways, growing it alongside such traditional tubers as taro and yams.[10] While Indigenous Pacific Islanders moved rats and chickens, perhaps as far as South America, they also moved South American nature in the other direction, flipping the dominant historical narrative of continental empires exerting powerful influence over disconnected and powerless islands and Islanders. Long before Chile annexed Rapa Nui (Easter Island), the ancestors of the Rapanui may have come to Chile. That Pacific peoples moved their own natures throughout the ocean, but also moved other peoples' natures across the ocean, demonstrates that the early Pacific was a world made and maintained by powerful Indigenous actors.

In the early modern era, the sweet potato diaspora continued. Spanish conquistadores brought potatoes to the Philippines, and from there, thanks to the Manila Galleon trade then dumping Spanish-American silver in Ming China (1368–1644), Spanish-American sweet potatoes got dumped on the Asian continent, rapidly making their way into the interior of China, where they played no small role in an agricultural revolution that displaced Indigenous peoples and contributed to rapid population growth and environmental change. In an early article on Pacific environmental history, John McNeill wrote, in 1994, that there "was no Magellan exchange" comparable to the Atlantic Ocean's Columbian exchange. This is true—Pacific nature did not need Ferdinand Magellan to set it into motion. When he arrived on the scene in the early sixteenth century, many centuries of transoceanic ecological exchanges had already occurred. Although Polynesian transoceanic trading and the exchange of natures certainly quieted down in the centuries leading up to James Cook's famous voyages of the 1760s and 1770s, Pacific nature was still on the move throughout the early modern period. It is easy to think of the Manila Galleons as a story of Spanish empire linked with Chinese empire. But as historians are starting to uncover, the galleons did not just move goods, but also Pacific peoples. Chinese migrant workers, and Indigenous Filipino laborers, came to New Spain (Mexico) as early as the sixteenth century. These laborers were involved in the movement of Pacific natures, a process already set in motion by Pacific Islanders.[11]

A Breadfruit Diaspora

Magellan's circumnavigation certainly expanded the spread and distribution of migrant natures; Captain James Cook's late eighteenth-century voyages provided the next great push. After Cook's death in 1779, Oceania was reanimated by new transoceanic migrations—human and non-human—in which Indigenous peoples again were central to effecting change in both local and global environments. Just as Polynesian peoples moved chickens and potatoes in the pre-Cookian world, in the post-Cookian world they moved sea otter furs, breadfruit, sandalwood, whale oil, and salmon. When Captain William Bligh of the ill-fated *Bounty* came to Tahiti for breadfruit in the late 1780s, Tahitian knowledge and Tahitian labor played a central role in shaping the coming breadfruit diaspora. Even Fletcher Christian and his mutineers were joined by Tahitian women who powerfully shaped the mutineers' desires and aspirations. Tahitian breadfruit eventually made its way to colonial Jamaica to feed African-descended slaves. This was, on the one hand, a tale of British ecological imperialism, moving nature from one periphery to another. But African and Tahitian peoples were also part of the story. When Christian and his fellow mutineers fled to Pitcairn Island, the Tahitians who traveled with them re-created Pitcairn Island as a "neo-Tahiti," planting yams, sweet potatoes, and other crops. This was a continuation of Polynesian-induced ecological exchanges in the Pacific. The colonization of uninhabited islands by Pacific peoples was a story that had begun thousands of years earlier and continued right up until the brink of the nineteenth century.[12]

Breadfruit continued to pop up in strange places. When German warships forced the Samoan leader Mālietoa Laupepa into exile in the late 1880s, he was held for a time on the coast of West Africa in German-occupied Cameroon. His captors fed him "bread, and tea, and rice, and bananas," but Laupepa was most pleased, on a walk near the prison, to discover a breadfruit tree in an English garden. He offered to purchase fruit from the tree's owner, but was met with the reply, "I am not going to sell breadfruit to you people." Instead, the tree's owner allowed Laupepa to take as many fruits as he wished at no cost. As Pacific peoples traveled the world's oceans in the nineteenth century, sometimes voluntarily and sometimes as forced migrants, they not only carried nature with them, but also brought ideas, associations, desires, and nostalgias for natures left behind, including breadfruit. A plantation in Jamaica and a tree in Cameroon became part of the story of the Pacific's many diasporas.[13]

A Salmon Diaspora

In the long nineteenth century, Native Hawaiians were recruited en masse to work on European and Euro-American ships. Perhaps as many as ten thousand Hawaiians left Hawai'i prior to 1876 to work on ships at sea and in foreign lands.[14] Many traveled to the Northwest Coast of North America to hunt sea otters. The furs of these animals were sent to Guangzhou, the great emporium of Qing China (1644–1911), where they were sold at a great profit. Toiling alongside the Columbia River, Hawaiian migrant workers developed a taste for a local food source: salmon. The fish were almost always on the move, hatching in the rivers, feeding in the great ocean, then returning to their riparian homelands to reproduce and die. In the early nineteenth century, the Hudson's Bay Company, which employed hundreds of Hawaiian men, salted and barreled salmon for export to distant places including Honolulu. Two centuries later, Hawaiians still eat *lomilomi* ("massaged") salmon as a popular dish. The reason salmon became part of Hawaiian cuisine is related to this history of Polynesian migrant labor and the way Indigenous migrant workers bridged two worlds, Hawai'i and the salmon-spawning Northwest Coast of North America.[15]

By the late nineteenth century, canned salmon had entered Polynesian foodways both near and far. Native workers consumed it on Hawaiian sugar plantations, linking their bodies with a diasporic ecology of salmon and sugar, just as the cane they were cutting was likely consumed by Hawaiian brothers and sisters living and working in San Francisco and in Sacramento.[16] Meanwhile, in Samoa, itinerant writer Robert Louis Stevenson thought it "curious" that "the common food of one race should be the delicacy of the other." As he sat on the coast of Tutuila eating pig, taro, and *miki* (coconut sauce)—all Samoan foods—his Native comrades lay "upon their sides, eating tinned salmon from home [North America]." But this was not strange. Centuries earlier, Polynesian peoples had incorporated the South American sweet potato into their diasporic foodways. In the nineteenth century, Pacific peoples encountered American shores for a second time and brought home yet another American nature: salmon.[17]

Diasporic Stories and Ideas

The nineteenth-century whaling industry presents yet another theater of Polynesian movements and transoceanic migrations. Hawaiian men worked on American whaling ships in prodigious numbers; as many as three thousand likely worked aboard American vessels at the industry's peak in the 1840s and 1850s. In Hawaiian-language newspapers, these migrant workers wrote letters home describing the things they had seen. Most notably, their stories and songs of distant and foreign environments circulated throughout a workers' diaspora. Whales, ice, snow, gale force winds, and contact with Inupiat peoples in Alaska became part of a Hawaiian national (and transnational) geography as migrant workers' experiences were translated into print and oral media that not only returned to Honolulu from distant points of production but also recirculated upon the ocean's currents—to California, to guano islands, and wherever Hawaiian-language newspapers were sold.[18]

Not only did Hawaiian migrant workers experience sea otters and cattle in Alta California and whales and ice in the Arctic, but they also experienced nature that was on the move. Wherever whales moved, whalemen followed. Hawaiian whale workers came to know the ocean through the act of following prey across the waves, throughout the seasons, and in the momentary chase of the hunt. As guano miners, Hawaiians encountered seabirds on the move: seabirds that flew thousands of miles away to capture energy from the ocean and then bring that energy back in the form of feces—guano—to nesting islands. Pacific nature moved both because people moved it but also because of the agency of whales and birds themselves, as well as ocean waves and wind currents. Pacific workers intimately knew and reported upon these various movements and migrations.[19]

In an age of increasing scientific racism, European and Euro-American employers of Pacific Islander labor contributed to misunderstandings of indigenous movement and mobility. Employers contended that Polynesian men, for example, were only fit for work in tropical and maritime environments and were unsuited for work in the cold or on land. Employers sought to bind Pacific peoples in place: to fix their "nature" in situ as "tropical" and brand them as "amphibious" peoples.[20] Unsurprisingly, Pacific peoples fought back. Hawaiian workers, told that they were unsuitable for work in the Arctic whaling industry, actually proved themselves better than their "Yankee" counterparts, according to one US Navy admiral. Similarly, European-descended peoples in the California Gold Rush wrote to newspapers asserting that Hawaiian migrant gold miners were unfit for the Sierra Nevada environment, but Hawaiian working-class authors responded. They refuted these claims, asserting their own narratives of success (and adventure and perseverance) in the gold fields. By the late nineteenth century, even the sovereign Kingdom of Hawaiʻi was convinced that Native subjects were unfit for labor in a modern, globalizing, capitalist economy. The Hawaiian state sought to save the Indigenous people yet replace them with foreign contract laborers. Once again, Indigenous workers fought back. They struggled against essentialist discourses that regarded their bodies as weak and immobile, and they struggled—through words and sometimes through fists—with rival workers at worksites near and far. They were defending not only their pride and their masculinity, but also an expansive history of Indigenous exploration, discovery, and power in the Pacific World—a narrative of Indigenous peoples on the move, moving nature with them, and facilitating the rearrangement of the world rather than falling victim to the rearrangement dreams of others.[21]

Twenty-First-Century Diasporas

In the late nineteenth century, Pacific Islander circulations and Indigenous diasporic formations—especially those based on extractive, maritime industries such as whaling and the fur trade—were in decline. But these were replaced by new circulations: the coerced migration of Chinese contract workers (the "coolie" trade) to Hawaiʻi, California, Peru, and elsewhere, and the forced migration of Melanesians ("blackbirding") to the cotton plantations of Queensland, Samoa, and beyond. New labor regimes opened up new worlds of Indigenous mobility and encounters with local and foreign natures. Chinese migrants encountered Hawaiian sugarcane and California minerals. Melanesian migrants encountered Australian cotton. Indian "coolie" workers encountered Fijian cotton. Banabans encountered phosphate mining. Pacific peoples continued to move, and Pacific nature moved in turn.[22]

When Epeli Hauʻofa wrote his essay "Our Sea of Islands" nearly thirty years ago, his motivation was to decolonize the way that Pacific peoples saw themselves. Following a long twentieth century of European, Asian, and American empire in the Pacific, Hauʻofa sought to convince Pacific Islanders that they could make it on their own despite the impression that their newly independent countries were like thousands of tiny islands in a far sea.

Decolonization necessitates reimagining Oceania as an interconnected "sea of islands," he wrote. Hau'ofa turned to history—the history of epic trans-Pacific migrations, island settlements, and inter-island exchanges—to show that Pacific peoples have always been on the move and have always been influential agents of historical change. Pacific peoples were global actors in the second millennium, and they will continue to be so in the third millennium.[23]

In this chapter, I have built upon Hau'ofa's idea. Whereas he sought to demonstrate a continuum from migrations past to the movements and circulations of the present, I have similarly sought to show how a thousand-year history of diasporic lives, of people and nature in motion, has led to a diasporic present. Migrations of human and non-human natures continue apace in the twenty-first century and show no signs of letting up.[24] This is revealed by the spread of yet one more Polynesian nature: tuna.

Pago Pago, capital of American Samoa, has one of the Pacific's most prized harbors. In the nineteenth century, one writer described it as the "safest port" in Samoa; another wrote that its "shelter cannot be equaled in the world." Robert Louis Stevenson wrote in the 1890s that the "tongue of water sleeps here in perfect quiet." The harbor's "colour is green like a forest pool," surrounded on all sides by "woody mountains."[25] Today Pago Pago's harbor is ringed on one side by huge metal containers stacked up to fifty feet high along the wharf with Danish and German names splashed upon their sides. On the other side of the harbor stand crumbling tuna canneries guarded by a painted statue of Charlie the Tuna. The US Navy administered American Samoa as a military colony for the first half of the twentieth century. Then, following closure of the US naval installation at Fagatogo in 1951, commercial shipping and industrial fishing rose to become the centerpiece of a new post-military-industrial economy. Van Camp (also known as Chicken of the Sea) established a tuna cannery in Pago Pago in 1954. StarKist followed in 1963. American Samoa became the center of a new global tuna diaspora (see figure 2.2).[26]

Five decades later, Pago Pago in the late 2000s was described as "one of the largest sites of canned tuna production in the world." The StarKist and Chicken of the Sea canneries then employed approximately five thousand workers, representing almost half of the territory's total workforce. Thousands of men and women were migrant workers from the neighboring state of Samoa. Samoan nationals came to Pago Pago to make wages of over three dollars an hour; their remittances flooded home with the smell of tuna-derived wealth. In American Samoa, the cannery workers spent part of their wages each month on housing, food, and entertainment. One study even suggested that tuna workers and their expenditures accounted for 80 percent of the territory's private-sector economy. In this tuna diaspora, millions of

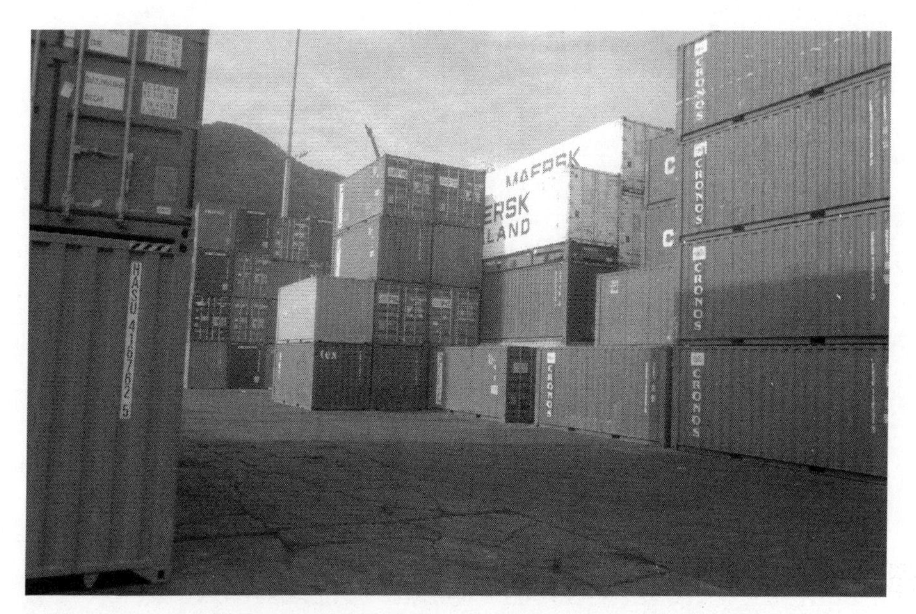

Figure 2.2. Twenty-First-Century Shipping Containers, Pago Pago Harbor, American Samoa, 2015. Photograph by the author (Rosenthal).

fish caught in the waters off of Samoa, Fiji, Niue, and other Pacific Islands were carried to Pago Pago's canneries, from which they ended up in American children's lunchboxes—including my own—in the 1980s and 1990s.[27] On the other hand, while the Samoan group is the ecological homeland of this tuna diaspora, economically, their role and rewards are peripheral to the global industry. Increasing labor costs in the territory, along with the threat of multilateral "free trade" agreements between the United States and tuna-canning countries, such as Ecuador and Thailand, have pressured companies such as Chicken of the Sea to leave American Samoa, as they did in 2009.[28]

Pacific peoples have never been the sole agents moving Pacific nature. Foreigners have brought colonialism and capitalism and set people and nature in motion in their own ways. But the Pacific's many diasporas continue, and Indigenous people are part of this twenty-first-century process. Diners in Brooklyn today can choose between either a basic tuna fish sandwich, made with canned tuna and mayonnaise, or eat Hawaiian-style poke at a new hip restaurant in Williamsburg. Not only does tuna move—from the hands of Pacific Ocean fisherfolk to Samoan cannery workers to California longshoremen to Jewish delicatessen owners in Brooklyn—but so do Polynesian ways of thinking about tuna: Hawaiian poke, Samoan oka, Chicken of the Sea, have your pick. The labor of Polynesian workers is embodied in these goods. American Samoa's congressional representative reminds us—as we chow down on Pacific Ocean tuna—that "the U.S. tuna industry was built on the backs of Samoans and our workers are among the best in the world." Tuna's stories are Polynesian in nature, and yet they are diasporically spread out across the world's many ships, seas, and stomachs.[29]

Epeli Hau'ofa noted in the 1990s that "at seaports and airports" the "delectables" of the world are moved by Polynesian hands. While living in Honolulu in 2015, I kept a P.O. box at my local post office where I regularly witnessed Hawaiians shipping coconuts to the US mainland and Micronesian residents sending coolers jam-packed with goods, sealed with packing tape, to places such as Saipan and Guam. In Samoa, one ship a week and five flights a day shuttle family members and their goods across the International Date Line between Samoa and American Samoa, separated by just seventy miles. Today's customhouses reveal what nineteenth-century customhouses also showed: Polynesia is more than a sea of "many islands"; it is a sea of many diasporas. From canoe plants to sweet potatoes to breadfruit, salmon, and tuna, Pacific history is the history of powerful Indigenous peoples effecting ecological change and environmental transformations both at home and abroad.

Conclusion

In this chapter, I have sought to provide a brief overview of a *longue durée* environmental history of the Pacific, focusing specifically on Indigenous peoples and natures in motion. While environmental historians have adequately catalogued the movements, mobilities, and migratory natures of elite European and Euro-descended peoples over time, there remains a tendency to neglect non-elite and non-white peoples as somehow locked in place, as in situ. These are the people "in nature," fighting back, yet somehow stuck in space and time. In Oceania, Pacific scholars such as Epeli Hau'ofa and Katerina Martina Teaiwa, among others, have railed against this racist discourse, recognizing how such narratives are a perpetuation of a centuries-old "Western" conception of Islands and Islanders as isolated and as passive victims of global economic change and colonization. This view not only deprives Islanders of their agency, but also of their history.

I have shown not only that Pacific peoples moved nature—such as sweet potatoes, breadfruit, and salmon—but that Pacific peoples have also moved through, in, and with natures—such as with whales and seabirds—and how disparate environments both near and far have shaped Pacific Islander movements. While I argue that Indigenous peoples have been at the forefront of ecological exchanges in the Pacific Ocean for one thousand years, I have not shied away from the fact that European explorers and colonizers have also influenced these Indigenous relationships with the natural world. For example, resource extraction and labor migration, propelled by changes in political economy and diplomacy, were key factors in the expansion of Hawai'i's diaspora in the nineteenth century. Although Indigenous Islanders have effected great ecological change across the ocean for over a millennium, some of these changes have been harmful to both humans and nature. Many of the processes examined in this chapter resulted in the decline of plant and animal species, the spread of disease epidemics, and the devaluation of Native lands. These ecological transformations, in turn, occasioned yet more human migrations and the creation of yet new diasporic ecological formations. Capitalism and globalization have increasingly shaped the nature of human and non-human diasporas; yet anti-colonial frameworks such as Hau'ofa's and Teaiwa's also point toward the continued resistance of Pacific peoples to foreign domination.

Today, in the twenty-first century, billions of people bounce around from place to place, living neither here nor there. This may be especially true of marginalized peoples: the poor, the displaced, the dispossessed; Indigenous peoples and migrant workers. The story of the Pacific's many diasporas is a reminder for twenty-first-century humans that migration and economic and ecological globalization are nothing new. Indigenous peoples have been and remain mobile actors in global contexts, notably as workers, and they have exercised an ability to effect transformations in environments near and far. Similar stories continue to unfold around the world, and these tales rightly deserve our attention.

Not all environmental historians will be inspired to unfix that which seems fixed: to see people and nature in motion. But to not do so is to ignore a key element of the human experience as well as a huge swath of humanity who, right now, as I write these words, are on the move, moving between and among natures and moving nature with them. When Epeli Hau'ofa wrote "we are the ocean," he referred not only to a contemporary understanding of Pacific Islander lives, like seeds, spread far and wide across the ocean, but also to an incredible history, spanning one thousand years, of Indigenous relationships with oceanic nature—a world of many diasporas.

Notes

1. Epeli Hau'ofa, "Our Sea of Islands," in *A New Oceania: Rediscovering Our Sea of Islands*, ed. Eric Waddell, Vijay Naidu, and Epeli Hau'ofa (Suva, Fiji: School of Social and Economic Development, the University of the South Pacific, in association with Beake House, 1993), 2–16, quote on 11; Katerina Martina Teaiwa, "Our Sea of Phos-

phate: The Diaspora of Ocean Island," in *Indigenous Diasporas and Dislocations*, ed. Graham Harvey and Charles D. Thomson, Jr. (Burlington, VT: Ashgate, 2005), 169–191, quotes on 177–178, 189; also, Epeli Hau'ofa, *We Are the Ocean* (Honolulu: University of Hawai'i Press, 2008); Katerina Martina Teaiwa, *Consuming Ocean Island: Stories of People and Phosphate from Banaba* (Bloomington: Indiana University Press, 2015). Elsewhere, Hau'ofa writes of "natural landscapes [as] maps of movements, pauses, and more movements"; see Epeli Hau'ofa, "Pasts to Remember," in *Remembrance of Pacific Pasts: An Invitation to Remake History*, ed. Robert Borofsky (Honolulu: University of Hawai'i Press, 2000), 453–471, 466.

2. For a synoptic overview, see Adam McKeown, "Movement," in *Pacific Histories: Ocean, Land, People*, ed. David Armitage and Alison Bashford (New York: Palgrave Macmillan, 2014), 143–165.

3. On the colonial demarcation of Oceania, see J. S. C. Dumont D'Urville, "Sur les îles du Grand Océan" [1832], trans. Isabel Ollivier, Antoine de Biran, and Geoffrey Clark, "On the Islands of the Great Ocean," *Journal of Pacific History* 38, no. 2 (2003), 163–174; Matt K. Matsuda, *Pacific Worlds: A History of Seas, Peoples, and Cultures* (New York: Cambridge University Press, 2012), 3.

4. On disease histories, see Madley in this volume; also, Seth Archer, *Sharks upon the Land: Colonialism, Indigenous Health, and Culture in Hawai'i, 1778–1855* (New York: Cambridge University Press, 2018).

5. Despite its etymological connections with seeds and spores, "diaspora" has yet received scant attention in environmental history, with the notable exception of African diasporic environmental histories. See Judith A. Carney, *Black Rice: The African Origins of Rice Cultivation in the Americas* (Cambridge, MA: Harvard University Press, 2001); David Eltis, Philip Morgan, and David Richardson, "Agency and Diaspora in Atlantic History: Reassessing the African Contribution to Rice Cultivation in the Americas," *American Historical Review* 112, no. 5 (2007): 1329–1358. Also see Dianne D. Glave, "African Diaspora Studies," *Environmental History* 10, no. 4 (October 2005): 692–694.

6. For a unidirectional thesis of ecological change, see Alfred Crosby, *Ecological Imperialism: The Biological Expansion of Europe, 900–1900* (New York: Cambridge University Press, 1986). Recent work in Pacific environmental history has challenged this model, including Jennifer Newell, *Trading Nature: Tahitians, Europeans, and Ecological Exchange* (Honolulu: University of Hawai'i Press, 2010); and Edward D. Melillo, *Strangers on Familiar Soil: Rediscovering the Chile-California Connection* (New Haven, CT: Yale University Press, 2015). That Indigenous peoples are important actors in the history of globalization, see Eric Wolf, *Europe and the People without History* (Berkeley: University of California Press, 1982); James Clifford, *Routes: Travel and Translation in the Late Twentieth Century* (Cambridge, MA: Harvard University Press, 1997); and the many essays in Harvey and Thomson, eds., *Indigenous Diasporas and Dislocations*.

7. On Polynesian migrants and the things they carried, see Patrick V. Kirch, *On the Road of the Winds: An Archaeological History of the Pacific Islands before European Contact* (Berkeley: University of California Press, 2000); Alan Ziegler, *Hawaiian Natural History, Ecology, and Evolution* (Honolulu: University of Hawai'i Press, 2002), chs. 24–25.

8. "Roots" and "routes" is from Clifford, *Routes*. "Portmanteau biota" and "neo-Europes" are from Crosby, *Ecological Imperialism*, although I have replaced "neo-Europes" with "neo-tropics." Earlier scholars referred to the Polynesians' canoe plants and animals as "transported landscapes"; see J. R. McNeill, "Of Rats and Men: A Synoptic Environmental History of the Island Pacific," *Journal of World History* 5, no. 2 (1994): 299–349, esp. 304.

9. This research is collected in Terry L. Jones, Alice A. Storey, Elizabeth A. Matisoo-Smith, and Jose Miguel Ramirez-Aliaga, eds., *Polynesians in America: Pre-Columbian Contacts with the New World* (Lanham, MD: AltaMira, 2011). The most relevant chapters are Alice A. Storey, Andrew C. Clarke, and Elizabeth A. Matisoo-Smith, "Identifying Contact with the Americas: A Commensal-Based Approach," 111–138; and, Alice A. Storey, Daniel Quiroz, and Elizabeth A. Matisoo-Smith, "A Reappraisal of the Evidence for Pre-Columbian Introduction of Chickens to the Americas," 139–170.

10. Ziegler, *Hawaiian Natural History, Ecology, and Evolution*, 318–321, 330; Storey, Clarke, and Matisoo-Smith, "Identifying Contact with the Americas," 125–127. Also see Tim Denham, "Ancient and Historic Dispersals of Sweet Potato in Oceania," *Proceedings of the National Academy of Sciences* 110, no. 6 (2013): 1982–1983; Caroline Roullier, Laure Benoit, Doyle B. McKey, and Vincent Lebot, "Historical Collections Reveal Patterns of Diffusion of Sweet Potato in Oceania Obscured by Modern Plant Movements and Recombination," *Proceedings of the National Academy of Sciences* 110, no. 6 (2013): 2205–2210.

11. McNeill, "Of Rats and Men," 313. On the sweet potato (and other New World crops such as corn and peanuts) in Ming and Qing China, see Peter C. Perdue, *Exhausting the Earth: State and Peasant in Hunan, 1500–1850* (Cambridge, MA: Harvard University Press, 1987); John F. Richards, *The Unending Frontier: An Environmental History of the Early Modern World* (Berkeley: University of California Press, 2003), 112–147; and, William T. Rowe, *China's Last Empire: The Great Qing* (Cambridge, MA: Belknap Press of Harvard University Press, 2009), 91–96. On early transpacific labor migrations, see Matsuda, *Pacific Worlds*, 55–63, 114–126.

12. David Mackay, *In the Wake of Cook: Exploration, Science & Empire, 1780–1801* (New York: St. Martin's Press, 1985); Trevor Lummis, *Pitcairn Island: Life and Death in Eden* (Brookfield, VT: Ashgate, 1997), 53–54; Newell, *Trading Nature*, ch. 5. Also see Richard H. Grove, *Green Imperialism: Colonial Expansion, Tropical Island Edens, and the Origins of Environmentalism, 1600–1860* (New York: Cambridge University Press, 1995), 222–239, 316–328; Alan Bewell, "Traveling Natures," *Nineteenth-Century Contexts* 29, no. 2–3 (2007): 89–110.

13. No. 261, "Letter from Ex-King Malietoa," in Great Britain. House of Commons. *Accounts and Papers. Volume 40. State Papers.* Vol. 86 (London: H.M. Stationery Office, 1889), 221–222; Robert Louis Stevenson, *Vailima Papers and a Footnote to History* (New York: Charles Scribner's Sons, 1925), 188.

14. Gregory Rosenthal, *Beyond Hawai'i: Native Labor in the Pacific World* (Oakland: University of California Press, 2018).

15. On the Northwest Coast fur trade, see James R. Gibson, *Otter Skins, Boston Ships, and China Goods: The Maritime Fur Trade of the Northwest Coast, 1785–1841* (Montreal: McGill-Queen's University Press, 1992); Arrell Morgan Gibson (with John S. Whitehead), *Yankees in Paradise: The Pacific Basin Frontier* (Albuquerque: University of New Mexico Press, 1993), esp. 93–101; Anya Zilberstein, "Objects of Distant Exchange: The Northwest Coast, Early America, and the Global Imagination," *William and Mary Quarterly* 64, no. 3 (2007): 591–620. On trade with Guangzhou, see Paul Van Dyke, *The Canton Trade: Life and Enterprise on the China Coast, 1700–1845* (Hong Kong: Hong Kong University Press, 2005); Rowe, *China's Last Empire*, 122–148. On salmon, see Jean Barman and Bruce Watson, *Leaving Paradise: Indigenous Hawaiians in the Pacific Northwest, 1787–1898* (Honolulu: University of Hawai'i Press, 2006), 47, 99; Lissa K. Wadewitz, *The Nature of Borders: Salmon, Boundaries, and Bandits on the Salish Sea* (Seattle: University of Washington Press, 2012), 36–38.

16. G. E. Beckwith to S. Savidge, September 26, 1864, "Haiku, Maui 1864 July-Sept," in Castle & Cooke Business Papers, 1850–1915, Hawaiian Mission Children's Society Library, Honolulu, Hawai'i. On Hawaiian sugarcane entering the port of San Francisco, see Louis J. Rasmussen, *San Francisco Ship Passenger Lists: Volume I* (Colma, CA: San Francisco Historic Records, 1965), 52, 96; Louis J. Rasmussen, *San Francisco Ship Passenger Lists: Volume II* (Colma, CA: San Francisco Historic Records, 1966), 6, 9, 12.

17. Stevenson, *Vailima Papers*, 64.

18. On Arctic whaling, see Susan A. Lebo, "Native Hawaiian Seamen's Accounts of the 1876 Arctic Whaling Disaster and the 1877 Massacre of Alaskan Natives from Cape Prince of Wales," *Hawaiian Journal of History* 40 (2006): 99–129; John R. Bockstoce, *Furs and Frontiers in the Far North: The Contest among Native and Foreign Nations for the Bering Strait Fur Trade* (New Haven, CT: Yale University Press, 2009), esp. 276–278, 312–313, 327; Bathsheba Demuth, *Floating Coast: An Environmental History of the Bering Strait* (New York: W. W. Norton & Co., 2019). On Native-language diasporic print cultures, see Pier M. Larson, *Ocean of Letters: Language and Creolization in an Indian Ocean Diaspora* (New York: Cambridge University Press, 2009); M. Puakea Nogelmeier, *Mai Pa'a i Ka Leo: Historical Voice in Hawaiian Primary Materials, Looking Forward and Listening Back* (Honolulu: Bishop Museum Press/Awaiaulu, 2010).

19. Gregory Rosenthal, "Life and Labor in a Seabird Colony: Hawaiian Guano Workers, 1857–1870," *Environmental History* 17, no. 4 (2012): 744–782.

20. On the racialization of Pacific Islander bodies, see Paul D'Arcy, *The People of the Sea: Environment, Identity, and History in Oceania* (Honolulu: University of Hawai'i Press, 2006), 30–33; Bronwen Douglas and Chris Ballard, eds., *Foreign Bodies: Oceania and the Science of Race, 1750–1940* (Canberra: ANU E-Press, 2008); Gary Y. Okihiro, *Pineapple Culture: A History of the Tropical and Temperate Zones* (Berkeley: University of California Press, 2009), 1–14.

21. "He mau wahi olelo no ka holo ana o na kanaka maoli i Kaliponia [A few words on the going of Hawaiians to California]," *Ka Hae Hawaii*, April 17, 1861; T. B. Kamipele, "No Kalifonia Mai [From California]," *Ka Hae Hawaii*, July 3, 1861; Henry Erben, Jr., "Remarks of Comm. Erben, upon passage from San Francisco to Honolulu, in the U.S.S. Tuscarora," [Nov. 1874], and "Report concerning Honolulu" from a midshipman to Commodore Henry Erben, Jr., January 8, 1875, Box 1: Correspondence & Papers, 1848–1875, Henry Erben, Jr. Papers, New York Historical Society, New York. On the discursive dispossession of Indigenous peoples, see Patrick Brantlinger, *Dark Vanishings: Discourse on the Extinction of Primitive Races, 1800–1930* (Ithaca, NY: Cornell University Press, 2003). On masculinity and memory in Hawai'i, see Ty P. Kāwika Tengan, *Native Men Remade: Gender and Nation in Contemporary Hawai'i* (Durham, NC: Duke University Press, 2008). On exploration, see David A. Chang, *The World and All the Things Upon It: Native Hawaiian Geographies of Exploration* (Minneapolis: University of Minnesota Press, 2016). On the rabble-rousing role of Pacific workers in the nineteenth century, see Gregory Rosenthal, "Workers of the World's Oceans: A Bottom-Up Environmental History of the Pacific," *Resilience: A Journal of the Environmental Humanities* 3, no. 1/2/3 (2016): 290–310.

22. H. E. Maude, *Slavers in Paradise: The Peruvian Labour Trade in Polynesia, 1862–1864* (Stanford, CA: Stanford University Press, 1981); Adam McKeown, *Chinese Migrant Networks and Cultural Change: Peru, Chicago, Hawaii, 1900–1936* (Chicago: University of Chicago Press, 2001); Gerald Horne, *The White Pacific: U.S. Imperialism and Black Slavery in the South Seas* (Honolulu: University of Hawai'i Press, 2007); Philip A. Kuhn, *Chinese among Others: Emigration in Modern Times* (Lanham, MD: Rowman & Littlefield, 2008),

ch. 5; Nicholas Thomas, *Islanders: The Pacific in the Age of Empire* (New Haven, CT: Yale University Press, 2010), esp. 196–199, 213.

23. On decolonizing methodologies and frameworks, see Haunani-Kay Trask, *From a Native Daughter: Colonialism and Sovereignty in Hawai'i* (rev. ed.; Honolulu: University of Hawai'i Press, 1999); Linda Tuhiwai Smith, *Decolonizing Methodologies: Research and Indigenous Peoples* (Dunedin, NZ: University of Otago Press, 1999); Noelani Goodyear-Ka'ōpua, Ikaika Hussey, and Erin Kahunawaika'ala Wright, eds. *A Nation Rising: Hawaiian Movements for Life, Land, and Sovereignty* (Durham, NC: Duke University Press, 2014).

24. On contemporary migrations and diasporas, see Grant McCall and John Connell, eds., *A World Perspective on Pacific Islander Migration: Australia, New Zealand, and the USA* (Kensington, Australia: Center for South Pacific Studies, University of New South Wales, 1993); Paul Spickard, Joanne L. Rondilla, and Debbie Hippolite Wright, eds., *Pacific Diaspora: Island Peoples in the United States and across the Pacific* (Honolulu: University of Hawai'i Press, 2002). For reflections on Hawaiian diaspora, specifically, see Rona Tamiko Halualani, *In the Name of Hawaiians: Native Identities and Cultural Politics* (Minneapolis: University of Minnesota Press, 2002).

25. Beachcomber, "Unfortunate Apia," *The Samoa Times and South Sea Advertiser*, April 13, 1889; Stevenson, *Vailima Papers*, 47; also see Hervey W. Whitaker, "Samoa: The Isles of the Navigators," *Century* 38, no. 1 (1889): 12–25, esp. 14–15.

26. Rachel A. Schurman, "Tuna Dreams: Resource Nationalism and the Pacific Islands' Tuna Industry," *Development and Change* 29, no. 1 (1998): 107–136, esp. 120; Liam Campling and Elizabeth Havice, "Industrial Development in an Island Economy: US Trade Policy and Canned Tuna Production in American Samoa," *Island Studies Journal* 2, no. 2 (2007): 209–228, esp. 213.

27. Schurman, "Tuna Dreams," 108–109; Campling and Havice, "Industrial Development," 211, 213–215.

28. Campling and Havice, "Industrial Development," 219–223; David B. Cohen, "Ravaged Samoa," *Forbes,* October 1, 2009.

29. "U.S. Tuna Industry 'Built on the Backs of Samoans,'" *Samoa Observer,* April 17, 2013; Ligaya Mishan, "Family Recipes from a Hawaiian Kitchen," *New York Times*, December 19, 2013.

THREE

Chinese Resource Frontiers, Environmental Change, and Entrepreneurship in the South Pacific, 1790s–1920s

James Beattie

Over the nineteenth century, several million Chinese exploited resource frontiers opening up around the Pacific. European imperialism and American westward expansion helped generate these new extractive zones. As goldminers, plantation laborers, and railroad navvies, as market gardeners, merchants, and in myriad other occupations, several million Chinese people—coupled with China's demand for raw materials—contributed to the great acceleration of environmental change in the seas and countries of the nineteenth-century Pacific.[1]

In exploring the environmental dimensions of what Henry Yu has coined the "Cantonese Pacific," I first survey how China's market demand created Pacific migrant ecologies.[2] Next, I explore the nature of explosive, boom-and-bust Pacific resource frontiers, notably goldmining. Goldmining, I show, stimulated Chinese migration by opening new resource frontiers, encouraging capitalist accumulation and new colonial investment, and, longer term, linking Pacific places and resources. I then examine Chinese entrepreneurship in the British Empire by focusing on a Chinese merchant—Chew Chong (c. 1830–1920)—whose enterprises opened up new local-level migrant ecologies between New Zealand, Australia, and China.

The chapter frames Pacific resource demand, and later Chinese enterprises like market gardening and agricultural investment, as examples of ecocultural networks—interlinked labor flows, migrant connections, and capital systems that transformed Pacific environments and made nature into commodities. Unlike other models, an ecocultural framework attempts to integrate cultural and material motivations and impacts related to environmental exchange and transformation. It focuses on the different components (such as technology and capital infrastructure), ecologies, and peoples brought together to form a network to commodify nature. It demonstrates the feedback loops of specific environmental exploitation on social systems, technologies, and ecologies, while also recognizing the sometimes-short-term duration of ecocultural networks. In a period dominated by the development of nationalism and the nation-state, an ecocultural approach can highlight the sub-national-level connections fashioned by non-state actors and the linkages between different locales. In the period under discussion, an ecocultural approach underlines the linkages of new energy regimes derived from fossil fuels with the resources of the so-called biological old world of the Pacific in which sun-powered photosynthesis released a series of energy transfers from plant to animal.[3] Examining the role of Chinese merchants and Chinese markers in developing Pacific ecocultural networks complicates the dominant narrative of Europeans as being the main drivers of Pacific resource exploitation.[4]

Creating "China's Pacific": Trade and Environmental Change, 1790s–1860s

As Ryan Tucker Jones has observed, from the eighteenth century to the 1910s, China's demand and Europe's desire to meet it transformed the Pacific "into the world's larder," by expanding China's environmental footprint from Southeast Asia into the wider Pacific.[5] The epicenter of China's Pacific was Canton, a bustling, polyglot trading port that engulfed American silver, English and Indian textiles, Pacific sealskins, furs, and sandalwood, and many other commodities besides.[6]

The Pacific dimension of the Canton trade challenges China's reputation as one of the world's great land empires. For the peoples of southern coastal China—today's provinces of Fujian and Guangdong—their frontiers were not the lands and waters, mountains and steppes of inland China, as it was for most other Chinese, but the bodies of oceans and islands, their peoples and resources, lying immediately to their south and accessible via Nan Hai (South Sea) and Nanyang (the Southern Ocean). Although China's Nanyang trade and the migration of its people to those countries began many centuries earlier, it particularly flourished in the eighteenth century.[7] In this period, Chinese junks were plying Nanyang waters, selling manufactured goods, and returning with raw materials and food. The placement of Chinese traders at the center of this flourishing exchange complicates the traditional picture of Yankee or European traders dominating China's foreign commerce; in reality, "the largest number of merchants to take [to] the seas [from the late seventeenth century] . . . were Chinese."[8]

In the eighteenth century, the processing, shipment, and labor requirements of China's monocultures linked different parts of China as well as Chinese regions with Nanyang countries.[9] Ecological simplification and the shift away from food crops to cash crops contributed to China's nineteenth-century socio-ecological crisis: interconnected problems of land degradation, population, government instability, and infrastructural collapse.[10]

By the eighteenth century, European powers were becoming more active in Asia. The English East India Company (EIC), after muscling out rivals, had established permanent trading bases, then settlements, in India (Bengal and Madras). Like Nanyang residents, in this period Europeans and North Americans were importing Chinese manufactured goods in large volumes—tea, porcelain, silks, and other luxury items. They largely paid for them with New World silver, which reached China through the Canton trading system (1757–1842). Carefully controlled by Chinese authorities, the Canton system intentionally restricted foreign trade to one port alone, to be conducted with a limited number of Chinese merchants only at certain times of the year.[11]

The limitations of this system, coupled with a ballooning balance-of-payments deficit, spurred European and American traders from the late eighteenth century to look to Pacific resources to send to Chinese markets. Pacific resource exploitation set up often transitory ecocultural networks lasting only so long as a resource did—and that was usually relatively short, given the rapacious demands of the period, and the smallness of islands and their limited resources.[12] Trade rapidly depleted resources, leading to a range of ecological and related social effects. Islanders commonly gained new technologies, as well as economic advantages and travel opportunities. Invariably, they experienced high mortality rates through exposure to new diseases, and a host of related social and economic effects.[13] Trading encounters also often laid the basis for an island's later colonization. Sealing, sea cucumber collecting, and sandalwood cutting illustrate the interrelated effects on people and nature of China's Pacific resource demand.

The hunting to near extinction of the southern fur seal (*Arctocephalus forsteri*) hinged on Britain's establishment in Australia in 1788 of the Botany Bay Penal Colony. This gave American and British traders a base in the Southern Ocean from which to exploit seals and other marine resources. Within a short time, a pattern for the exploitation of seals had emerged. Vessels would drop off sealers in remote locations—often for months at a time—to catch and prepare skins. There they would engage in an orgy of killing. Contemporaries describe rocks stained red with the blood of seals, and seals—unaccustomed to human contact—seemingly allowing themselves to be clubbed to death. In 1802, French Commodore Baudin aptly described hunters' activities in Bass Strait (off Australia's southeastern coast) as "a destructive war carried on against" seals. "You will soon hear that they have entirely disappeared," he warned.[14]

Baudin's warnings went unheeded. By 1802, sealers had destroyed Bass Strait rookeries. Sydney-based sealers then turned their attention to New Zealand and, soon afterward, to the small islands lying to its south, west, and east, as they gradually exhausted one rookery after another. The large number of skins collected explains the rapid decline of southern Pacific seal populations. In the first decades of the 1800s, a Sydney newspaper recorded, "it was not uncommon for a vessel to obtain in a short trip from 80,000 to 100,000 skins."[15] Sealing gangs arrived on the Antipodes Islands in 1804. By 1809 crews had skinned as many as a quarter of a million seals (this figure counted only seals skinned, not killed).[16] Sealers operating in New Zealand and its outlying islands, Rhys Richards estimates, killed some seven million fur seals before 1833. Slaughter peaked twice, in 1804–1809/10, and 1823–1826/7. Most sealskins—air-dried—went directly to London or Canton, as well as going from London to Canton. (The figures also ignore "illegal" [i.e., non-EIC] trade, pre-1833, the year it lost its monopoly on eastern trade.[17]) In China, "frontier products like fur became markers of elite Chinese fashion," and immensely popular.[18] Fur was "high value, low weight, easy to carry, and durable and so remarkably well suited for long-distance trade."[19] In Canton, furriers made garments from seals' thick fur, also using the guard hairs removed in this process to make cheap felt hats. Oversupply meant the bottom fell out of the Canton market in the early 1830s, with many traders subsequently turning to the London market, even though the China market later picked up.[20]

In New Zealand, both Māori and foreign crews participated in killing seals and preparing skins.[21] The lengthy onshore stays of sealers—and their occasional marooning—commonly fostered relationships with local peoples. (Indeed, my daughters' Māori ancestry dates to this period of contact in 1810s southern New Zealand.) Contact through sealing brought new iron tools, animals (e.g., pigs), and food crops (especially the potato), which together revolutionized diet and contributed to changing settlement patterns among southern New Zealand Māori. It also brought unwanted migrant ecologies, such as disease, as well as paving the way for later colonization.[22]

Sea cucumbers and sandalwood, often exported together, experienced similar pulses of boom and bust to sealing. Sea cucumber—or *bêche-de-mer* (*Holothuroidea* spp.)—is a marine animal eaten and used in Chinese medicine.[23] Prior to European interest in it, Chinese traders had obtained the dried delicacy from Indonesia, India, and the Philippines. Macassan traders also visited northern Australia during the monsoon, collecting and curing *bêche-de-mer*—a trade of considerable size. In 1803, for example, Matthew Flinders estimated that the sixty Macassan vessels he encountered were carrying about six million cured sea cucumbers.[24] Europeans and New Englanders began to enter the lucrative sea cucumber trade in the late eighteenth century, often also peddling other commodities like shells and sandalwood.[25]

The sea cucumber trade varied by timing and extent, but it encompassed much of the Pacific at one stage or another, as traders exhausted one source after another. Fijian trade boomed from 1828 to 1835, averaging thirty-five to seventy tons per annum, and enjoyed another spurt from 1842 to 1850.[26] In Noumea, trans-shipments to China via Sydney peaked from 1864 to 1873.[27] Relying on trans-shipment from Queensland (Australia), New Guinea's trade crested between 1873 and 1885 but diminished thereafter.[28]

Aside from its effects on Pacific sea cucumber populations, the Chinese trade drew in considerable local labor supplies and consumed a large volume of timber.[29] The involvement of Pacific peoples in this trade was paramount, since its preparation required long onshore time and considerable labor resources. *Bêche-de-mer* traders constructed large drying houses—American-built ones commonly measured 100–120 feet in length and 20 feet in width. Larger operations employed as many as 200 people to source sea cucumber, 100 to cut firewood, and perhaps 50 to keep fires burning.[30] Because its drying required constantly lit fires, timber consumption rapidly depleted scarce island forests. For example, *bêche-de-mer*–induced deforestation on western Viti Levu (Fiji) further disrupted "an ecosystem already in a delicate state of balance [and] made conditions for tree regeneration difficult."[31] In Fiji alone, R. G. Ward estimates the trade consumed one million cubic feet of timber.[32]

Traders commonly exported sandalwood (*Santalum* spp.) alongside sea cucumber. Chinese carpenters made furniture and chests from sandalwood, while its aromatic oils found use in perfumes, medicines, and incense. With China's Indian supplies dropping in the nineteenth century, traders had turned to Pacific resources to meet Chinese demand. Traders and local peoples systematically worked out this resource on a number of islands: moving from Fiji (1804–1816) and then Marquesas (1814–1820) to "Hawaii (1811–31), where an efficient royal monopoly expedited depletion, and lastly to Melanesia, especially the New Hebrides (1841–65)."[33]

Sandalwood-getters commonly cut and killed young trees, effectively consigning their enterprise to the short term.[34] Like *bêche-de-mer*, sandalwood extraction required a large labor force, its control enriching some Native leaders. For example, King Kamehameha I's (c. 1736–1819) control of Hawai'i's sandalwood supply enabled him to consolidate power and contributed to complex social, political, and demographic changes occurring within Hawai'ian society. Labor redistribution from local agricultural production to sandalwood extraction and food production for trade contributed to famine in 1811–1812, while diseases introduced by traders appear to have led to considerable depopulation.[35]

Colonization, Goldmining, and Migration, 1840s–1870s

While China's resource demand in the Pacific from the latter eighteenth century has attracted scholarly attention, much less well examined is the role of Chinese as agents of environmental change, and it is on this topic the rest of the chapter focuses.

The unequal treaties imposed by Britain on China in the wake of the Opium Wars (1839–1842; 1856–1860) unintentionally provided the mechanism for large-scale Chinese migration.[36] The Opium Wars not only swept aside the old trading order by opening China to foreign interests, but the establishment of British Hong Kong (ceded in 1842) connected Chinese migrants and capital with international shipping networks and newly opening commodity frontiers around the world. Originally envisaged as an entry point into the lucrative China market and Chinese coastal trade, Hong Kong's future was changed by California's gold rush (1848). This signal event transformed Britain's new possession "into Asia's leading Pacific gateway . . . linking North America and Asia."[37]

By 1939—nearly 100 years after its establishment—an astonishing 6.3 million Chinese had embarked from Hong Kong, with 7.7 million returning via the island.[38] Hong Kong served as a gateway to the Pacific and elsewhere, including Britain's new possession of Singapore. Chinese merchants and then laborers established new migrant ecologies in the expanding European colonial spheres. In 1840s Singapore, for example, Chinese shifting cultivators grew pepper and gambier for commercial production, using horticultural practices that maximized productivity but left land exhausted and useless within a few years. The extent of this destruction, likely coupled with racist sentiments, moved one European observer to liken the Chinese planter to a locust, in leaving behind a trail of desolation. Land exhaustion in Singapore encouraged Chinese to develop plantations in Peninsular Malaya.[39]

After 1848, commodities, people, and ideas flowed from Hong Kong to California, and then to a series of other Pacific rushes (e.g., Victoria, Australia; Otago, New Zealand; Yukon, Canada). Chinese institutions, families, and businesses made use of these thickening networks of trade, transportation, and communication wrapping around the Pacific to move people, goods, and money to and from South China.[40] Chinese merchants controlled Pacific labor migration, from recruitment and transportation to accommodation and working conditions. Their multinational firms linked kin in different parts of the Pacific, and often also controlled a good's production and marketing. Although Chinese were commonly sojourners, permanent settlements developed around the Pacific, notwithstanding anti-Chinese legislation later in that century.[41]

Chinese migrated and worked claims in groups, bound together by common threads of kinship, place, and language. Most Cantonese arrived uninvited in the goldfields (only Otago Province invited Chinese miners, guaranteeing them equality before the law and providing Chinese-speaking constables). Even if, as in Otago, they might have initially received a reasonably warm welcome, most Chinese experienced hostility, and sometimes even violence, at one stage or another.[42] Chinese initially took up alluvial goldmining. It appealed because it was relatively low tech, highly labor intensive, and cheap, requiring in effect only picks and shovels, pans and cradles. Although some Chinese later participated in hydraulic sluicing and quartz mining, these activities required more capital than alluvial mining, and so were not as common.[43]

In following successive Pacific rushes, Chinese goldmining polluted waterways, washed away soil, removed vegetation, and irrevocably changed landscapes.[44] In this respect, they transformed nature into commodities and caused environmental problems, just like Europeans and other groups did.[45] For example, on Otago's Round Hill goldfield, Chinese mining contributed to polluting and sludging up ninety-one-hectare Lake George/Uruwera and Whakapatu Bay.[46] While in this respect Chinese impacts mirrored those of other nationalities, in some cases Chinese technology and networks, and hence their environmental impacts, differed from Europeans'. For example, the environmental footprint of Chinese miners' demand for goods and food—rice and other foods, opium, and traditional medicines—reached China.[47] Later, Chinese also developed specialized ecocultural networks reliant upon connections with China (discussed later in this chapter).

With respect to water management in Pacific environments, Chinese miners utilized skills honed over generations of living in the Pearl River Delta to build sophisticated water races on goldfields, skills recognized by private companies and colonial governments, which employed Chinese teams to build races and sludge channels. Chinese also adapted technology or developed their own ingenious solutions to particular mining problems. For example, Cantonese adapted the traditional waterwheel from their homeland, using it to raise or lower water levels in mining areas, and to bring water uphill, something copied by Europeans.[48] Some utilized other techniques different from Europeans. Victorian miners obtained very small grains of gold by using much longer sluicing boxes than those used by Europeans. It is difficult to ascertain with certainty the differential impacts of Chinese technology, especially given the prevalence of technology transfer between regions, countries, and different groups, and the parallel development of technology, which might also have multiple introductions into a region. A case in point is the technological adaptation of a mobile dredge by Choie Sew Hoy (c. 1836–1901). Utilizing Chinese and Euro-American engineering, his design sparked Otago's major 1890s dredging boom. Subsequently adapted, the design was used around the world.[49]

Post-Mining Industries, 1870s Onward

When mining ended in one area, Chinese moved on to other newly opening Pacific goldfields, returned home (if they had made enough money), or moved into other occupations. As an example of the last, Chinese market gardeners supplied goldfields and later other settlements with fresh produce. Chinese truck laborers transformed Californian agriculture, leasing land and providing settlements with fresh vegetables, until the impacts of Chinese exclusion acts limited their activities.[50] Despite similar racially exclusionary policies in Australasia, Chinese market gardeners dominated this industry as the main suppliers of fresh produce well into the twentieth century.[51] Chinese also succeeded as fruit growers, introducing a variety of new edible and ornamental species from South China into Australasia.[52] In horticulture, Chinese utilized the highly intensive manuring and watering methods of their homeland, which meant that, as a European writer observed of Queensland Chinese, "as gardeners they are preeminent."[53] Chinese market-gardening success provoked complaints (and sometimes attacks) from European rivals, but this did not seem to diminish the popularity or importance of Chinese-grown foods in colonial Australasia.

As plantation workers in northern Queensland, Chinese helped to transform environments and simplify ecologies through deforestation and the introduction of monocultures. Chinese merchants pioneered Queensland's banana trade by controlling investment, labor, production, sales, and marketing, just as on Hawai'i they had developed its early sugar industry.[54] Chinese workers labored on major railway projects and in Pacific phosphate mines (see figure 3.1). By cutting, blasting, and digging their way through and over the Sierra Nevada Mountains, Chinese laborers played a major role in the construction of the Transcontinental Railroad, which linked markets and centers of production and enabled goods, capital, and people to flow along a west-east axis.[55] As phosphate miners, they extracted minerals that fertilized the late nineteenth-century agricultural booms of Australasia and Britain.[56] Here, as plantation workers, and throughout the Pacific, they endured often appalling conditions.[57]

Figure 3.1. Central Pacific RR—Snow Sheds P.2, Utah State Historical Society Classified Photo Collection, Image No. 00560. Used by permission, Utah State Historical Society.

Traditional cultural ideas and belief systems offered Chinese interpretive frameworks through which to view overseas environments. For example, some Cantonese interpreted overseas landscapes in light of traditional Chinese geomancy, identifying places efficacious for settlement or modifying their dwellings' situations accordingly. Likewise, cosmological geographies, mediated by religious rituals, as well as by visitations of ghosts, friends, and ancestors, enabled Chinese to both transcend the geographical divide between their homeland and place of residence outside China, and bridge the world of the living and the dead. Materially, self-help organizations oversaw transfers to China of the remains of those who died overseas.[58]

Whether indentured laborers or merchants, Cantonese yoked their extensive migrant networks onto colonial systems. In places like New Zealand, as I show, some naturalized citizens utilized colonial financial and legal apparatuses to develop land and other resources, and to set up ecocultural networks linking resources, people, and markets in one part of the world with those elsewhere.

The Fungus King: Chinese Entrepreneurship and Environmental Change in Taranaki, New Zealand, 1880s–1910s

> To the first settlers who faced the wilderness with determination and hope as their only capital the little Chinese pedlar . . . came as a general benefactor.
>
> —"*The Chinese Benefactor,*" Hawera & Normanby Star, *July 5, 1923*

Chew Chong's story illustrates how a Chinese person with initiative could succeed in a British colony. Chew Chong directed profits from selling New Zealand fungus in China to pioneer a regional North Island dairy industry. Born in Canton around 1828, Chew Chong left China for household service in Singapore in the 1830s, before trying his luck as a goldminer and merchant in Victoria, Australia, from 1856.[59] He arrived in Otago, New Zealand, in 1868, then traveled to the North Island, settling in Taranaki in 1870.

Chong identified a potential commodity growing abundantly on rotting logs in Taranaki's undeveloped forests: edible tree fungus, Wood Ear (*Auricularia polytricha*), prized in Chinese cuisine and medicine. Realizing its potential value, Chew Chong secured a market for this product, leaving instructions (and money) with a local storekeeper to purchase as much fungus as possible.[60] Over the next decades, fungus exports generated significant revenue for Chew Chong and others involved in the industry; from 1880 to 1920, New Zealand fungus exports totaled £401,551.

Chew Chong established an ecocultural network marrying colonial labor and resources with Chinese investment, business connections, and markets. He collected fungus from poor white bush settlers and Māori. Once dried, he shipped it to Dunedin for export to China via Sydney. This operation took place in conjunction with other Chinese merchants. In New Zealand, this included Dunedin-based Choie Sew Hoy (1836–1901), Auckland-based Chan Dah Chee (c. 1850–1931), as well as possibly a handful of Europeans. It also tapped into Sydney merchant and shipping connections with China, and relied upon Canton merchants, who took over its sale and distribution in China.

Chong's enterprise stimulated Taranaki's stagnant economy, moving it from a barter system to a cash economy. Regional investment and agricultural development lay in the doldrums after decades of intermittent warfare between settlers and some Māori tribes. Most settlers simply lacked the capital to convert forest to farmland. Until Chong's arrival in 1870 and his payment in cash, a newspaper explained, many had struggled even to pay their annual rates, and "but for the fungus . . . would have had ruin staring them in the face."[61] In particular, Māori, desperately poor following war and land confiscation, benefited from the money earned through its collection. Increases in payments for dried fungus chart the industry's growth. Collectors initially received payments of "only a halfpenny per lb." This rose in the 1890s to "as high as threepence and fourpence per lb."[62] Chew Chong's model stimulated other North Island bush settlements to start collecting fungus.

The fungus grew abundantly on mahoe (*Melicytus ramiflorus*), pukatea (*Laurelia novae-zelandiae*), and tawa (*Beilschmiedia tawa*), and required careful preparation. Collectors had to pick it in dry weather (difficult in the region's damp climate) and avoid destroying its roots, otherwise the organism dies. Next, they dried it. A newspaper report recommended placing fungus on iron sheets, in the sun. It noted the need to keep the fungus "clean and free from sand, dirt, or grit—a very necessary precaution as the fungus is used chiefly in the concoction of soups for human consumption." Unlike other commodities sent to China such as *bêche-de-mer*, which required drying over a fire, any artificial heating diminished the fungus's value by corrupting its taste. And even when "thoroughly dried and bagged," care was still required to ensure it remained moisture free.[63]

Chew Chong—styled "the Fungus King of New Zealand" by one colonial newspaper—pioneered a largely sustainable industry, provided overharvesting did not occur and the forest remained.[64] Profits from fungus collecting enabled Chew Chong to open three stores in Taranaki.[65] Each bought dried fungus and cocksfoot seed (sown in the soils of newly fired forest) locally and on-sold them. Chew Chong's stores sold a variety of goods to settlers, including imported Chinese silks, fancy goods, camphor boxes, preserved ginger, and even Chinese export oil paintings.[66]

From Forest to Farmland: "Chew Chong Establishes the Factory System"

In the late 1880s, Chew Chong helped to transform and industrialize South Taranaki's landscape by pioneering the refrigerated butter industry, which he financed with profits made from fungus exports.[67] The new industry effectively destroyed the earlier because dairying accelerated the conversion of forests to farmland (see figure 3.2).

In contrast to an economically buoyant North Taranaki in the 1880s, South Taranaki was languishing because of environmental barriers posed by the area's wetlands and forests, coupled with difficulties of transportation and the lack of a well-paying industry to stimulate development. Amid

Figure 3.2. An illustration of the impacts of converting forest to dairy farms. Unknown photographer, Dairy Farm on Auroa Road, South Taranaki (1890), collection of Puke Ariki, New Plymouth, PHO2008-267. Reproduced by permission of Puke Ariki, New Plymouth.

South Taranaki's forested landscape, settlers eked out family farms in small clearings that at best provided a living, not a livelihood. Most settlers kept eight to ten cows, supplementing income with fungus collecting and grass seed growing (cocksfoot).[68]

Chew Chong's experience of buying poor quality milk for butter-making spurred him to drive for improvements to the whole industry through the introduction of dairy factory production to South Taranaki. In concert with developing transportation networks and growing global demand, Chew Chong introduced technological innovations enabling him to produce high-quality butter. In 1887, he opened three butter factories, as well as creameries.[69] Chew Chong's Jubilee Butter Factory cost £3,700 to build and equip, boasting the latest imported technology, as well as his own ingenious inventions.[70] Two Danish cream separators enabled processing of 150 gallons of milk an hour.[71] A new type of butter churn and air cooler, both of which Chew Chong patented, further improved production.[72] Chew Chong built the air cooler himself by dynamiting a tunnel to channel water over one hundred meters underground. This drove an eight-horsepower waterwheel, which powered the machinery and cooled the butter.[73] A few years later, he improved processes further, by installing a steam engine and

Hall's refrigerating machine, "probably the first freezing machine in a New Zealand butter factory."[74] Chew Chong's efforts illustrate a pattern writ large across much of the nineteenth-century world: the application of fossil fuel technology to drive for improvements in efficiency, transportation, and, ultimately, the exploitation of nature.[75]

As well as investing in new technology, Chew Chong took full control of milk production, sales, and marketing. He introduced the key innovation—now characteristic of all modern dairy farming—of share milking to ensure a reliable supply of milk, as well as owning 200 stock. Chew Chong also custom designed the packaging of his new product, in easily stackable one-pound parchment-wrapped blocks. He named it "Jubilee" Butter for Victoria's Jubilee, a clever marketing ploy to British customers wanting to buy empire-made products.

Chew Chong was able to develop Taranaki's dairy industry only because refrigerated shipping enabled milk products to reach British consumers at a time when Danish butter production was tailing off. (In time, New Zealand butter usurped Danish butter's role as the primary supplier for the UK market.)[76] Although difficulties of transportation remained in Taranaki, Chew Chong also benefited from recent government investment in port infrastructure and railways, developments he cannily made use of—five minutes' walk took one from his Jubilee Factory to Eltham railway station.[77] Eltham's line linked producers with international and national markets via the ports of either Patea or New Plymouth. From 1885, businesses could rail products directly to Wellington and Auckland consumers and from their ports to overseas consumers.

By shaping the region's natural endowments through the application of technology, the butter produced from Chew Chong's factories was far superior to the salty, poor quality product made by settlers on bush farms. At the 1889 Dunedin Exhibition, his butter "gained two certificates and a silver cup for the best half-ton available for export."[78] Government inspectors and experts praised his factories and their products.[79] The quality of his products earned him more money than milled butter. Chew Chong's first shipment realized "a hundredweight more than milled butter in the same shipment. Passed onto the farmer this meant more income and [the] possibility of better conditions."[80] Once his factories became established, farmers' incomes increased. Chew Chong paid twopence a gallon to suppliers in 1888, threepence the next year. "In short," a correspondent wrote in 1894, "the factory system realized the district, bettered the condition of the people, and placed them in a position they would never otherwise have achieved in so short a space [of time]."[81]

"Islands of Log Littered Grasslands": Environmental Change and the Dairy Industry

Dairy factories and creameries owned by Chew Chong acted as bridgeheads of social, economic, and environmental change in South Taranaki.[82] Dairying moved deforestation inland, drawing it "from the coastal belt in the west, Hawera in the south and from the mountain road (main road from Hawera to Stratford) and railway in the East."[83] It also provided the impetus and capital for settlers to take up most of South Taranaki's bush districts, which they did from 1893 to 1920, a frontier charted by burnt forests, introduced pastureland, and rising European populations and livestock.[84] Demand created by dairying increased land prices six-fold to 1900, stimulating improvements to transportation and further reinvestment in the industry.[85]

Chew Chong's enterprise contributed to deforestation in two main ways. First, it stimulated conversion of forests into grassland and, second, it provided demand for Taranaki timber to supply butter boxes. For example, in establishing his Jubilee Factory, Chew Chong engaged contractors to fell bush and sow in grass seed several acres near his factory.[86] Dairying also drove other environmental modification. Chew Chong's creamery, located on the edge of swampland inland from Eltham, encouraged the area's draining, deforestation, and development, a process all but complete by 1920.[87]

Dairying had far-reaching ecological consequences.[88] Much of the rich mosaic of life associated with the region's dense forest disappeared, replaced by introduced pastureland, and eventually, in some parts, by a grass monoculture. Dairy cows altered soil composition. Effluent runoff reached streams. Fertilizers, excavated from Nauru and Ocean Island by Chinese navvies and dumped onto Taranaki's soils to maintain grass growth, leached into soils and waterways.[89] Deforestation, meanwhile, exposed some farms and orchards to storms screaming in from the coast, bringing salt-laden water that stripped orchards, gardens, and grasslands.[90] Even the hastily introduced shelterbelts of introduced plants to replace the trees they had removed caused later problems—many of these introductions (such as boxthorn and barberry) later became weedy species.

Faced with stiff competition from newer co-operatives, Chew Chong retired in the early 1900s, but not before earning respect from Māori and settlers for his entrepreneurial nature and role in kick-starting South Taranaki's economy. He received an illuminated address and eighty-five gold sovereigns as thanks from prominent citizens for his civic-mindedness, and an honorary chieftainship in recognition of his role in helping struggling Māori.[91] His ventures share some characteristics with other Chinese Pacific enterprises. In developing the fungus industry, Chew Chong yoked knowledge of China,

Figure 3.3. Chew Chong and family, unknown photographer, c. 1903, collection of Puke Ariki, New Plymouth, PHO2004-292. Reproduced by permission of Puke Ariki, New Plymouth.

and Chinese networks and markets, with New Zealand resources and labor. In contrast, his dairying enterprises utilized Australian and British markets, rather than Chinese ones, and also drew on colonial and family labor. Despite increasingly stringent anti-Chinese immigration laws and amid growing racism from the late 1800s, Chew Chong still moved within the upper echelons of Taranaki society. He served as a founding member of the New Plymouth Chamber of Commerce and supported several civic projects. (Unlike other colonies, New Zealand permitted aliens to purchase land in this period.) Fluency in English and Chinese, as well as marriage to a European woman (Elizabeth Whatton), further smoothed his passage into colonial society (see figure 3.3).[92] Other Chinese merchants in Australasia negotiated between colonial and Chinese worlds much as Chew Chong did, enabling them to forge successful careers reliant on developing colonial and Chinese resources and markets, sometimes in partnership with European investors.[93]

Conclusion

Demand from Chinese markets (1790s onward) and the later activities of Cantonese people (1840s onward) accelerated the exploitation and commodification of Pacific resources. Moving resource frontiers created ecocultural networks connecting people, places, and natures through environmental exchanges and their associated technologies, labor systems, and capital exchanges. Sealing, sea cucumber collecting, and sandalwood-getting chart China's impact, as European traders and local peoples scoured the oceans and islands of the Pacific for products for the China market, with resulting social, demographic, and environmental impacts.

From the mid-nineteenth century, millions of Cantonese took advantage of opportunities presented by British imperialism and American expansion to seek their fortune on a series of Pacific goldfields. As well as goldmining, Chinese took up other occupations, such as laboring and market gardening, each of which brought significant environmental impacts. Cantonese networks spanned the Pacific. Ideas, organisms, and technologies traveled with workers, not necessarily in a linear fashion, because resource commodification connected different local places in South China and the Pacific. Exchanges and interactions with different groups also took place, from technology to intermarriage. Like Chew Chong and Choie Sew Hoy, some Chinese merchants married colonial labor and resources with Chinese investment, business connections, and markets. Profits from one enterprise, when invested in others, fashioned new migrant ecologies, some of which overlapped with Indigenous and colonial systems.

For far too long the ethnocentric bias of writers has masked the story of the Chew Chongs and many other less well-known Chinese who traveled to the Pacific. This chapter argues that it is time for historians to give a voice back to the Chinese and to recognize their agency in the creation of a nineteenth-century Cantonese Pacific and in the fashioning of sometimes distinct migrant ecologies.

Notes

1. Figure calculated from Adam McKeown, "Global Migration, 1846–1940," *Journal of World History* 15, no. 2 (2004): 158.

2. Henry Yu, "The Intermittent Rhythms of the Cantonese Pacific," in *Connecting Seas and Connected Ocean Rims: Indian, Atlantic, and Pacific Oceans and Chinese Seas Migrations from the 1830s to the 1930s*, ed. Donna R. Gabaccia and Dirk Hoerder (Leiden, Netherlands: Brill, 2011), 393–414.

3. James Beattie, Edward D. Melillo, and Emily O'Gorman, eds., *Eco-cultural Networks and the British Empire: New Views on Environmental History* (New York: Bloomsbury, 2015); J. Beattie, E. O'Gorman, and E. Melillo, "Rethinking the British Empire through Eco-cultural Networks: Materialist-Cultural Environmental History, Relational Connections and Agency," *Environment and History* 20, no. 4 (2014): 561–575.

4. Overwhelmingly, environmental historians have stressed the centrality of European capital and colonization in driving Pacific environmental change, especially from the

mid-nineteenth century. Exceptions include Robert Hellyer, "The West, the East, and the Insular Middle: Trading Systems, Demand, and Labour in the Integration of the Pacific, 1750–1875," *Journal of Global History* 8 (2013): 391–413; and Edward Dallam Melillo, *Strangers on Familiar Soil: Rediscovering the Chile-California Connection* (New Haven, CT: Yale University Press, 2015). For example, Paul Star posited that a handful of wealthy Europeans ("biota barons") had a disproportionate environmental impact in colonial New Zealand. Paul Star, "New Zealand's Biota Barons: Ecological Transformation in Colonial New Zealand," *ENNZ: Environment and Nature in New Zealand* 6 (2011): 1–12.

5. Ryan Tucker Jones, "The Environment," in *Pacific Histories: Ocean, Land, People*, ed. D. Armitage and A. Bashford (Basingstoke, UK: Palgrave Macmillan, 2014), 130.

6. David Igler, *The Great Ocean: Pacific Worlds from Captain Cook to the Gold Rush* (New York: Oxford University Press, 2013), 30.

7. On earlier patterns of migration, note Diana Lary, *Chinese Migration: The Movement of People, Goods, and Ideas over Four Millennia* (Lanham, MD: Rowman & Littlefield, 2012).

8. Robert B. Marks, *Tigers, Rice, Silk, & Silt: Environment and Economy in Late Imperial South China* (New York: Cambridge University Press, 1998), 163.

9. Marks, *Tigers*, chs. 5–10.

10. Mark Elvin, "Three Thousand Years of Unsustainable Growth: China's Environment from Archaic Times to the Present," *East Asian History* 6, (1993): 7–46; E. N. Anderson, "Agriculture, Population and Environment in Late Imperial China," in *Environment, Modernization and Development in East Asia: Perspectives from Environmental History*, ed. Ts'ui-jung Liu and James Beattie (Basingstoke, UK: Palgrave Macmillan, 2016), 31–58.

11. J. R. McNeill, "Of Rats and Men: A Synoptic Environmental History of the Island Pacific," *Journal of World History* 5, no. 2 (1994): 319.

12. On Chinese-Australasian-Indian-British trade and environmental change, see Beattie, "Thomas McDonnell's Opium: Circulating, Plants, Patronage, and Power in Britain, China and New Zealand, 1830s–1850s," in *The Botany of Empire in the Long Eighteenth Century*, ed. Sarah Burke Cahalan and Yota Basaki (Washington, DC: Dumbarton Oaks/Harvard University Press, 2017), 163–188; Beattie, "Plants, Animals and Environmental Transformation: New Zealand/Indian Biological and Landscape Connections, 1830s–1890s," in *The East India Company and the Natural World*, ed. Vinita Damodaran and Anna Winterbotham (Basingstoke, UK: Palgrave Macmillan, 2014), 219–248.

13. The best overviews are McNeill, "Of Rats and Men"; Jones, "Environment."

14. Commodore Baudin to Governor King, 1802, *Historical Records of New South Wales*, vol. 5, p. 832, cited in Robert McNab, *Murihiku: A History of the South Island of New Zealand and the Islands Adjacent and Lying to the South, from 1642 to 1835* (Wellington: Whitcombe and Tombs Limited, 1909), 256.

15. *Sydney Gazette*, July 29, 1829, cited in McNab, *Murihiku*, 261.

16. Richard Ellis, *The Empty Ocean: Plundering the World's Marine Life* (Washington, DC: Island Press, 2003), 175–176.

17. Rhys Richards, "New Market Evidence in the Depletion of Southern Fur Seals, 1788–1833," *New Zealand Journal of Zoology* 30, no. 1 (2003): 1–9.

18. Jonathan Schlesinger, *A World Trimmed with Fur: Wild Things, Pristine Places and the Natural Fringes of Qing Rule* (Stanford, CA: Stanford University Press, 2017), 18. See also 25–33, ch. 4.

19. Schlesinger, 131.

20. Ellis, *Empty Ocean*, 176.

21. Jim McAloon, "Resource Frontiers, Environment and Settler Capitalism, 1769–1860," in *Making a New Land: Environmental Histories of New Zealand*, ed. Eric Pawson and Tom Brooking, 2nd ed. (Dunedin, NZ: Otago University Press, 2013), 76.

22. Potato-growing encouraged more sedentary living, in place of a system of seasonal resource use. Athol Anderson, *Welcome of Strangers: An Ethnohistory of Southern Maori 1650–1850* (Dunedin, NZ: Otago University Press, 1998).

23. Pacific peoples also consume it, in a variety of forms.

24. C. Conand, *The Fishery Resources of Pacific Island Countries. Part 2. Holothurians* (Rome: Food and Agriculture Organization of the United Nations, 1990), 14.

25. For a fascinating discussion of sea cucumber's exchange value, see Edward D. Melillo, "Making Sea Cucumbers out of Whales' Teeth: Nantucket Castaways and Encounters of Value in Nineteenth-Century Fiji," *Environmental History* 20, no. 3 (2015): 449–474.

26. C. Campbell, *A History of the Pacific Islands* (Berkeley: University of California Press, 1989), 64.

27. "Table 3: *Bêche-de-mer* shipments via Noumea from 1862 to 1887," Conand, *Fishery Resources*, 20.

28. Conand, *Fishery Resources*, 15.

29. Micronesians and some Polynesians would eat the body wall raw or with lime juice, while Fijians might cook it in coconut milk and Papua New Guineans grill it. Conand, *Fishery Resources*, 18.

30. R. G. Ward, "The Pacific *Bêche-de-mer* Trade with Special Reference to Fiji," in *Man in the Pacific: Essays on Geographical Change in the Pacific Islands*, ed. R. G. Ward (Oxford: Clarendon Press, 1972), 107–108.

31. G. Ross Cochrane, "Problems of Vegetation Change in Western Viti Levu, Fiji," in *Settlement and Encounter: Geographical Studies Presented to Sir Grenfell Price*, ed. Fay Gale and Graham H. Lawton (Melbourne: OUP, 1969), 120.

32. Ward, "Pacific *Bêche-de-mer*," 118.

33. McNeill, "Of Rats and Men," 322.

34. Dorothy Shineberg, *They Came for Sandalwood: A Study of the Sandalwood Trade in the South-West Pacific* (New York: Cambridge University Press, 1967), 2.

35. Ross H. Cordy, "The Effects of European Contact on Hawaiian Agricultural Systems—1778–1819," *Ethnohistory* 19, no. 4 (1972): 410; O. A. Bushnell, *The Gifts of Civilization: Germs and Genocide in Hawaii* (Honolulu: University of Hawaii Press, 1993).

36. Qing attempts to break British opium imports into China gave the British the excuse they had long waited for to attack China. Ironically, as John McNeill notes, opium's rise might well have saved further Pacific Island species from extinction. McNeill, "Of Rats and Men," 325–326.

37. Elizabeth Sinn, *Pacific Crossing: California Gold, Chinese Migration, and the Making of Hong Kong* (Hong Kong: Hong Kong University Press, 2013), 2.

38. Sinn, *Pacific*, 2.

39. Jeyamalar Kathirithamby-Wells, *Nature and Nation: Forests and Development in Peninsular Malaysia* (Singapore: Singapore University Press, 2005), 37–38.

40. Madeline Y. Hsu, *Dreaming of Gold, Dreaming of Home: Transnationalism and Migration between the United States and South China, 1882–1943* (Stanford, CA: Stanford University Press, 2000).

41. For an introduction to the topic, see Paul D'Arcy, "The Chinese Pacifics: A Brief Historical Review," *Journal of Pacific History* 49, no. 4 (2014): 396–420; John Fitzgerald, *Big White Lie: Chinese Australians in White Australia* (Sydney: UNSW Press, 2007); Adam

McKeown, *Chinese Migrant Networks and Cultural Change: Peru, Chicago, and Hawaii, 1900–1936*, 2nd ed. (Chicago: University of Chicago Press, 2001); Adam McKeown, *Melancholy Order: Asian Migration and the Globalization of Borders* (New York: Columbia University Press, 2008).

42. McKeown, *Chinese Migrant Networks*; Jean Gittens, *The Diggers from China: The Story of the Chinese on the Goldfields* (Melbourne; London; New York: Quartet Books, 1981); James Ng, *Windows on a Chinese Past: How the Cantonese Goldseekers and Their Heirs Settled in New Zealand*, vol. 1 (Dunedin, NZ: Otago Heritage Books, 1993).

43. Joanna Boileau, "Chinese Market Gardening in Australia and New Zealand, 1860s–1960s: A Study in Technology Transfer" (PhD diss., University of New England, 2014).

44. J. Beattie, "'Hungry Dragons': Expanding the Horizons of Chinese Environmental History—Cantonese Gold-Miners in Colonial New Zealand, 1860s–1920s," *International Review of Environmental History* 1 (2015): 103–145; Fei Sheng, "Environmental Experiences of Chinese People in the Mid-Nineteenth Century Australian Gold Rush," *Global Environment: A Journal of History and Natural and Social Sciences* 7–8 (2012): 99–127; Corey Ross, "The Tin Frontier: Mining, Empire, and Environment in Southeast Asia, 1870s–1930s," *Environmental History* 19 (2014): 454–49.

45. On which topic, see J. Beattie, "Eco-cultural Networks in Southern China and Colonial New Zealand, 1860s–1910s," in *Eco-cultural Networks and the British Empire: New Views on Environmental History*, ed. James Beattie, Emily O'Gorman, and Edward Melillo (New York: Bloomsbury, 2015), 151–179.

46. *Otago Witness*, October 7, 1882, 11.

47. Beattie, "Eco-cultural Networks in Southern China."

48. Christopher Davey, "The Origins of Victorian Mining Technology, 1851–1900," *The Artefact* 19 (1996): 54; Nicol Allan MacArthur, "Gold Rush and Gold Mining: A Technological Analysis of Gabriel's Gully and the Blue Spur, 1861–1891" (MA diss., University of Otago, 2014), 39. On its use in Southeast Asia, see Ross, "Tin Frontier," 458.

49. Beattie, "'Hungry Dragons': Expanding the Horizons of Chinese Environmental History," 103–145.

50. Sucheng Chan, *This Bittersweet Soil: The Chinese in California Agriculture, 1860–1910* (Berkeley: University of California Press, 1986).

51. Warwick Frost, "Migrants and Technological Transfer: Chinese Farming in Australia, 1850–1920," *Australian Economic History* 42 (2002): 113–131; Lei Guang and James Gerber, eds., *Agriculture and Rural Connections in the Pacific, 1500–1900* (Burlington, VT: Ashgate/Variorum, 2006); Stanin Zvonka, "From Li Chun to Yong Kit: A Market Garden on the Loddon, 1851–1912," *Journal of Australian Colonial History* 6 (2004): 15–34; Beattie, "Eco-cultural Networks in Southern China"; J. Beattie, "'The Empire of the Rhododendron': Reorienting New Zealand Garden history," in *Making a New Land: Environmental Histories of New Zealand*, 241–257, 365–367.

52. Beattie, "Eco-cultural Networks in Southern China."

53. E. Thorne, *The Queen of the Colonies, or, Queensland as I Knew It by an Eight Years Resident* (London: Sampson, Lowe, Marston, Searle and Rivington, 1876), 114.

54. Sophie Couchman, "The Banana Trade: Its Importance to Melbourne's Chinese and Little Bourke Street, 1880s–1930s," in *Histories of the Chinese in Australasia and the South Pacific*, ed. P. Macgregor (Melbourne: Museum of Chinese Australian History, 1995), 75–90; C. F. Yong, "The Banana Trade and the Chinese in NSW and Victoria, 1901–1921," *ANU Historical Journal* 1, no. 2 (1964): 28–35; Wai Jane Char, "Chinese Merchant-Adventurers and Sugar Masters in Hawaii: 1802–1852," *Hawaiian Journal of History* 8 (1974): 3–25.

55. Iris Chang, *The Chinese in America: A Narrative History* (New York: Viking, 2003), 53–64; Huang Annian, ed., *The Silent Spikes: Chinese Laborers and the Construction of North American Railroads*, trans. Zhang Juguo (Beijing: China Intercontinental Press, 2006).

56. Gregory T. Cushman, *Guano and the Opening of the Pacific World: A Global Ecological History* (Cambridge: Cambridge University Press, 2013).

57. The experience of Chinese sojourners is powerfully evoked by Maxine Hong Kingston's family memoir/novel, *China Men* (London: Vintage, 1989 [1980]).

58. Beattie, "Eco-cultural Networks in Southern China."

59. "Mr Chew Chong," *Otago Daily Times*, February 25, 1890, 3.

60. "A Market for Fungus," *Evening Post*, August 19, 1922, 12.

61. "The Chinese Benefactor," *Hawera & Normanby Star*, July 5, 1923, 15.

62. "A Market for Fungus," *Evening Post*, August 19, 1922, 12.

63. "Bush Farming," *Auckland Star*, March 7, 1923, 12.

64. "Compliment to a Chinese," *Feilding Star*, January 30, 1911, 2.

65. James Ng, *Windows on a Chinese Past*, vol. 3 (Dunedin, NZ: Otago Heritage Books, 1993–1999).

66. Helen Wong, *In the Mountain's Shadow: A Century of Chinese in Taranaki, 1870–1970* (n.p.: Helen Wong, 2010), 24.

67. Section title from "The Chinese Benefactor," *Hawera & Normanby Star*, July 5, 1923, 15.

68. G. I. Rawson, "The Evolution of the Rural Settlement Pattern of Lowland South Taranaki, 1860–1920" (MA thesis, University of Canterbury, 1967), 53.

69. A creamery served as a subsidiary plant "in areas where there were insufficient cows" for a dairy factory "but enough to make their contribution worthwhile." They were usually located no more than ten miles from the factory, beyond which it was uneconomic to transport cream. Over time, as an area developed, farmers converted creameries to dairy farms. Rawson, 52.

70. H. J. Andrew, *The History of Eltham, N.Z. Cradle of the Dairy Export Industry* (Eltham, NZ: Eltham Borough Council, 1959), 25.

71. "Mr Chew Chong," *Otago Daily Times*, February 25, 1890, 3.

72. "Patents, Designs and Trade-marks. Third Annual Report of the Registrar," *Appendix to the Journals of the House of Representatives*, 1892 Session I, H-02.

73. "The Chinese Benefactor," *Hawera & Normanby Star*, July 5, 1923, 15.

74. James Ng, "Chew Chong," from the *Dictionary of New Zealand Biography* in *Te Ara: The Encyclopedia of New Zealand*, updated October 30, 2012, http://www.TeAra.govt.nz/en/biographies/2c17/chew-chong.

75. J. R. McNeil, *Something New under the Sun: An Environmental History of the Twentieth Century* (New York: W. W. Norton, 2000); Daniel R. Headrick, *Humans versus Nature: A Global Environmental History* (Oxford: Oxford University Press, 2020), chs. 8–9.

76. Tom Brooking and Eric Pawson, *Seeds of Empire: The Environmental Transformation of New Zealand* (London: I. B. Taurus, 2011). Chew Chong also exported to Australia.

77. "The Jubilee Butter Factory," *Bush Advocate*, March 27, 1890, 3.

78. "Obituary," *Hawera & Normanby Star*, October 8, 1920, 4.

79. "Mr Chew Chong," *Otago Daily Times*, February 25, 1890, 3.

80. Rawson, "Evolution," 53.

81. *Hawera & Normanby Star*, February 23, 1894, cited in Rawson, "Evolution," 53.

82. The quotation in the subhead is from G. C. Petersen, *Forest Homes* (Wellington: A. H. and A. W. Reed, 1956), cited in Rawson, "Evolution," 90.

83. Rawson, "Evolution," 29.

84. Rawson, "Evolution," 28.

85. Andrew, *The History of Eltham*, 26.

86. Estimates based on description of area converted. Andrew, *History of Eltham*, 26.

87. Rawson, "Evolution," 57. Chew Chong's creamery was taken over by the Eltham Dairy Company in 1902. "Eltham Jubilee Supplement," *Taranaki Daily News*, March 6, 1959, 12.

88. Bruce Clarkson, "Ecological Heritage in the Taranaki Region," in *Making Our Place: Exploring Land-Use Tensions in Aotearoa New Zealand*, ed. Jacinta Ruru, Janet Stephenson, and Mick Abbott (Dunedin, NZ: Otago University Press, 2011), 115–128. See also Beattie and Bruce Clarkson, *Ecological Change in the Taranaki Ringplain Since the Last Ice Age* (draft book manuscript).

89. Brooking and Pawson, *Seeds of Empire*.

90. Rollo Arnold, *Settler Kaponga, 1881–1914: A Frontier Fragment of the Western World* (Wellington: Victoria University Press, 1997), 145–146.

91. Ng, "Chew Chong."

92. Rawson, "Evolution," 58.

93. Beattie, "'Hungry Dragons'"; Beattie, "Dragons in the South Pacific: Chinese Migration, Resource Exploitation and Environmental Change" (draft manuscript).

Four

The Third Vector

Pacific Pathogens, Colonial Disease Ecologies, and Native American Epidemics North of Mexico

Benjamin Madley

The epidemics that Native Americans endured following the arrival of newcomers are formative events in North American history. In what are now the continental United States and Canada, diseases, colonialism, and violence reduced the Indigenous population from perhaps 18,000,000 or more to some 365,137 by about 1900.[1] Pathogens served as invaders' invisible allies, cutting down untold numbers of Native American people. Diseases helped to make invasion and, ultimately, the modern United States and Canada, possible.

Our understanding of the epidemics endured by Indigenous peoples north of the Rio Grande River is shaped by a paradigm in which these epidemics are generally conceptualized as moving overland from east to west and from south to north. This Atlantic-World-focused understanding is based upon substantial evidence. Epidemics scythed through the Indigenous populations of Mexico in the sixteenth century, and some of these outbreaks moved north via overland routes.[2] Seventeenth-century epidemics then tore through New England and west to the Great Lakes.[3]

There was, however, a third direction by which exotic microbes arrived in North America: from the Pacific. As the historian David Igler has shown, pathogens traversed the world's largest ocean as part of the "global exchange that transformed the Pacific . . . the spread of epidemic diseases from trading vessels to native communities."[4] This chapter will explore how Pacific pathogen vectors likely introduced four extremely lethal epidemics to California between 1827 and 1844, how colonial disease ecologies magnified their impact, and why exploring Pacific World migrant ecologies can help us to better understand Native American histories.[5]

Colonial Disease Ecologies

Diphtheria, influenza, malaria, measles, plague, scarlet fever, smallpox, typhus, whooping cough, and other "Old World" pathogens devastated Native American populations. Many have argued that this happened largely because American Indians had never before been exposed to such diseases and thus had limited biological resistance to them. In 1976, the historian Alfred Crosby coined the phrase "virgin soil epidemics" to describe such outbreaks, defining them as "those in which the populations at risk have had no previous contact with the diseases that strike them and are therefore immunologically almost defenseless."[6] For Crosby, "virgin soil epidemics" were theoretically inevitable: "In theory, the initial appearance of these diseases is as certain to have set off deadly epidemics as dropping lighted matches into tinder is certain to cause fires."[7] Such immunological determinism largely blames Native American population declines on Indigenous immune systems while muting both contingency and the roles of colonialism, violence, and other factors.

Recent scholarship has complicated the "virgin soil epidemics" equation to consider how colonialism amplified the lethality of "Old World" pathogens in the Western Hemisphere.[8] Native American people often encountered these pathogens when their bodies and immune systems were most compromised, that is in the context of colonial disease ecologies. Colonialism routinely involved invasion, violence, and the loss of clothing, food, housing, medicine, resources, tools, and community members. Dispossession, displacement, further violence, and institutionalized oppression routinely followed. Meanwhile, newcomers introduced non-indigenous flora and fauna. These invasive organisms destroyed the plants, animals, agricultural systems, and environments upon which Indigenous peoples built their civilizations. Large-scale hunting, fishing, fur trapping, logging, and mining further stressed traditional food systems and economies. In short, invasion often transformed or destroyed Indigenous ecologies. These were the conditions in which Native American people often encountered Eastern Hemisphere pathogens: without adequate access to customary foods, clean water, housing, or clothing let alone medical care, all of which made them more likely to contract and less likely to recover from sicknesses. By making so many people ill, epidemics also sometimes catalyzed the abandonment of traditional land management practices, crops, fishing nets, bows, and traps. Epidemics thus contributed to malnutrition and even starvation that either killed people outright or made them vulnerable to successive microbial assaults. "Old World" diseases also sometimes rendered Indigenous people sterile, created widows and widowers, suppressing birth rates, increasing infant mortalities, and ultimately thwarting Indigenous demographic recovery. Scholar John Ware has argued simply: "If Native mortality was higher, it was because Natives were subjected to many additional stresses."[9] Colonial disease ecologies were a major stress.

Beginning in 1492, colonial disease ecologies began spreading. They transformed much of the Americas, fostering conditions in which Eastern Hemisphere pathogens generated epidemics among Indigenous communities. The numbers are staggering. The epidemiologist and microbiologist Francis Black conservatively estimated that "approximately 56 million people died as a result of European exploration in the New World [and] most died of introduced diseases."[10] The Indigenous peoples of California endured multiple microbial assaults, including four introduced between 1827 and 1844, all of which seem to have arrived from the Pacific at a time when colonialism was radically transforming California environments (see figure 4.1).

The 1827-1828 Epidemic

During 1827–1828, an epidemic raged along California's coast. The Kentuckian adventurer James Pattie was there and later wrote that some time before December 20, 1828, "the small pox began to rage on the upper part of the coast, carrying off the inhabitants by hundreds."[11] According to a 1973 study by the medical historian Rosemary Valle, this epidemic first appeared at Southern California's Mission San Gabriel in 1827, near present-day Los Angeles, before expanding up and down the coast, with Pacific shipping helping to spread the disease. The epidemic was devastating. Valle reported a 76 percent death rate increase in California's Franciscan missions during the epidemic, with higher than normal death rates at seventeen of them.[12]

The devastation wrought by this and other epidemics was—in part—a product of colonial disease ecologies that had been spreading since Spaniards first began colonizing California in 1769. Exotic plants overran the land in and around the missions, competing with the native species that California Indians harvested. Introduced animals were even more damaging. By 1821, the imported animal population had surged to 1,469 goats, 1,633 pigs, 2,011 mules, 19,830 horses, 149,730 cattle, and 191,324 sheep.[13] As the historian John Ryan Fischer observed, "Spanish livestock . . . exacted a toll on resources."[14] Indeed, these quadrupeds reshaped entire ecosystems. The ethnoecologist M. Kat Anderson explained, "Grazing was among the activities that caused the greatest damage. Coastal prairies, oak savannahs, prairie patches in coastal redwood forests, and riparian habitats, all rich in plant species diversity and kept open and fertile through centuries of Indian

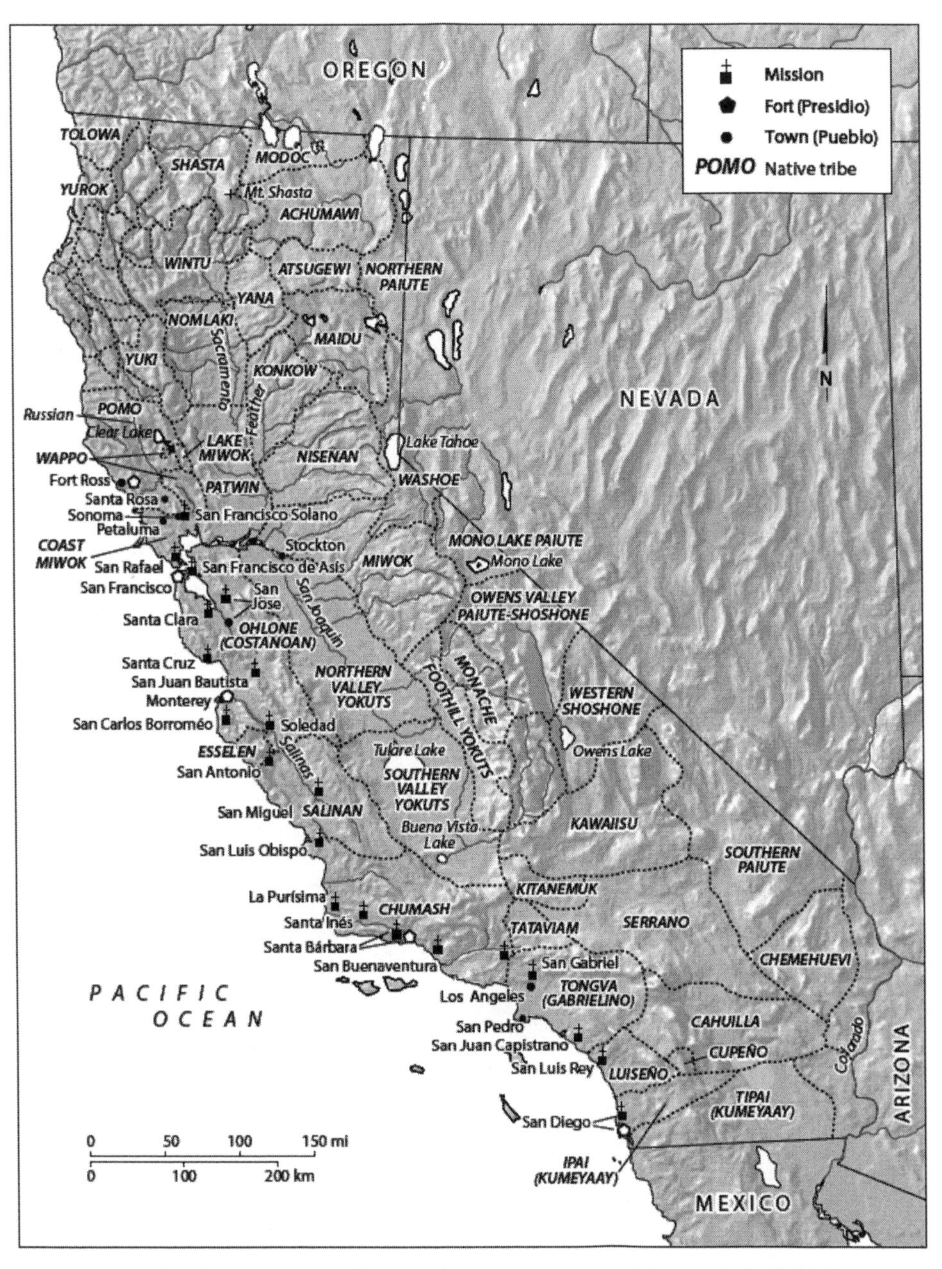

Figure 4.1. California missions, forts, and towns, 1769–1844. Map made by Bill Nelson.

Figure 4.2. José Cardero, "Vista del Presidio de Monte Rey," 1791–1792. Drawing on paper, ink, wash, pencil. Robert B. Honeyman, Jr. Collection of Early Californian and Western American Pictoral Material, BANC PIC 1963.002:1310—FR. Courtesy of the Bancroft Library, University of California, Berkeley.

burning, became grazing land for vast herds of cattle, sheep, goats, hogs, and horses owned by the Spanish missions." Imported animals devoured plants central to many California Indian diets and competed for forage with game animals that California Indians hunted. Anderson emphasized, "Overgrazing eliminated native plant populations [and] brought about extensive and irreversible changes in many ecosystems."[15] New colonial policies also changed ancient land management regimes (see figure 4.2).

Spanish officials banned the traditional California Indian land management practice of burning grasslands. These bans decreased the yields of customary hunting and gathering, both of which relied upon pyrogenic land management.[16] Imported plants and animals, in concert with these new restrictions, all tore at the fabric of traditional California Indian economies, decreased nutritional intake, and made California Indian people more vulnerable to pathogens and less able to recover from illnesses.

Meanwhile, the concentration of thousands of Native Americans in California's missions created local disease ecologies that facilitated pathogen transmission. As one eyewitness wrote of California mission Indians under Mexican rule, "confinement behind infected walls was very harmful to them": concentration, poor ventilation, and unsanitary conditions led to illnesses, while a lack of medical care providers mitigated against recoveries.[17]

A Mexican smallpox vaccination program, in which Pattie reported playing the lead role, may have limited the spread of smallpox—which causes painful, oozing sores.[18] Smallpox inoculations and vaccinations were not new to California. As early as 1798, the military surgeon at Monterey distributed instructions to California missions and presidios on how to administer smallpox inoculations.[19] Spaniards in California began vaccinating against the disease in 1817, if not earlier, using vaccine lymph brought by the Spaniard José Verdia. Over the next twelve years, ships brought vaccine lymph to California from as far across the Pacific as Lima, Peru. Several of these vessels were Russian, suggesting that lymph might also have come to California on ships sailing from Russian Alaska or even Siberia.[20] Either way, Pattie claimed to have personally vaccinated 23,500 people—both American Indians and non-Indians—from San Diego to Russia's colonial outpost at Fort Ross, north of San Francisco, by July 1829. Sherburne F. Cook, however, challenged Pattie's narrative as "unreliable and highly exaggerated."[21]

Pattie may have misidentified the epidemic. Valle insisted it was measles, and there is evidence to support her claim.[22] In 1831, California governor Emanuel Victoria reported: "Smallpox is almost unknown, nevertheless vaccination has been indifferent . . . practiced only during the last few years by amateurs."[23] Some might have confused measles and its full body rash with the pustules that cover the bodies of smallpox victims. Whatever the malady, it likely came to California via a Pacific vector, given that in the late 1820s ships were California's main connection to the non-Indigenous world.

The 1830–1833 Malaria Epidemic

In late February 1829, the United States brig *Owyhee* moored in the lower Columbia River at Fort Vancouver, across the water from what is now Portland, Oregon. The *Owyhee* had just completed a six-month-long voyage from Boston, sailing south around Cape Horn and then north across the equator to the Pacific Northwest.[24] On March 3, her sister ship, the *Convoy*, arrived.[25] In April, both sailed further north seeking furs in the Strait of Juan de Fuca and southeastern Alaska.[26] In August, the *Owyhee* returned to the Columbia and overwintered there.[27] Meanwhile, in September, the *Convoy* sailed southwest to Hawai'i and Tahiti before returning to the Columbia in March 1830.[28] During their Pacific peregrinations, both ships visited multiple ports, taking on food, water, furs, trade goods, and perhaps a group of microscopic stowaways: *Plasmodium* protozoans.[29]

In July or August 1830, after the two ships' crews had been trading on the Columbia and beyond, an epidemic erupted at Fort Vancouver in Chinookan territory.[30] Malaria—with its debilitating fatigue, fevers, chills, vomiting, and death—now spread through the lower Columbia River Valley, among both Chinookan-speaking Native Americans and newcomers.[31] In an October 11 letter from the "Entrance of the River Columbia," the Scottish botanist David Douglas reported how "a dreadfully fatal intermittent fever broke out in the lower parts of this river about eleven weeks ago, which has depopulated the country." He explained, "Villages, which had afforded from one to two hundred effective warriors, are totally gone; not a soul remains! The houses are empty . . . while the dead bodies lie strewed in every direction on the sands of the river." Douglas added, "I am one of the very few persons among the Hudson Bay Company's people who have stood it, and sometimes I think, even I have got a *great shake*, and can hardly consider myself out of danger."[32]

The *Owyhee* or the *Convoy* likely brought either a person infected with the *Plasmodium* protozoans that cause malaria or the female *Anopheles* mosquitos that spread the protozoans between humans.[33] Multiple sources—both Native and non-Native—suggest that one of these ships introduced the epidemic.[34] Still, as Cook observed, malaria might have come from "the *Dryad*, the *Isabella*, or the *Vancouver*, all British ships, which . . . dropped anchor in the Columbia during the years 1829 and 1830."[35] It is also possible that a combination of vessels brought the pestilence. Whichever ship transported the protozoans to the lower Columbia, they seem to have come via the Pacific.

Over time, they apparently migrated south, carried by whining mosquitos and traveling humans.[36] Fur trappers headed south to California from Fort Vancouver, as well as California's four existing species of Anopheles mosquitos, likely assisted the migration. As the historian Linda Nash emphasized, "the temperate climate and long, hot summers of California were conducive to an epidemic outbreak, as they fostered multiple cycles of mosquito reproduction."[37] The consequences for many California Indians would be disastrous, in part because of the changing environmental conditions created by newcomers.

California Indian communities faced the disease in the context of expanding and intensifying colonial disease ecologies. By 1832, California's twenty-one missions had perhaps 420,000 cattle, 320,000 sheep, goats, and hogs, and 60,000 horses and mules.[38] Meanwhile, Mexican officials granted lands for twenty new private ranchos between 1822 and 1832.[39] According to Anderson, cattle and sheep ate "native plants such as California bromes, blue wild rye, and clovers, which produced seeds, grains, and greens used as food by the Indians." These plant populations thus "shrank dramatically." Newcomers also cut down large numbers of oaks, diminishing the supply of acorns, a dietary staple of many California Indian communities.[40] In addition,

Figure 4.3. Emanuel Wyttenbach, *Rodeo*, no date. Pen and ink drawing. Courtesy of the California History Room, California State Library, Sacramento, California.

as Fischer noted, exotic flora "displaced native plants like chia (*Salvia columbariae*) and purple needlegrass (*Nassella pulchra*) [whose seeds] had served as an important food source for California Indians."[41] Mission cultivation of wheat and other crops also wrought significant changes to the land.[42] Russian farming and ranching likewise modified northwestern California's Fort Ross region. Colonialism's ecological impact was greatest in the coastal zone between San Diego and Fort Ross, but as invasive flora and fauna pushed inland, the new ecologies created by colonialism affected more and more California Indian people, helping to set the stage for catastrophe (see figure 4.3).

On December 2, 1832, the Hudson's Bay Company fur trapping expedition leader John Work wrote of Northern California Indians, possibly Maidus, "There appears to be some sickness resembling an ague prevailing among them."[43] On August 6, 1833, he observed: "Some sickness prevails among the [Konkow or Maidu] Indians on feather [*sic*] river. The villages which were once so populous and swarming with inhabitants when we passed that way in Jany and Febry last seem now almost deserted & have a desolate appearance."[44] The illness also infected Work's expedition. By August 20, sixty-one expedition members were "ill, a good many of them attacked with trembling fits."[45] Moving north to the upper Sacramento River Valley, Work observed: "the villags [*sic*] seem almost wholly depopulated."[46] The malady seems to have been new to Northern California Indians. A "trapper" explained that along the Sacramento, "it was a fever of the remittent class. . . . We were informed by the Indians that they have no traditions of any similar scourge in past time." It was "the 'malaria of the marshes.'"[47]

By the summer of 1833, the malady had torn through California's Sacramento and San Joaquin River valleys.[48] Tens of thousands of California Indians inhabited these food-rich regions, and the impact was awful.[49] On March 26, 1837, the United States Navy officer William A. Slacum reported: "The ague and fever, which commenced on the Columbia in 1829," moved south, deep into Oregon, while "still further south, in Tularez, near St. Francisco, California, entire villages have been depopulated."[50] Recollecting that summer, Colonel Philip L. Edwards described "an intermittent fever" among his trapping party, which "prevailed with such mortality that the few survivors of a village sometimes fled from their homes." Yet Edwards also emphasized Indigenous survival: "Still, the Indians in this [Sacramento River] valley are numerous."[51] Later, in the early 1870s, "an old [Yokuts] Indian, named Chuchuka," told the ethnographer Stephen Powers "that many years ago there was a terrible plague, which raged on both sides of the Fresno [River], destroying thousands of people. According to his account, it was a black-tongue disease." Powers added, "Abundant evidences of his truthfulness have been discovered in those localities, in the shape of human bones."[52] Indeed, in 1879 the former trapper J. J. Warner recollected how, "late in the summer of 1833, we found the valleys depopulated." Warner explained,

> From the head of the Sacramento, to the great bend and slough of the San Joaquin, we did not see more than six or eight live Indians; while large numbers of their skulls and dead bodies were to be seen under almost every shade tree, near water, where the uninhabited and deserted villages had been converted into graveyards; and, on the San Joaquin river, in the immediate neighborhood of the larger class of villages, which, the preceding year, were the abodes of a large number of those Indians, we found not only many graves, but the vestiges of a funeral pyre.[53]

In total, perhaps 20,000 to 50,000 Central Valley California Indians died in 1833.[54] The epidemic may also have taken the lives of Indigenous people living to the south.

The Luiseño Indian man Pablo Tac of Mission San Luis Rey, on the Southern California coast, may have described this epidemic. In about 1835, Tac wrote, "In Quechla not long ago there were 5,000 souls, with all their neighboring lands. Through a sickness that came to California 2,000 souls died, and 3,000 were left."[55] What Tac did not emphasize—perhaps because he was writing in Rome while studying for the priesthood and surrounded by Catholic clergy—was that the loss of 40 percent of his people was a result of contact with non-Indians and that the epidemic, likely amplified by colonial disease ecologies, smashed his world. It deprived him of friends, relatives, and much of his community's social fabric. Like other epidemics that arrived from the Pacific, this biological scourge devastated California Indian communities. Sadly, it would not be the last.

The 1837–1840 Miramontes Smallpox Epidemic

In 1812, agents of the Russian-American Company established Fort Ross Colony on the coastal bluffs north of San Francisco. It served as a base from which to supply and support sea otter hunting along California's coast. By the 1830s, dozens of vessels a year plied the Pacific sea-lane between the Russian-American Company headquarters at Sitka, Alaska, in Tlingit territory, and the company's southernmost outpost in Northern California, in Kashaya Pomo territory.[56] As Igler observed, "trading vessels provided the primary means for diseases to travel around the Pacific's vast waterscape, reach isolated island populations, and strike various communities along the American coastline."[57] One such ship carried the deadly cargo that triggered a major California epidemic.

In 1835, smallpox broke out in Sitka. According to the anthropologist Robert T. Boyd, the *Variola* virus likely arrived there via one of four Pacific vectors: on the *Diana* from China, on the *America* from Kamchatka, on the Russian-American Company bark *Sitka* from Latin America, or via the distribution of "contaminated vaccine," which itself likely arrived overwater from the Pacific.[58] In September 1836, the epidemic spread south to the Hudson's Bay Company's Fort Simpson, in Tsimshian territory, within what is now British Columbia.[59]

The disease next seems to have sailed to California with the tide. Boyd suggested that fur-trading vessels, such as the *Diana*, *Joseph Peabody*, *Llama*, or *La Grange*, might have carried the virus to northwestern California.[60] Yet it may have arrived via another vessel or vessels. In February 1837, the *Llama* sailed through the Golden Gate into San Francisco Bay from Honolulu, Hawai'i.[61] Either it or the *Sitka*, which anchored at Fort Ross from Alaska, may have brought the lethal virus.[62] Whichever ship carried it, the virus landed at Fort Ross. According to Cook, "Late in 1837 [the Mexican] General Mariano S. Vallejo sent to Fort Ross a corporal of cavalry named Ignacio Miramontes to bring back a cargo of cloth and leather goods for the troops stationed at Sonoma. When Miramontes and his men returned they also brought with them the smallpox."[63] Catastrophe followed.

As elsewhere in Native America, smallpox proved extremely contagious and lethal. According to the Mexican official J. Fernandez, the epidemic "propagated with an incredible velocity," first affecting the people of Sonoma in Pomo territory.[64] It also spread south and east, likely along established trade routes, bombarding Indigenous communities already struggling with the material conditions imposed by Mexican colonialism. According to the colonist and historian E. Cerruti, "the smallpox spread with such virulence that it almost exterminated all of the Indians of the valleys of Sonoma, Petaluma, Santa Rosa, Russian River, Clear Lake, Sugano[?], the Tulares, and extended to the skirts of Mount Chasta or Shasta." Cerruti added that it "also invaded the missions."[65] If accurate, the Miramontes smallpox epidemic infected Miwok, Nomlaki, Patwin, Pomo, Wappo, Wintu, and Wintun communities, and perhaps others as well.

The impact was cataclysmic. According to a survivor named Caskibell, ten to twenty members of his Russian River Indigenous community died each day during the worst days, while "nearly all died" in some tribes.[66] Recollecting the epidemic's impact at Mission Santa Cruz, the Ohlone/Yokuts man Lorenzo Asisara explained: "*la peste de viruelas* [the plague of smallpox] finished the Indian population" of Ohlone and Yokuts people there.[67] The Mexican rancher José Manuel Salvador Vallejo later estimated that the epidemic killed "upwards of sixty thousand."[68] Others reported higher death tolls. According to Fernández the number was approximately 100,000.[69] Cerruti insisted that it killed "200 whites, 3000 mestizos and mission Indians, and 100,000 free Indians."[70] This too may have been an underestimate. Juan Bautista Alvarado, California governor from 1837 to 1842, recollected that the "viruelas Miramontes" killed three-fifths of California Indians during 1838, 1839, and 1840, or "two or three hundred thousand Indians."[71] Whatever the precise numbers, the epidemic killed many thousands. Four years later, California would suffer another major smallpox outbreak, once again likely introduced via maritime vectors into environments increasingly transformed by colonialism.

The 1844 Smallpox Epidemic

Under Mexican rule, California ranchos multiplied and expanded as Mexican officials privatized mission lands and issued land grants further afield. Mexican authorities issued twenty titled land grants between 1822 and 1832, ninety between 1832 and 1836, and approximately six hundred ninety between 1836 and 1846.[72] Such grants were large: typically ten to twenty thousand acres. Supplying the thriving hide and tallow trade, some of these new ranches—with their cattle, other animals, and grain and grape production—pushed into previously uncolonized areas, such as the Sacramento River Valley. Such ranchos rapidly transformed local environments, undermining traditional Indigenous economies and food security. Ultimately, Mexican-era ranchers came to own some ten million acres, or roughly 10 percent of California.[73] Thus, by the mid-1840s, expanding and intensifying colonial disease ecologies made more California Indians more vulnerable to imported pathogens than ever before. These diseases also helped newcomers to seize and transform the land, further undermining Indigenous subsistence strategies and feeding back into colonial disease ecologies. It was a vicious cycle.

On January 27, 1844, the Mexican government schooner *California* sailed north from Mazatlán for California. During the voyage, a trio of Native Hawaiian crewmembers came down with smallpox.[74] All three recovered.[75] Yet, when the *California* reached San Pedro on February 23, its crew or passengers apparently brought the *Variola* virus ashore in Southern California.[76] On May 17, P. H. Reid wrote, "The number of smallpox cases is three and I think the smallpox is of the type *Variola vacinae* which is relatively milder and less contagious."[77] Later that month, when the *California* anchored further north in Monterey Bay, the passenger and United States Consul Thomas Larkin sent apparently infected laundry from the ship to the pueblo of Monterey.[78] Larkin thus inadvertently introduced the virus into his hometown. He and Mr. Hartwell then vaccinated some three hundred people.[79] Still, according to the historian Steven Hackel, smallpox killed "a minimum of 125 of the town's residents, at least eighty-five of whom were Indian. As many as hundreds more succumbed without their deaths' being recorded in local Catholic Church records."[80] The outbreak also took a deadly toll on California's Central Coast. On August 22, 1844, Father José J. Jimeno reported: "the horrible plague of the smallpox finished the poor [Mission La] Purisima, very few Indians are left and the infirmity continues there."[81] This epidemic also spread into California's interior.

From the white adobe buildings and dusty streets of Monterey, the disease apparently spread north to San José. There, the stockman David Kelsey "went to see a sick Indian who had the small-pox." In August, Kelsey carried the virus to what is now Stockton, on the San Joaquin River.[82] In or near Stockton, the epidemic "broke out among the tribes and caused a great mortality among them." Thus, "hundreds died." Moreover, "one after another died until those who survived fled from the vicinity."[83] The disease also spread to the Miwok people of what is now Amador County, killing an unknown number there.[84] It may also have extended further north. While visiting Clear Lake in 1851, treaty-making expedition member George Gibbs blamed Indigenous population losses there on "the ravages of the small-pox, at no very remote period. Some old Indians, who carry with them the marks of the disease, state it positively; and it is reported, by native Californians, that over 100,000 perished of this disease in the valleys drained by the Sacramento and the San Joaquin."[85] Perhaps referring to this epidemic, in 1877, Powers wrote: "About 1846 there was an epidemic among [the Wintun] which produced fever and raging thirst."[86] Clearer than the total death toll is the fact that the 1844 smallpox outbreak arrived via the Pacific.

Conclusion

Of the four California epidemics considered in this chapter, measles or smallpox very likely arrived via the Pacific in 1827, malaria apparently came from Oregon via the Pacific in 1832, and smallpox arrived on ships in both 1837 and 1844. Understanding the origins of these epidemics does not change their cataclysmic outcome. We cannot bring back the dead. Yet, epidemics, in combination with colonial invasion and violence, are crucial to understanding Native American and Western Hemisphere histories. Epidemics killed untold numbers of Indigenous people while facilitating colonial invasions and, ultimately, transforming the hemisphere. Thus, it is vital to understand how connected they were to Pacific World migrations—both human and non-human—and the making of our modern world.

New studies of Pacific World migrant ecologies, and their profound impacts on Indigenous peoples, demand our attention. The history of pathogens in interaction with colonial disease ecologies and Indigenous peoples has been extensively studied within the framework of the Atlantic World.[87] However, they have yet to be studied with such rigor in the Pacific World. Indeed, the historical movement of organisms throughout and across this vast oceanic space remains poorly understood and undertheorized. Pathogens did arrive in California via Pacific vectors and arrived in zones transformed by newcomers into colonial disease ecologies with cataclysmic consequences for Indigenous people. California Indians survived. Yet, these epidemics were formative events in California history.

Notes

The author thanks Paul Kelton, Jacob Lahana, Edward D. Melillo, Preston S. McBride, Christopher Parsons, Joshua L. Reid, Martin Rizzo-Martinez, and Jeremiah Sladeck for their help with this chapter.

1. For some pre-contact population estimates for the region north of the Rio Grande River, see Douglas H. Ubelaker, "Prehistoric New World Population Size: Historical Review and Current Appraisal of North American Estimates," *American Journal of Physical Anthropology* 45, no. 3 (November 1976): 661–665, table 2, 664; Henry Dobyns, *Their Number Become Thinned: Native American Population Dynamics in Eastern North America* (Knoxville: Univer-

sity of Tennessee, 1983), 42; Russell Thornton, *American Indian Holocaust and Survival: A Population History since 1492* (Norman: University of Oklahoma Press, 1987), 32, 60; William Denevan, ed., *The Native Population of the Americas in 1492*, 2nd ed. (Madison: University of Wisconsin Press, 1992), xxviii. Census takers reported 237,196 American Indians in 1900 in the United States and 127,941 "Indian and Eskimo" people in Canada in 1901. United States Bureau of the Census, *Indian Population in the United States and Alaska, 1910* (Washington, DC: Government Printing Office, 1915), 10; Canada Yearbook in Social Science Federation of Canada and Statistics Canada, *Historical Statistics of Canada* (Ottawa: Statistics Canada, 1983), Table A125: "Origins of the Population, Census Dates, 1871 to 1971."

2. Noble David Cook, *Born to Die: Disease and New World Conquest, 1492–1650* (New York: Cambridge University Press, 1998), 95–140, 154–165.

3. David S. Jones, *Rationalizing Epidemics: Meanings and Uses of American Indian Mortality since 1600* (Cambridge, MA: Harvard University Press, 2004), 21–67; Dean R. Snow and Kim M. Lanphear, "European Contact and Indian Depopulation in the Northeast: The Timing of the First Epidemics," *Ethnohistory* 35, no. 1 (Winter 1988): 15–33; Richard White, *The Middle Ground: Indians, Empires, and Republics in the Great Lakes Region, 1650–1815* (New York: Cambridge University Press, 1991), 217–218, 229–230, 275.

4. David Igler, "Diseased Goods: Global Exchanges in the Eastern Pacific Basin, 1770–1850," *American Historical Review* 109, no. 3 (June 2014): 693–719.

5. British literature scholar Alan Bewell coined the phrase "colonial disease ecologies" in Bewell, *Romanticism and Colonial Disease* (Baltimore: Johns Hopkins University Press, 2003), 4. However, Bewell did not elaborate on its meaning or impacts on Indigenous peoples.

6. Alfred Crosby, "Virgin Soil Epidemics as a Factor in the Aboriginal Depopulation in America," *William and Mary Quarterly* 33, no. 2 (April 1976): 289.

7. Ibid., 290.

8. See, for example, David S. Jones, "Virgin Soils Revisited," *William and Mary Quarterly* 60, no. 4 (October 2003): 703–742, revised as David S. Jones, "Death, Uncertainty, and Rhetoric," in *Beyond Germs: Native Depopulation in North America*, ed. Catherine M. Cameron, Paul Kelton, and Alan C. Swedlund (Tucson: University of Arizona Press, 2015), 16–49.

9. John Ware, "Forward," in Catherine M. Cameron, Paul Kelton, and Alan C. Swedlund, eds., *Beyond Germs: Native Depopulation in North America* (Tucson: University of Arizona Press, 2015), viii.

10. Francis L. Black, "Why Did They Die?," *Science* 258 (December 11, 1992): 1739.

11. James O. Pattie and Timothy Flint, eds., *The Personal Narrative of James O. Pattie of Kentucky, during an Expedition from St. Louis, through the Vast Regions between That Place and the Pacific Ocean, and Thence Back through the City of Mexico to Vera Cruz, during Journeyings of Six Years . . .* (Cincinnati: John H. Wood, 1831), 202 (hereafter Pattie, *Personal Narrative*).

12. Rosemary K. Valle, "James Ohio Pattie and the 1827–1828 Alta California Measles Epidemic," *California Historical Quarterly* 52, no. 1 (Spring 1973): 28–36.

13. Robert Archibald, *The Economic Aspects of the California Missions* (Washington, DC: Washington Academy of American Franciscan History, 1978), 179–181.

14. John Ryan Fischer, *Cattle Colonialism: An Environmental History of the Conquest of California and Hawai'i* (Chapel Hill: University of North Carolina Press, 2015), 44.

15. M. Kat Anderson, *Tending the Wild: Native American Knowledge and the Management of California's Natural Resources* (Berkeley: University of California Press, 2005), 76. According to Fischer, "the impact of introduced animals on endemic animals is unclear" during California's mission period. See Fischer, *Cattle Colonialism*, 50.

16. Henry T. Lewis, *Patterns of Indian Burning in California: Ecology and Ethnohistory* (Ramona: Ballena Press, 1973), v–xlviii; Jan Timbrook, John R. Johnson, and David D. Earle, "Vegetation Burning by the Chumash," *Journal of California and Great Basin Anthropology* 4, no. 2 (Winter 1982): 163–186; Steven W. Hackel, *Children of Coyote, Missionaries of Saint Francis: Indian-Spanish Relations in Colonial California, 1769–1850* (Chapel Hill: University of North Carolina Press, 2005), 337–338 (hereafter Hackel, *Children of Coyote*).

17. Antonio María Osio, Rose Marie Beebe, and Robert M. Senkewicz, eds. and trans., *The History of Alta California: A Memoir of Mexican California* (Madison: University of Wisconsin Press, 1996), 123–124.

18. Pattie, *Personal Narrative*, 211–217.

19. Rosemary Keupper Valle, "Prevention of Smallpox in Alta California during the Franciscan Mission Period (1769–1833)," *California Medicine* 119, no. 1 (July 1973): 74, 75.

20. Hubert H. Bancroft, *The Works of Hubert Howe Bancroft,* 39 vols. (San Francisco: The History Company, 1888), 34:632.

21. Pattie, *Personal Narrative*, 211–220; S. F. Cook, "Smallpox in Spanish and Mexican California, 1770–1845," *Bulletin of the History of Medicine* 7, no. 2 (February 1939): 183.

22. Valle, "Prevention of Smallpox," 76.

23. Victoria in Cook, "Smallpox," 183.

24. S. F. Cook, "The Epidemic of 1830–1833 in California and Oregon," *University of California Publications in American Archaeology and Ethnology* 43, no. 3 (May 10, 1955): 308 (hereafter Cook, "Epidemic of 1830–1833"); F. W. Howay, "The Brig Owhyhee in the Columbia, 1829–30," *Oregon Historical Society Quarterly* 35, no. 1 (March 1934): 1.

25. Howay, "Brig Owhyhee in the Columbia," 16.

26. David Igler, *The Great Ocean: Pacific Worlds from Captain Cook to the Gold Rush* (New York: Oxford University Press, 2013), 66; Howay, "Brig Owhyhee in the Columbia," 13–14.

27. Howay, "Brig Owhyhee in the Columbia," 14.

28. Ibid., 15, 19; Samuel Eliot Morison, "Columbia River Salmon Trade, 1830," *Oregon Historical Quarterly* 28, no. 2 (June 1927): 113.

29. Cook, "Epidemic of 1830–1833," 308; Igler, *Great Ocean*, 66.

30. David Douglas to W. J. Hooker, October 11, 1830, in W. J. Hooker, *The Companion to the Botanical Magazine; Being a Journal, Containing Such Interesting Botanical Information as Does Not Come within the Prescribed Limits of the Magazine; With the Occasional Figures,* 2 vols. (London: Printed by Edward Couchman for the proprietor Samuel Curtis, 1836), 2:146, 148; John McLeod in Cook, "Epidemic of 1830–1833," 308.

31. Cook, "Epidemic of 1830–1833," 303.

32. Douglas to Hooker, October 11, 1830, in Hooker, *Companion to the Botanical Magazine*, 2:146, 148. Italics in original.

33. Cook, "Epidemic of 1830–1833," 308.

34. For Native American voices, see William A. Slacum to John Forsyth, Secretary of State, March 26, 1837, in William A. Slacum, "Memorial of William A. Slacum, Praying Compensation for His Services Obtaining Information in Relation to the Settlements on the Oregon River," Senate Document No. 24, 25th Cong, 2nd Sess., 1837, p. 8 (hereafter Slacum, "Memorial"); Charles Wilkes, *Narrative of the United States Exploring Expedition,* 5 vols. (Philadelphia: C. Sherman, 1844), 5:148; Robert T. Boyd, *Coming of the Spirit of Pestilence: Introduced Infectious Diseases and Population Decline among Northwest Indians, 1774–1874*

(Seattle: University of Washington Press, 1999), 110. For non-Native voices, see Cook, "Epidemic of 1830–1833," 308.

35. Cook, "Epidemic of 1830–1833," 308.

36. Ibid.; Kenneth Thompson, "Insalubrious California: Perception and Reality," *Annals of the Association of American Geographers* 59, no. 1 (March 1959): 57.

37. Linda Nash, *Inescapable Ecologies: A History of Environment, Disease, and Knowledge* (Berkeley: University of California Press, 2006), 22, footnote 11.

38. Charles F. Lummis, *The Spanish Pioneers and the California Missions* (Chicago: A. C. McClurg & Co., 1929), 332. The Mexican official, landowner, and historian Antonio María Osio estimated 435,000 cattle in Osio, Beebe, and Senkewicz, trans. and ed., *History of Alta California*, 119–120. According to the Franciscan historian Zephyrin Englehardt, the twenty-one California missions possessed only 151,180 cattle, 137,971 sheep, 1,711 goats, 1,164 swine, 14,522 horses, and 1,575 mules as of December 31, 1832. See Zephyrin Engelhardt, *The Missions and Missionaries of California: Upper California*, 4 vols. (San Francisco: The James H. Barry Company, 1913), vol. 3, part 2: 653.

39. David Hornbeck, Phillip Kane, and David L. Fuller, *California Patterns: A Geographical and Historical Atlas* (Mountainview, CA: Mayfield Publishing Company, 1983), 58.

40. Anderson, *Tending the Wild*, 82.

41. Fischer, *Cattle Colonialism*, 48.

42. Anderson, *Tending the Wild*, 76.

43. John Work and Alice Bay Maloney, eds., "Fur Brigade to the Bonaventura: John Work's California Expedition of 1832–1833 for the Hudson's Bay Company," *California Historical Society Quarterly* 22, no. 3 (September 1943): 212.

44. John Work and Alice Bay Maloney, eds., "Fur Brigade to the Bonaventura: John Work's California Expedition of 1832–1833 for the Hudson's Bay Company (Concluded)," *California Historical Society Quarterly* 23, no. 2 (June 1944): 131–132.

45. Ibid., 134.

46. Ibid., 133.

47. "Trapper," n.d., in Cook, "Epidemic of 1830–1833," 305.

48. Cook, "Epidemic of 1830–1833," 310.

49. Ibid., 316–320.

50. Slacum to Forsyth, March 26, 1837, in Slacum, "Memorial," 16. For additional primary source accounts of the epidemic, see Cook, "Epidemic of 1830–1833," 316–19, and Peter Ahrens, "John Work, J. J. Warner, and the Native American Catastrophe of 1833," *Southern California Quarterly* 93, no. 1 (Spring 2011): 1–32.

51. Philip L. Edwards, *California in 1837: Diary of Col. Philip L. Edwards Containing an Account of a Trip to the Pacific Coast* (Sacramento: A. J. Johnston & Co., 1890), 30–31.

52. Stephen Powers, "The California Indians No. IX.—The Yocuts," *The Overland Monthly and Out West Magazine* 11, no. 2 (August 1873): 111.

53. Warner in Frank T. Gilbert, *History of San Joaquin County, California* (Oakland, CA: Thompson & West, 1879), 11–12. Warner first described the epidemic in "The Indian Pestilence in 1833," *Los Angeles Star*, August 23, 1874, 2.

54. Cook, "Epidemic of 1830–1833," 322; Cook, "Historical Demography," 92.

55. Pablo Tac and Minna and Gordon Hewes, eds. and trans., "Indian Life and Customs at Mission San Luis Rey: A Record of California Mission Written Life by Pablo Tac, an Indian Neophyte (Rome, c. 1835)," *The Americas* 9, no. 1 (July 1952): 98.

56. Igler, "Diseased Goods," 703.

57. Ibid., 716.

58. Boyd, *Coming of the Spirit of Pestilence*, 117.

59. James R. Gibson, *Otter Skins, Boston Ships, and China Goods: The Maritime Fur Trade of the Northwest Coast, 1785–1841* (Montreal: McGill-Queens University Press, 1992), 276.

60. Boyd, *Coming of the Spirit of Pestilence*, 130.

61. Hubert H. Bancroft, *The Works of Hubert Howe Bancroft*, 39 vols. (San Francisco: A. L. Bancroft & Co., 1886), 21:105; Hubert H. Bancroft, *The Works of Hubert Howe Bancroft*, 39 vols. (San Francisco: A. L. Bancroft & Co., 1884), 27:341–342.

62. Boyd, *Coming of the Spirit of Pestilence*, 130.

63. Cook, "Smallpox," 184.

64. J. Fernandez, "Cosas de California," 1874, BANC MSS C-D 10, Bancroft Library, Berkeley, California, 48–49.

65. E. Cerruti, "Establecimientos Rusos de California," 1875, BANC MSS C-D 75, Bancroft Library, Berkeley, California, 8.

66. No Author, *An Illustrated History of Sonoma County, California* (Chicago: The Lewis Publishing Company, 1889), 213.

67. Lorenzo Asisara in José Maria Amador, "Memorias sobre la Historia de California por José María Amador natural del Pais que nació el año de 1781 y vive hoy cerca del pueblito de Whiskey Hill," 1877, BANC MSS C-D 28, Bancroft Library, Berkeley, California, 112. Historian Martin Rizzo-Martinez kindly brought this source to my attention.

68. José Manuel Salvador Vallejo, "Notas Historicas sobre California por El Mayor Salvador Vallejo," 1874, BANC MSS C-D 22, Bancroft Library, Berkeley, California, 45.

69. Fernández in Cook, "Smallpox," 185.

70. Cerruti, "Establecimientos Rusos de California," 8.

71. Juan B. Alvarado in Cook, "Smallpox," 185. Juan Bautista Alvarado, *Historia de California*, 5 vols. (1876), BANC MSS C-D 4, Bancroft Library, Berkeley, California, 4:165–166.

72. Hornbeck, Kane, and Fuller, *California Patterns*, 58.

73. Hackel, *Children of Coyote*, 389, footnote 34.

74. *California*, 1844 (February-April) in Adele Ogden, "Trading Vessels on the California Coast: 1786–1848," 1979, BANC MSS 80/36c, Bancroft Library, Berkeley, California, 1104.

75. Cook, "Smallpox," 188.

76. *California*, 1844 in Ogden, "Trading Vessels on the California Coast," 1104–1105.

77. P. H. Reid in Cook, "Smallpox," 189.

78. *California*, 1844 (February–April), in Ogden, "Trading Vessels on the California Coast: 1786–1848," 1104; Cook, "Smallpox," 188.

79. Thomas O. Larkin to Messrs. Stearns & Temple, June 25, 1844, in "Abel Stearns Papers," MSS Stearns Papers, Box 39, Huntington Library, California.

80. Hackel, *Children of Coyote*, 422.

81. F. José J. Jimeno to F. Narciso Duran, August 22, 1844, DLG-0549-L01, Santa Bárbara Mission-Archive Library, Santa Barbara, California, 1–2.

82. No Author [Jesse D. Mason], *History of Amador County, California, with Illustrations and Biographical Sketches of Its Prominent Men and Pioneers* (Oakland, CA: Thompson & West, 1881), 35; See also George H. Tinkham, *A History of Stockton from Its Organization up to*

the Present Time Including a Sketch of San Joaquin County . . . (San Francisco: W. M. Hinton & Co. Printers, 1880), 107.

83. Tinkham, *History of Stockton from Its Organization up to the Present Time Including a Sketch of San Joaquin County* . . . , 25, 66, 67. Mention of 1845 is almost certainly a typographical error.

84. S. F. Cook, "The Conflict between the California Indian and White Civilization, II. The Physical and Demographic Reaction of the Nonmission Indians in Colonial and Provincial California," *Ibero-Americana* 22 (February 10, 1943): 18.

85. George Gibbs, "Journal of the Expedition of Colonel Redick M'Kee, United States Indian Agent, through Northwestern California," in Henry R. Schoolcraft, *Information Respecting the History, Conditions and Prospects of the Indian Tribes of the United States,* 6 vols. (Philadelphia: Lippincott, Grambo & Co., 1853), 3:107.

86. Stephen Powers, *Contributions to North American Ethnology,* 8 vols. (Washington, DC: Government Printing Office, 1877), 3:232.

87. For some major works on the history of disease in the Atlantic World, see David S. Jones, "Disease in the Atlantic World," *Oxford Bibliographies,* https://www.oxfordbibliographies.com/view/document/obo-9780199730414/obo-9780199730414-0079.xml.

Five

Sentiment and Gore
Whaling the Pacific World

Lissa Wadewitz

American whalers began entering the Pacific in the 1790s, but the height of the industry in that vast ocean was between the 1830s and 1860s. In those decades, thousands of mixed-race crews penetrated the far reaches of the Pacific World in pursuit of whales and other animals.[1] They caught, tormented, maimed, and killed thousands of marine animals, but whales were the most prized. Indeed, the indiscriminate hunting of whales contributed to the decline of both the whale populations and the industry as a whole. These events also had significant, albeit uneven, environmental, cultural, and economic repercussions for people located closest to the most popular hunting grounds.[2]

From our twenty-first-century perspective, the violence that pervaded the whaling industry of these decades is striking.[3] This is admittedly due to the nature of our contemporary relations with animals. Most Americans today are far less likely to have experience killing animals as did our nineteenth-century counterparts. Still, the sheer volume of animal blood shed is a critical part of this history that deserves more attention from historians. Somewhat surprisingly, many seafarers of this era in fact expressed a level of wonder and sentimentality toward whales that was at odds with both their role as market hunters and the gore that typified the industry.[4] This chapter seeks to assess how American debates regarding both human–non-human relations and the aesthetics of nature appear to have filtered into the Pacific whaling fleet and influenced whalers' behavior. The evidence, although predominantly produced by literate Euro-Americans, further suggests that the unique social, cultural, and economic relations that emerged on board nineteenth-century vessels also factored into whalers' interactions with the leviathans they pursued across the Pacific World.

Signs of Sentimentality

Whalers' behavior toward many creatures suggests that they viewed most animals as sources of food, potentially life-threatening, and/or fully expendable. They caught porpoises and turtles for food variety, toyed with albatrosses and other birds, and took great pleasure in catching and tormenting sharks.[5] But whales, so much larger and distinct from terrestrial fauna, were different. Many seafarers recognized the intelligence and human-like qualities of whales, and they frequently described their encounters with a sense of awe and humility. It seems that discussions about the human place in nature and the proper treatment of animals that permeated American popular culture in the 1800s influenced such reactions. Given the aims of the commercial whaling industry, it is easy to overlook the capacity of whalers to express appreciation for the mammals they pursued, but the sentiments are there.

Like both land-based sports and market hunters, whalers quickly learned the habits of their prey. They could identify spouting patterns from great distances and noted different species' habits and personalities. They also observed and marveled at the whales' curiosity and intelligence. Sailors soon realized, for instance, that over time, whales could identify whaling boats and gave them wide berth. After striking a large male whale in the summer of 1852, novice whaler Enoch Cloud noted, "He was well acquainted with a boat it seems! He allowed no chance to get a lance at him & after running 'til night we cut the line & let him go!"[6] Mariners further imbued whales with other human qualities, such as the ability to taunt or feel emotions. According to another seasoned hand, "'Whales has [*sic*] feelings as well as any body. They don't like to be stuck in the gizzards, and hauled alongside, and cut in, and tried out in them 'ere boilers no more than I do.'"[7]

Whalers regularly also remarked upon the sociability of the whaling groups they encountered. As the Reverend Henry Cheever observed, "It is evident that the societies of these great sea monsters seldom go to war, but live together in cordial and happy amity, and render each other all the help in their power when in distress."[8] In the spring of 1840, Francis Olmsted witnessed the lancing of a pilot whale that had been playing about the ship's bow. When the boats towed the bleeding animal to the vessel, they were "accompanied by all his companions spouting and foaming around the boats like attendant tritons. So affectionate are these poor fish, that when one of their number is struck by the whaler, the school continues around the sufferer, appearing to sympathize with him in his agonies." "Even when dead," he continued, "they do not desert him, and it was not until a long time after the victim had been hoisted upon deck, far from their sight, that they abandoned him."[9] That same season Olmsted also watched a large school of sperm whales feeding and playing on the open ocean. However, "let one of the school become alarmed at the approach of danger, and with a flourish of his flukes, well understood, the alarm is instantly communicated to the others, though scattered for several miles over the ocean, and they betake themselves to precipitate flight."[10] Whalers recognized that these creatures had both a sense of community and forms of communication.

Even seasoned seafarers were not immune to the enchanting characteristics of whales. Arctic whaler Herbert Aldrich was surprised to learn that several captains in the fleet had heard whales sing. "I at first took this for a sophomoric joke, slyly intended for me to bite at, so I kept quiet," he remarked. "But one day there was a rehearsing of experiences, and I found that the masters really believed that whales do sing."[11] William Whitecar was similarly impressed by a sperm whale breach in the 1850s: "I was struck with the greatness of the Creator's works in this, to us, almost unknown element."[12] Seaman and memoirist Richard Henry Dana's description of a school of whales he encountered off Cape Horn in the 1830s echoes the experiences of other men at sea:

> We were surrounded by shoals of sluggish whales and grampuses, which the fog prevented our seeing . . . heaving out those peculiar lazy, deep, and long-drawn breathings which give such an impression of supineness and strength. Some of the watch were asleep, and the others were perfectly still, so that there was nothing to break the illusion, and I stood leaning over the bulwarks, listening to the slow breathings of the mighty creatures.[13]

Tellingly, whalers often recorded the touching maternal behavior of female whales. By the 1840s, for instance, American whalers found that attacking gray whales on their birthing grounds was a highly effective way to kill female whales. According to one whaler, the calf's mother "will not readily desert her offspring, and in her extreme solicitude for her young, is a frequent victim. The taking of one of a school, almost always ensures the capture of another, for his [or her] comrades do not immediately abandon the victim, but swim around him, and appear to sympathise with him in his sufferings."[14] A whaling captain's wife recorded a similarly haunting scene in Magdalena Bay off the coast of California in 1846: "A plenty of boats stove every day and they all say these are the worst whale to strike they ever saw. The only way they can get fast [harpoon] is to chase the calf till it gets tired out then they fasten to it and the whale [the mother] will remain by its side and is then fastened too."[15] In his remarks on the behavior of the right whale, Reverend Cheever noted that "its immediate recourse is to flight, except when it has young to look out for, and then it is bold as a lion, and manifests an affection which is itself truly affecting."[16]

Many of these attitudes and observations suggest that cultural movements back in the states may have factored into American whalers' attitudes toward their prey. The expanding animal welfare movement that emerged out of the Second Great Awakening of the early 1800s seems to especially be at work here. Children's literature and Sunday school curricula increasingly presented anthropomorphized stories of animals with human-like feelings and animal families that showcased maternal love.[17] This attention to animal families in popular culture may have prompted Euro-American whalers in particular to pay closer attention to these behaviors among the whales they encountered. Given that providing a "voice" to "dumb animals" was becoming a central tenet of this movement by the mid-nineteenth century, it is possible that witnessing various forms of intra-whale communication and human-like qualities added to the ability of whalers to appreciate the leviathans in their midst.[18]

Although predominantly engaged in by Euro-American elites, the rise of Romanticism and debates regarding the meaning of nature may have further influenced how some whalers perceived the mammals they pursued. In particular, the concept of the sublime as articulated by Edmund Burke and Immanuel Kant in the mid-to-late 1700s and then perpetuated in the works of painters and writers like Thomas Cole and Henry David Thoreau in the 1800s seems to animate the whalers' language and reactions to seeing whales firsthand.[19] Both Cole and Thoreau especially grappled with the natural world's sublimity, that is, the ability of nature to evoke both terror and awe simultaneously. After experiencing a violent thunderstorm in the Catskill Mountains in the 1820s and venturing to the White Mountains of New Hampshire, for instance, Cole observed that "man may seek such scenes and find pleasure in the discovery, but there is a mysterious fear [that] comes over him and hurries him away. The sublime features of nature are too severe for a lone man to look upon and be happy."[20] Thoreau had a similarly sobering experience climbing Maine's Mt. Katahdin in the 1840s. As historian Roderick Nash notes, "The line between the sublime's delightful horror and genuine terror was thin."[21] Coming face-to-face with an angry eighty-ton sperm whale might have easily evoked such mixed feelings among American whalers as ruminations about the meaning of nature swirled through popular culture forms of this period.

The Hunt Proceeds

Whalers thus often admired the whales they encountered for various reasons, but, as the above examples demonstrate, this reverence did not interfere with the hunt for several reasons. As with land-based market hunters, some of these motivations are fairly straightforward given the demands of the industry and the economic goals of individual workers. But to fully grasp whalers' ability to both revere whales and then indiscriminately kill them requires a more in-depth examination of the ways in which ideas about the human place in nature and human–non-human relations were changing over the course of the nineteenth century and what those ideas meant for life aboard American whaling vessels as they increasingly expanded across the Pacific Ocean.

Perhaps the most compelling driver of whaler behavior was the desire to make money. When whalers signed onto a voyage, they agreed to both a specific position on board and a set "lay," or percentage of the ship's total profits at the conclusion of the trip. A captain might receive 1/12th of the cut and officers 1/25th, but the average unskilled crewman usually earned just 1/200th of the final profits.[22] Because the length of the voyage and the crew members' pay were both directly related to how much whale oil and baleen a ship accumulated, everyone kept careful track of the vessel's total supplies.[23] Mary Lawrence, a whaler's wife, worried about the strain on her husband when his ship experienced a bout of bad luck. "What long faces greet my eyes. Everybody is discouraged," she wrote in her journal in the summer of 1858.[24] The need to return home with a specific amount of oil and bone in the hold so as to make the voyage worthwhile could provide tremendous incentive for whalers to harden their hearts against the animals around them.

The fact that chasing and harpooning whales was also a highly dangerous activity most certainly compelled these men to feel animosity toward their prey.[25] Every time whalers engaged in the chase, they were acutely aware they were putting their lives at risk. Indeed, stories about angry whales smashing boats and summarily tossing men into the sea pervaded whaling lore (see figure 5.1). Whalers regularly broke bones, suffered from exposure, and, of course, lost their lives in pursuit of these creatures.[26] As one whaler remarked in the 1850s, "It gives a faint idea of the monstrous size of the terrible animals with which we have to deal! But . . . to be an eyewitness of their amazing strength & agility . . . to be seated in a frail boat, in the middle of the ocean, exposed to their fury . . . there is no sport connected with any part of it!"[27]

Chasing whales could also be an incredibly frustrating experience, particularly if the whales were elusive or made off with expensive equipment. Such encounters would no doubt have heightened the whalers' desire to kill their quarry. Enoch Cloud's ship ran into a spate of such bad luck in 1852. Although the crew pursued whales for nearly seventy successive days, more than half of those days were spent chasing whales for hours on end, with nothing to show for the effort.[28] Even if a harpooner managed to "fasten to" a whale, the animal often escaped by diving or, if dying, by sinking; many

Figure 5.1. A whale staving a boat. Painting by Charles Sidney Raleigh (1877). Courtesy of the New Bedford Whaling Museum. Catalog No. 2001.100.4329.

vessels lost thousands of dollars of equipment when this happened, which then affected the overall profit margins.[29]

Less obvious motivations likely emanated from the class tensions that pervaded most whaling ships in this period. Like most marine vessels, status and skill determined both wage rates and the strict hierarchy by which whalers were organized. The captain and the ship's officers tended to be experienced men who earned the ship's largest returns. Because they also generally hailed from more privileged backgrounds, officers often believed themselves to be of a higher social status than the rest of the crew. If the captain or an officer were also prone to corporal punishment and cruel behavior, as many were, these class/status tensions were often exacerbated further.[30] Harpooning a whale might thus affirm an officer's sense of authority and superiority. For their part, the crew sometimes deliberately defined themselves in opposition to the ship's officers, using these divisions to strengthen the bonds among the men of the forecastle.[31] Successfully killing the voyagers' primary prey could also be a way for "lowly" forecastle workers to assert power and highlight the value of their contributions to the ship.

Such class concerns likely often intersected with expressions of masculinity that could potentially unify the crew as well. Whalers' rituals and social interactions might build gender solidarity in the forecastle, for example. The historian Margaret Creighton argues that many men—especially younger men from New England—embarked on whaling voyages as a rite of passage toward independent manhood. According to Creighton, this expectation then fueled these whalers' actions toward one another, their attitudes toward women and sex, and their desire to create a collective masculine identity while on board ship.[32]

As American whalers more regularly entered the Pacific, racial tensions grew in overall intensity as well. Since finding adequate hands for the entire fleet became more difficult over time, whaling captains started leaving New England ports with mere skeleton crews, seeking to hire additional laborers en route to the Pacific.[33] The result was that American whaling crews grew strikingly diverse. Over three thousand African Americans worked aboard New Bedford whaling ships between 1803 and 1860, while one in six whaleships had at least one Native American on board, and some had as many as six or seven Native men per voyage.[34] After leaving New England, these ships made for the Azores, or "Western Islands," whose population was Portuguese, Catholic, and mixed-race. Next was the Cape Verdean archipelago, which had a darker-skinned population that white Americans often referred to as "Portuguese blacks." Once in the Pacific, whaling captains acquired additional islander crewmembers, particularly Hawaiians and Māori (see figure 5.2).[35]

Given that the crews were so diverse and the dominant racial discourse of the nineteenth-century United States so predicated on Euro-American men's sense of cultural and biological superiority, tensions between racial groups were bound to erupt. Historians have found that African American men, due to white racism, stereotypes about their abilities, and the fact that they were often hired to do "women's work," fared the worst in the American whaling fleet.[36] They had fewer opportunities overall, were often poorly treated, and were sometimes murdered.[37] Racial animosities were so pronounced on some vessels, living spaces were segregated by race.[38]

Despite instances of violence and inter-racial tensions, whaling still generally offered thousands of men of color a way to achieve some sense of freedom, self-respect, and a chance to advance in their careers (see figure 5.3). In the words of one African American boardinghouse master, while aboard American whaling vessels, "A coloured man is only known and looked upon as a man, and is promoted in rank according to his ability and skill to perform the same duties as the white man."[39] Native American men ironically benefited from stereotypes about their skills as hunters to advance to officer positions after the 1830s. According to historian Nancy Shoemaker, rank trumped race. "That a man of color as an officer had special privileges could have fueled

Figure 5.2. Portrait of Captain Amos Haskins, a Wampanoag Indian (1850s). Courtesy of New Bedford Whaling Museum. Item No. 00.231.27.

white foremast hands' resentment," she argues, "but ship rules protected and legitimated those privileges."[40]

Did this unfamiliar world of race relations affect whalers' relationships with or attitudes toward the whales of the Pacific World? It is difficult to say with certainty, but the lines between animal and human were definitely blurry, especially for Euro-Americans thrown into new situations and places inhabited by "exotic," dark-skinned people who were rumored to eat human flesh. Although debates about race, racial origins, and the link between the worlds of human and animal had a long history in Europe and the United States, Euro-American ideas about the proximity of non-whites to the animal kingdom intensified over the course of the 1800s.[41] That Americans also were

Figure 5.3. Japanese artist Usui Shozo's depiction of American whalemen in Japan in 1845 affirms the diversity of American whaling crews. Courtesy of the New Bedford Whaling Museum. Catalog No. 1986.36.

unsure about how to classify whales (fish or mammal?) merely added to the confusion about the accuracy of existing taxonomies of the natural world.[42] It is certainly possible that these uncertainties influenced how whalers perceived and acted toward the whales they so eagerly killed. Many Euro-American whalers in fact specifically referred to the new dark-skinned peoples they encountered as not only lesser human beings but as actual animals. Whaler George Blanchard, for instance, expressed repugnance for the women of Cape Verde: "Love could never nestle on the thick Black Lips of a Portugee niggar . . . Saving their faces (the best resemblance to which is their imitative companions of the woods, the monkeys) the young ladies . . . might rival the finest figures in our own Country. In purchasing one of these Animals, you don't buy a Pig in a Poke, you see your bargain."[43] With regard to some Pacific Islanders with whom his expedition was having trouble, mariner William Reynolds took these sentiments further when he fumed, "So that I regard the bloody fiends as I do the sharks, and would feel the same kind of inward joy in killing them in battle, as I exult in when one of those monsters of the sea is torn from his hold on life."[44] Likening human beings to lowly beasts no doubt gave Euro-Americans license to treat people of color both poorly and violently—if they could. But what if such behavior was not only unsanctioned, but punishable?[45]

The Pacific whaling grounds thus presented a racially mixed-up world where whites continually articulated their deeply held beliefs about Euro-American superiority and the clash between "savagery" and "civilization," despite the racially diverse reality of the whaling fleet and the actual power structures they lived under every day.[46] Unable to freely lash out at the men of color aboard their ships and seeing the special privileges afforded all officers regardless of skin color may have pushed Euro-American whalers to channel their frustrations at the animals they *did* have license to harm.

But what about whalemen of color? Despite a paucity of written records, available evidence suggests men of color may have felt compelled to join the hunt due to their frustrations with racist treatment or abuse. Although whalers of all ethnicities also saw the whale hunt as a way to distinguish themselves as men and prove their worth to their fellow crewmembers, men of color may have more acutely felt a need to appear brave, skilled, and thus worthy of their positions. Surrounded by potentially hostile white crewmates, some men of color may have more zealously pursued the whales they encountered as a result.[47]

Unlike animal slaughter involving smaller species, the whale hunt was inherently a collective enterprise. The whaleboat required all hands to act in concert as they determinedly rowed toward their quarry on the open water. That their target was revered as intelligent and powerful heightened the challenge. During the hunt, differences of race and class fell away out of necessity, and the goal at hand pitted man against leviathan.[48] As Enoch Cloud observed following a particularly arduous, but ultimately successful, whale chase in the winter of 1851, "It was the most terrible sight I ever witnessed. Three hearty cheers burst from the four boats as a stream of blood shot from her spout-holes, full 30 feet into the air! . . . And when I saw this, the largest & most terrible of all created animals bleeding, quivering, dying a victim to the cunning of man, my feelings were indeed peculiar!"[49] Such cohesiveness in the heat of battle does not mean, however, that issues of power and perceived social or biological difference were not continually at work in this watery world, affecting the ways these diverse peoples interacted and negotiated their roles on board ship.

Conclusion

Nineteenth-century whalers had an impact on many of the animal populations of the Pacific World, not just their primary prey.[50] Yet their attitudes and actions toward whales were based on complicated and evolving understandings of human–non-human relations and perceived ethno-racial hierarchies. Emerging debates regarding slavery, taxonomies of the natural world, the aesthetics of nature, and the proper treatment of animals also all appear to have combined and seeped into the whaling world of the 1800s, particularly for whalers of Euro-American descent. A closer examination of the relationships between whalers and whales, and the social, cultural, and economic tensions that increasingly infused the Pacific whaling fleet, thus adds nuance to our understanding of the daily experiences of the thousands of diverse workers whose muscles powered this vital industry. The wanton destruction of whales and its attendant environmental consequences are important pieces of the larger story of imperial ambition, cultural contact, and global exchange that characterized the nineteenth-century Pacific World.

Notes

Sincere thanks to Tom Mertes, Jennifer Seltz, Rachel St. John, Louis Warren, the volume editors and fellow authors, and the participants of the UC Davis Environment and Society Colloquium and the University of Oregon History Department Brownbag Series for valuable feedback on earlier drafts of this chapter. Thanks also to my able student research assistants, Annika Albrecht and Kara Skokan.

1. Nancy Shoemaker, *Native American Whalemen and the World: Indigenous Encounters and the Contingency of Race* (Chapel Hill: University of North Carolina Press, 2015), 12.

2. David Igler, *The Great Ocean: Pacific Worlds from Captain Cook to the Gold Rush* (New York: Oxford University Press, 2013), 99–128.

3. Igler, *The Great Ocean*, 105, 113, 116–117, 121–124.

4. John F. Reiger, *American Sportsmen and the Origins of Conservation*, 3rd rev. ed. (Corvallis: Oregon State University Press, 2001), 5–44.

5. Henry T. Cheever, *The Whale and His Captors; or, The Whaleman's Adventures, and the Whale Biography, as Gathered on the Homeward Cruise of the "Commodore Preble"* (1850; reprint Fairfield, WA: Ye Galleon Press, 1991), 97, 132; William B. Whitecar, *Four Years aboard the Whaleship Embracing Cruises in the Pacific, Atlantic, Indian and Antarctic Oceans in the Years 1855, '6, '7, '8, '9* (Philadelphia: J. B. Lippincott & Co., 1860), 36–37; Francis Allyn Olmsted, *Incidents of a Whaling Voyage to Which Are Added Observations on the Scenery, Manners and Customs, and Missionary Stations of the Sandwich and Society Islands* (Rutland, VT: Charles E. Tuttle Co., 1969), 91–94, 113; J. C. Mullet, *A Five Year's Whaling Voyage, 1848–1853* (Fairfield, WA: Ye Galleon Press, 1977), 12–13.

6. Enoch Carter Cloud, *Enoch's Voyage: Life on a Whaleship, 1851–1854*, ed. Elizabeth McLean (Wakefield, RI: Moyer Bell, 1994), 151 (quotation). See also, Cheever, *The Whale and His Captors*, 155; Stanton Garner, *The Captain's Best Mate: The Journal of Mary Chipman Lawrence on the Whaler Addison, 1856–60* (Hanover, NH: University Press of New England, 1966), 37, 38.

7. Quoted in Cheever, *The Whale and His Captors*, 126.

8. Ibid., 135.

9. Olmsted, *Incidents of a Whaling Voyage*, 155.

10. Ibid., 158.

11. Herbert L. Aldrich, *Arctic Alaska and Siberia or, Eight Months with the Arctic Whalemen* (Chicago: Rand, McNally, and Co., 1889), 3335.

12. Whitecar, *Four Years aboard the Whaleship*, 36.

13. Richard Henry Dana, *Two Years before the Mast: A Personal Narrative of a Life at Sea* (New York: Penguin Books, 1981), 69.

14. Charles Melville Scammon, *Journal aboard the Bark Ocean Bird on a Whaling Voyage to Scammon's Lagoon, Winter of 1858–1859*, ed. David A. Henderson (Los Angeles: Dawson's Book Shop, 1970), 27 (quotation); Olmsted, *Incidents of a Whaling Voyage*, 61.

15. Quoted in Igler, *The Great Ocean*, 122.

16. Cheever, *The Whale and His Captors*, 110 (quotation); Cloud, *Enoch's Voyage*, 55, 111–112, 257, 259.

17. Katherine C. Grier, "'The Eden of Home': Changing Understandings of Cruelty and Kindness to Animals in Middle-Class American Households, 1820–1900," in *Animals in Human Histories: The Mirror of Nature and Culture*, ed. Mary J. Henninger-Voss (Rochester, NY: Rochester University Press, 2002), 340–341. For a rare example of such a story focused on whales, see "Barbarity of Whale Fishing" (July 4, 1827) in *Youth's Companion*, ed. Lovell Thompson et al. (Boston: Houghton Mifflin Co., 1954), 1119–1121.

18. Grier, "'The Eden of Home,'" 348–349.

19. Edmund Burke, *A Philosophical Enquiry into the Origin of our Ideas of the Sublime and Beautiful* (New York: Garland Publishers, 1971); Immanuel Kant, *Critique of Judgement* (Mineola, NY: Dover Publications, 2005).

20. Quoted in Roderick Frazier Nash, *Wilderness and the American Mind*, 4th ed. (New Haven, CT: Yale University Press, 2001), 79.

21. Nash, *Wilderness and the American Mind*, 79.

22. Eric Jay Dolin, *Leviathan: The History of Whaling in America* (New York: W. W. Norton & Company, 2007), 270–271.

23. Cloud, *Enoch's Voyage*, 146, 149, 151, 265, 293; Igler, *The Great Ocean*, 122–123.

24. Garner, *The Captain's Best Mate*, 103, 115 (quotation), 121.

25. Alan Taylor makes similar arguments about colonial America; see his "'Wasty Ways': Stories of American Settlement," *Environmental History* 3, no. 3 (July 1998): 291–310.

26. Cheever, *The Whale and His Captors*, 100, 103–104, 148, 171, 186–188, 211–229; Cloud, *Enoch's Voyage*, 79, 80–81, 93, 127–129, 132–133, 145–146, 149, 251–252, 264; Olmsted, *Incidents of a Whaling Voyage*, 113–115; Garner, *The Captain's Best Mate*, 40, 62, 150; Margaret Creighton, *Rites & Passages: The Experience of American Whaling, 1830–1870* (Cambridge: Cambridge University Press, 1995), 63–68; Igler, *The Great Ocean*, 119–122.

27. Cloud, *Enoch's Voyage*, 263–264.

28. Ibid., 63–169, 90–91 (quotation); Olmsted, *Incidents of a Whaling Voyage*, 79–82, 102–104, 112–113.

29. Cheever, *The Whale and His Captors*, 133; Scammon, *Journal aboard the Bark Ocean Bird*, 30–33; Whitecar, *Four Years aboard the Whaleship*, 35–36; "Charles Town Nareganst Joseph Ammons," July 29, 1846, journal entry excerpted in *Living with Whales: Documents and Oral Histories of Native New England Whaling History*, ed. Nancy Shoemaker (Amherst: University of Massachusetts Press, 2014), 82.

30. Jeffrey Bolster, *Black Jacks: African American Seamen in the Age of Sail* (Cambridge, MA: Harvard University Press, 1997), 75, 79–82; Dolin, *Leviathan*, 255–260; Elmo Paul Hohman, *The American Whaleman: A Study of Life and Labor in the Whaling Industry* (New York: Longmans, Green and Co., 1928), 59–60.

31. Creighton, *Rites & Passages*, 116–138.

32. Ibid., 116–138.

33. Whitecar, *Four Years aboard the Whaleship*, 22, 40; Dolin, *Leviathan*, 221.

34. Shoemaker, *Native American Whalemen*, 13–14; Cloud, *Enoch's Voyage*, 38; James Farr, "A Slow Boat to Nowhere: The Multi-Racial Crews of the American Whaling Industry," *The Journal of Negro History* 68, no. 2 (Spring 1983): 159–170; Bolster, *Black Jacks*, 177, 220; David A. Chappell, *Double Ghosts: Oceanian Voyagers on Euroamerican Ships* (Armonk, NY: M. E. Sharpe, 1997). The number of Native American and African American mariners declined as the nineteenth century progressed.

35. Shoemaker, *Native American Whalemen*, 40–57, 48, 147; Joan Druett, *In the Wake of Madness: The Murderous Voyage of the Whaleship* Sharon (Chapel Hill, NC: Algonquin Books, 2003), 100–110; Dolin, *Leviathan*, 223–226; Hohman, *The American Whaleman*, 50–57. Some whalers claimed that owners/captains chose mixed-race crews so as to limit the chances of rebellion. See John Newton, *A Savage History: Whaling in the Pacific and Southern Oceans* (Sydney: New South Publishing, 2013), 186, and S. George Ellsworth, ed., *The Journals of Addison Pratt: Being a Narrative of Yankee Whaling in the Eighteen Twenties, A Mormon Mission to the Society Islands, and of Early California and Utah in the Eighteen Forties and Fifties* (Salt Lake City: University of Utah Press, 1990), 15–16.

36. Bolster, *Black Jacks*, 5; Chappell, *Double Ghosts*, 43–77; Creighton, *Rites & Passages*, 159; Stephen Reynolds, *The Voyage of the New Hazard: To the Northwest Coast, Hawaii, and China, 1810–1813*, ed. Judge F. W. Howay (Salem, MA: Peabody Museum, 1938), 11, 14 footnote 31, 15, 112.

37. Druett, *In the Wake of Madness*, 113–132; Chappell, *Double Ghosts*, 60, 68–69.

38. Bolster, *Black Jacks*, 177; Creighton, *Rites & Passages*, 185–187.

39. Quoted in Bolster, *Black Jacks*, 177; Creighton, *Rites & Passages*, 187.

40. Shoemaker, *Native American Whalemen*, 66.

41. See, for instance, James Belich, "Race," in *Pacific Histories: Ocean, Land, People*, ed. David Armitage and Alison Bashford (New York: Palgrave Macmillan, 2014), 263–281; Cristin Ellis, *Antebellum Posthuman: Race and Materiality in the Mid-Nineteenth Century* (New York: Fordham University Press, 2018), 23–60; Ibram X. Kendi, *Stamped from the Beginning: The Definitive History of Racist Ideas in America* (New York: Nation Books, 2016).

42. D. Graham Burnett, *Trying Leviathan: The Nineteenth-Century New York Case That Put the Whale on Trial and Challenged the Order of Nature* (Princeton, NJ: Princeton University Press, 2007).

43. Diary of George Blanchard, *Pantheon*, December 26, 1842, Private Collection, Tom Bullock Jr. Courtesy of Tom Bullock Jr., Jane Blanchard Bullock, and George Blanchard, quoted in Creighton, *Rites & Passages*, 158.

44. William Reynolds, *The Private Journal of William Reynolds: United States Exploring Expedition, 1838–1842*, ed. Nathaniel Philbrick and Thomas Philbrick (New York: Penguin Books, 2004), 241. Although Reynolds was not on a whaleship, this behavior appears to have been common across the merchant marine and the whaling fleet. See also Chappell, *Double Ghosts*, 63; Cloud, *Enoch's Voyage*, 176, 214–215.

45. For more on the intersections of race and wildlife, see Miles A. Powell, *Vanishing America: Species Extinction, Racial Peril, and the Origins of Conservation* (Cambridge, MA: Harvard University Press, 2016), 1–45. See also Josiah C. Nott and George R. Glidden, *Types of Mankind, or, Ethnological Researches Based upon the Ancient Monuments, Paintings, Sculptures, and Crania of Races* (Philadelphia: Lippincott, Grambo, and Co., 1854).

46. Shoemaker, *Native American Whalemen*, 78–79.

47. Bolster, *Black Jacks*, 166–170; Creighton, *Rites & Passages*, 73–74.

48. Bolster, *Black Jacks*, 82–93.

49. Cloud, *Enoch's Voyage*, 52–53.

50. See, for instance, Chappell, *Double Ghosts*, 78–156; Igler, *The Great Ocean*, 43–97; J. R. McNeill, "Of Rats and Men: A Synoptic Environmental History of the Island Pacific," *Journal of World History* 5, no. 2 (Fall 1994): 31225; Gregory Rosenthal, *Beyond Hawai'i: Native Labor in the Pacific World* (Oakland: University of California Press, 2018).

Six

Changes on the Plantation

An Environmental History of Colonial Samoa

Holger Droessler

With their tropical climate and ample rainfall, the Samoan Islands have sustained human inhabitants for three millennia. Surrounded by coral reefs, the Samoan archipelago consists of ten inhabited islands. The three biggest islands—Savai'i, Upolu, and Tutuila—were home to over 35,000 inhabitants at the end of the nineteenth century.[1] In contrast to other Pacific archipelagos, such as Hawai'i or the Philippines, Samoa lacks substantial areas of flat land that could be turned into large-scale plantations. The islands' volcanic origin accounts for their steep and rugged terrain, susceptible to soil erosion and coastal floods in the wake of cyclones and tsunamis. Only a third of Tutuila and the Manu'a islands have ground slopes of 30 percent or less.[2] Due to its narrow coastal plains, overall arable land in Samoa encompasses not more than a fifth of its total area. The largest island in the archipelago, Savai'i, is also the best suited for plantation agriculture because its central peak, the volcano Mt. Matāvanu, rises gradually from the coast. The rainforest that stretches over 280 square miles across Savai'i's volcanic center is home to most of Samoa's native species of flora and fauna. Despite Savai'i's topographical advantages, most large Euro-American plantations emerged in Upolu, the traditional center of population and commerce.

From the mid-nineteenth to the early twentieth century, the natural environment of Samoa changed dramatically, as Germany, Great Britain, and the United States vied for control over the islands. Over the course of a half century, long-standing Samoan farming practices came under increasing pressure by the expansion of Euro-American cash crop plantations. To maximize profits and control, Euro-American plantation owners and colonial officials introduced clearly defined and permanent lines into the Samoan natural and social landscapes. Putting fences and barbed wire around parcels of land, as Euro-American settlers did, challenged Samoan conceptions of space and time. Ultimately, Samoa's integration into the global capitalist economy at the turn of the twentieth century had profound environmental consequences. Colonization brought changes on Samoan plantations by alienating land, turning coconuts into cash crops, and introducing new plants and animals. Confronted with migrant ecologies, Samoans protected their sustainable farming practices and adapted to the new world of copra.

Relating Land and Language

Since first human settlement, Samoa's tropical climate has shaped patterns of living and laboring. On tiny Ofu Island in the eastern Manu'a chain, as on other Samoan islands, permanent human settlements emerged close to coastlines with access to the ocean.[3] Trading and fishing took place there, and the occasional breeze promised relief from the humid air. Samoa's location just south of the equator minimizes differences in temperature throughout the year, which ranges from 65°F to 105°F. Life in Samoa is dominated by a rainy season between November and April, followed by a dry season from May to October, with at times strong trade winds from the southeast.[4] Samoans developed their own nomenclature to demarcate five distinct seasons throughout the year: *vāipalolo* (more rain), *vāitoelau* (nice and cool trade winds), *vāituputupu* (when plants grow in abundance), *aununu* (hurricanes and cyclones), and *tuiefu* (when the sun becomes unbearable and the soil hard as rock).[5] To Samoans, wind, water, and sun are inextricably connected to the changing fortunes of the soil.

Samoans also divided their land into five different categories of ownership and use. First, village house lots were located close to the sea and typically included a small patch of taro plants and the family tombs. Village house lots were not used for full-scale agriculture, which was practiced on designated plantation lots, the second land category. These plantation lots were family owned and were located behind villages, hosting major stands of coconut and breadfruit trees, together with occasional taro and banana plots. On their plantation lots, Samoans cultivated food crops both for their own consumption and for sale. In this primary "coconut zone," irregularly spaced palm trees of varying ages overshadowed extensive thickets of scrub.[6] Third, family reserve sections, further uphill from the plantation lots, consisted of taro, yam, and banana plots, and were not under steady cultivation. Harvested crops had to be carried a long way back to the village on the narrow paths zigzagging through the dense forest.[7] Together with the plantation lots, the family reserve sections yielded the majority of food crops to Samoans.

When the denser secondary tropical rainforest in these higher-lying sections was cleared, the soil provided a base for taro production.

The other two categories of land—village lands and district lands—were also significant for Samoan food production. The village lands were located further uphill from the family reserve sections, but also included the reef and village common (*malae*) and the areas between village boundaries. Cultivation established use right on village lands, which the village council (*fono*) often subsequently confirmed. Samoans hunted pigs in the wilder village lands uphill, whereas communal fishing with large nets took place in the reef. District lands, finally, were primarily claimed by district councils to establish political boundaries between separate districts. There, Samoans hunted pigeons and pigs and also collected wild forest products.[8] Crucial to this Samoan taxonomy of land was the fact that ownership of land did not necessarily entail ownership of the crops the land yielded.

Social conceptualizations of agricultural land and of space, more broadly, are reflected in the Samoan language. For example, the word *vā* is used to describe "spaces in between" in several Polynesian languages, including Samoan. These spaces can be located between things like coconuts on a family field (*vānui*) or between people. *Teu le vā* (tend to the space between) functions as an important social imperative in Samoan society when parents admonish their children to pay respect to status hierarchies and their proper place within them. In general, social, political, and spatial boundaries in Samoan society derive from shifting relationships between points (*mata*) rather than from sharply defined boundaries, such as those set by colonial settlers around their plantations.[9] Hence, *matāmutia* literally means "grassy area" (*mutia*) made up of "points" (*mata*) and refers to the small taro plots next to Samoan homes. *Matāvao*, to pick another example of this point-field spatiality, describes the dynamic area between the edge of a Samoan plantation and the virgin bush.[10]

Producing Food Crops

Informed by these understandings of space, Samoans cultivated a variety of root and tree crops, from taro to coconuts and breadfruit. Due to the advantageous climate, planting could be done all year around. Taro (*talo*) was the most important food crop that most Samoan families relied on. Many of the bigger taro plantations were located on family reserve sections, further away from home. Taro planting and harvesting continued throughout the year, except during the drier months of July and August. On average, a half-acre taro plantation could yield more than 1,200 tubers a year for a family.[11] Taro was so important to Samoan life and culture that a family's inability to serve taro to guests indicated a fundamental breakdown in family and village relations. Several Samoan proverbs attested to the cultural significance of taro. The saying "Let each plant two taros in a particular spot" conveyed the importance of economic self-reliance and the advantage of having both household and plantation taro in times of need.[12] During droughts, Samoans cherished giant taro (*ta'amu*), a related species that could be grown with little initial investment.[13] Overall, taro and its bigger brother demonstrated the central aim of Samoan farming to ensure not only physical survival but also cultural continuity.

Like the taro, the coconut (*niu*) combined economic and cultural functions. As nourishment for body and mind, the coconut fed generations of Samoans and held their houses together. Coconut trees are among the most widespread plants in the South Pacific, providing Samoans and other Islanders with both calories and canoes (see figure 6.1). A medium-sized coconut yields more than 1,400 calories and is rich in iron, potassium, and saturated fat. Samoans often cultivated coconut trees close to their homes, using the different parts of the tree for different purposes.[14] Because growing coconut trees required little sustained attention, Samoans were fond of saying, "Give a coconut a day and it will give you a lifetime."[15]

A well-known story from Samoa and other parts of Polynesia explains the origins of the first coconut tree. According to oral tradition, a beautiful girl named Sina had a pet eel (*tuna*) who fell in love with her. Afraid, Sina ran away, but Tuna followed her to a pool in a neighboring village. Before village chiefs could kill Tuna, Sina granted him his last wish: cut his head off and plant it in the ground. From her planting grew the first coconut tree. The face of the eel—two eyes and a mouth—can still be seen in the three round marks of the husked coconut.[16] When Portuguese explorers brought back the first coconuts to Europe in the mid-sixteenth century, they called the fruit *coco*, or grinning face. The round form of the coconut with its three indentations, indeed, resembles a human skull, an association that influenced even Samoan Plantation Pidgin: "White man coconut belong him no grass he stop [The white man's head is bald]."[17]

The coconut's anthropomorphic appearance was matched by its great practical use for humans. Shells served as drinking cups and to carry water, the palm and midrib were used to make baskets, and coconut fiber was plaited into sennit by older Samoan men to build houses and canoes.[18] The husking and splitting of nuts, followed by the grating and squeezing of the meat inside, were arduous and time-consuming labor processes. As a consequence, the time invested into the preparation of a coconut tended to correlate with the special occasion or the status of the guests to be treated.[19] An average family coconut grove was less than one acre in size, but could yield up to sixty

Figure 6.1. Cocospalme, 1900. In Ländereien und Pflanzungen der Deutschen Handels- und Plantagengesellschaft der Südseeinseln zu Hamburg in Samoa, Hamburg Chamber of Commerce/Library of Commerce, 1964/546 Anhang, Figure 5.

nuts per tree per year.[20] Coconut trees took between six and eight years to mature, but some trees bore fruit for seventy years. Samoans did not plant coconut trees in a particular order or distance from one another, but they made sure to plant them close to taro and yam fields to have quick refreshment available for workers.[21] That way, Samoans knew that no spot on their islands was farther than half an hour from the nearest coconut, which could provide food and drink in times of need.[22] While coconut trees were owned by the families on whose ground they stood, passers-by had the right to pluck or pick up a few nuts to refresh themselves.[23] Fallen nuts, in particular, were often left free to be picked up by anyone who found them.[24]

To harvest the still-green fruits, Samoan men climbed up coconut trees that grew as tall as 100 feet. Using only a sling wrapped around their feet as support, they hugged the tree trunk with their arms and scaled the tree like a caterpillar. Once at the top of the tree, the climber plucked the nuts from their stems and dropped them onto the ground.[25] Mature coconuts could be more conveniently picked up from the ground and collected in baskets, usually made out of coconut leaf midribs.[26] As Samoans quickly found out, ripe coconuts also made better copra, the dried meat of the coconut.[27] Traditionally, young women carried the harvested fruits in two baskets, one on the back and one on the front of their bodies, connected with a stick across their shoulders.[28] Filled to the top, two baskets of coconuts could weigh up to 150 pounds. Young men then further processed the coconuts, making use of their individual components. First, the husk of the coconut was split off and removed by pounding the nut against a sharpened wooden stick (*mele'i*) that was rammed into the ground. Next, Samoans straddled a wooden scraping stool (*'ausa'alo*) to scrape the open coconut against the seashell-like part of a coconut shell fastened to the stool's point. The scraped-off pieces of the coconut kernel were then collected in a vessel or on a leaf placed below the stool. Finally, the scraped matter was poured into a strainer and its juice squeezed into a bowl for further mixing with other foodstuffs.[29]

In addition to taro and coconuts, Samoans also cultivated other food crops, such as bananas, yams, and breadfruit. The most prominent among them was the breadfruit (*'ulu*), which could be stored in pits to prepare for times when food was scarce. The breadfruit derives its English name from the texture of the cooked fruit, which resembles baked bread. Like yam, breadfruit was more susceptible to drought than taro or the coconut, which made it a less reliable part of the Samoan diet. Still, breadfruit trees are among the highest-yielding food plants in the world, yielding between 50 and 150 fruits per year in the South Pacific environment. Like the coconut, the breadfruit could not only be eaten but also provided excellent timber for the construction of durable houses. The breadfruit tree's lightweight wood, resistant to termites and shipworms, was ideal for housing structures as well as boats. Moreover, Samoans were able to tap latex from the trees, which they used to caulk fishing canoes. Finally, breadfruit leaves served as oven covers, food wrappers, and platters.

Samoans had millennia of experience with sustainable agriculture to produce food crops and other useful items. Accustomed to this non-capitalist mode of production, Samoans gradually grasped the new opportunities that presented themselves with the increasing presence of Euro-American missionaries and traders beginning in the 1830s. After the introduction of export-oriented agriculture by German traders in the 1860s, Samoans fought to maintain their economic and cultural autonomy as they adapted to the new world of copra.

Enclosing Land, Partitioning Islands

In the mid-nineteenth century, the German trading house Godeffroy & Co. began its business activities in the South Pacific, establishing its headquarters in Apia in 1857.[30] From Apia, Godeffroy expanded its trade in tropical fruit throughout Polynesia and into Melanesia and Micronesia. In its first years, Godeffroy relied on local Samoan producers to supply the increasingly valuable cash crops. In the mid-1860s, the young Godeffroy manager Theodor Weber took advantage of a long drought, a hurricane, and a pest plague, and bought twelve acres of land from starving Samoans to set up the first cotton plantation.[31] During the global cotton famine caused by the US Civil War in the mid-1860s, a few Samoans had worked for wages on these cotton plantations.[32] By 1868 when the Samoan cotton boom was over, the firm owned 2,500 acres, almost 1 percent of the total land area of Upolu.[33] After its reorganization into the *Deutsche Handels- und Plantagengesellschaft der Südseeinseln* (DHPG) in 1878, the German firm continued to prosper.[34]

By the 1870s, copra had become Samoa's main export to Europe and North America, where it was processed into a variety of products, including high-quality soap, margarine, and even dynamite.[35] Driven by growing demand for copra, the DHPG dramatically expanded its plantation holdings by purchasing land from Samoan titleholders (*matai*). Samoans were divided over these escalating land sales to outlanders. Some feared that foreign ownership would undermine long-standing ways of life based on subsistence agriculture, while others welcomed the considerable profits they reaped from the sales. These profits often came in the form of Western arms, which *matai* used to gain the upper hand over their rivals. The result was what outside observers called "civil war," which belied the active support that Euro-American traders and plantation owners provided to different sides of competing Samoan parties. A vicious cycle of selling land for arms ensued.

Throughout the late nineteenth century, violent conflicts among competing Samoan factions had put severe limits on the time and resources Samoans could devote to subsistence agriculture, often resulting in famines.[36] During the turbulent years of the tridominium (1889–1899), German diplomats in Apia regularly reported on the relationship between war and economic stagnation.[37] Consul Max Biermann noted in April 1894 that during Samoan wars subsistence production was interrupted, forcing many Samoans to consume their coconuts instead of selling them as dried copra. Even worse, the consul observed, Samoans had to be supplied with provisions from German plantations, which offered the only sources of food in times of war. As a result, Biermann concluded, the copra trade came to a halt, plantation output decreased, Samoan purchasing power declined, and imports and exports dropped.[38] Because continued warfare among Samoans—fueled by the competing interests of the colonial powers—had a negative impact on agricultural exports, the pressure to ensure political stability on the islands increased. After decades of proxy wars, international treaties, and hundreds of casualties, an international treaty divided the Samoan islands between Germany and the United States in December 1899, while Great Britain withdrew in exchange for concessions elsewhere. The European and American diplomats who agreed on Samoa's division in distant Berlin did not bother to consult the islands' inhabitants.

After political partition, German and American colonial officials and plantation owners also tried to impose boundaries on Samoa's economy. As a consequence, they expected Samoans to produce cash crops in ever-greater quantities. To be sure, Samoans had always produced a small surplus of food crops to have a reserve in case of emergencies or to host traveling parties and ceremonies.[39] Most Samoans continued their sustainable farming practices, but they increasingly began to sell their surplus crops to Euro-American traders. And occasionally, Samoans entered into wage contracts on larger Euro-American plantations to earn additional cash. Among other uses, Samoans saved their hard-earned money as insurance against environmental crises, such as cyclones, droughts, or pests.

While Samoans worried about a steady supply of food, colonial officials were preoccupied with increasing copra production. Throughout the nineteenth century and into the colonial era, Samoans remained, by far, the largest producers of copra. In 1896, Samoans produced as much as 80 percent of overall copra exports that year on their family plantations.[40] The remaining 20 percent was produced by the DHPG, the largest foreign trading company operating in Samoa at the time. Given Samoans' preponderant role in copra production and their general unwillingness to work on foreign plantations, Euro-American plantation owners, traders, and colonial officials sought to devise different means to increase agricultural output.

Driven by the economic interests of German traders and plantation owners, the German colonial administration lost little time in putting pressure on Samoan farmers. On August 31, 1900, only a few months into formal annexation, Governor Wilhelm Solf passed a regulation that required every Samoan head of family to plant fifty coconut trees a year. On average, it took around six thousand mature Samoan coconuts to produce a single ton of copra. The German administration appointed Samoan officials to inspect plantations on a regular basis and punish individuals who failed to meet their quota.[41] This stricter policy was hard to enforce but did lead to a considerable increase in the number of coconut trees in German Samoa. In 1908, there were 455,280 coconut trees on German plantations, more than 90 percent of which belonged to the DHPG.[42] Between 1900 and 1913, Samoans planted an astounding one million new coconut trees in Savai'i and Upolu (see figure 6.2).

Over the same time period, however, overall copra output did not increase significantly.[43] This was mainly due to the fact that most Samoans, while following the official dictate to plant new trees, did not substantially increase their workload and generally only produced and sold as many coconuts as they needed to survive and earn cash. Even so, Samoan stands of coconut trees covered three times the area of European copra plantations.[44] To protect the "natural fruit lands of Sāmoans," the Berlin Act of 1899 had prohibited the sale of all lands outside of the municipal district of Apia.[45] In November 1907, a regulation passed by the German colonial administration confirmed this ban in principle, but enlarged the area in which the sale of Samoan lands was allowed.[46] From then on, no Samoan lands were to be sold outside of the so-called "plantation district," an area of roughly seven square miles around Apia where most of the foreign-owned, large-scale plantations were located. In addition, every Samoan was guaranteed at least 3.2 acres of land to cultivate. Its good intentions notwithstanding, the regulation clearly benefited the largest landholder outside of the plantation district: the DHPG.[47] The German company now enjoyed a "virtual monopoly of land which other Europeans could buy."[48] A DHPG business report from 1907 duly noted that the company could now proceed to sell the majority of its uncultivated lands at a profit.[49]

Other strategies the colonial administrations pursued to increase agricultural production among Samoans included restrictions on Samoan *malaga* (visiting parties), the introduction of copra kilns, and head taxes. Yet despite these attempts by colonial administrations to force Samoans into wage labor on foreign plantations, the overwhelming majority of Samoans continued sustainable farming that offered greater control over their lives. Since Samoans owned most of the land on which coconut trees grew, their surplus production dominated the copra export market throughout the colonial era. Their vibrant subsistence and cash crop economy provided Samoans not only with an insurance against environmental disasters, but, equally important, with a strong foundation to protect their political and social self-

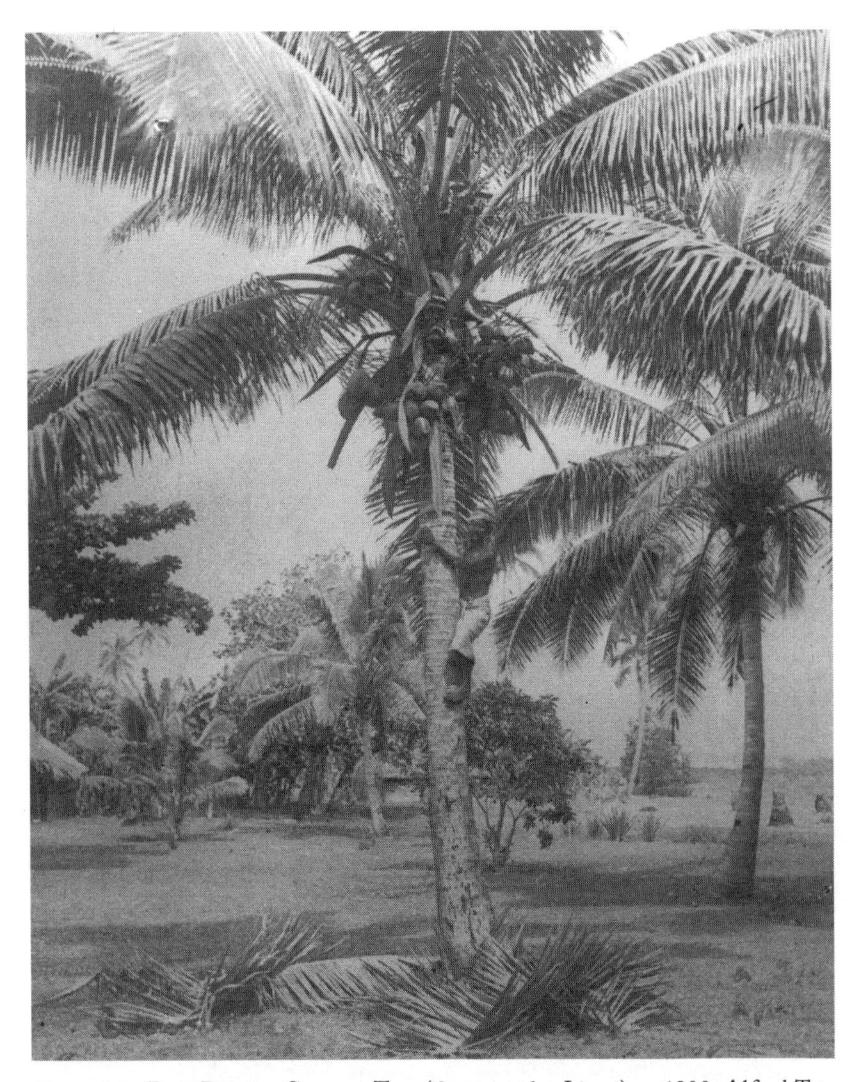

Figure 6.2. Fruit-Bearing Coconut Tree (*Cocos nucifera* Linné), c. 1900, Alfred Tattersall. In Ferdinand Wohltmann, *Pflanzung und Siedlung auf Samoa* (Berlin: Süsserott, Kolonialwirtschaftliches Komitee, 1904), 100. Bildarchiv der Deutschen Kolonialgesellschaft, Universitätsbibliothek Frankfurt am Main (042-0245-31).

determination against colonial demands. In addition, they continued taking fruit from commercial plantations and founded their own producer cooperatives to cut out Euro-American traders from the lucrative copra trade. Samoans were thus able to respond to the introduction of a large-scale plantation economy largely on their own terms.

Reordering Samoan Plantations

Despite considerable Samoan control, cash crop agriculture radically transformed the natural environment of the islands. The reordering of Samoan plantations took many forms. In the late 1890s, German plantation owners were so worried about Samoans stealing valuable cash crops that they secured large parts of their plantations with barbed wire.[50] Euro-American plantation owners also replanted coconut trees, which had been growing in uncontrolled fashion in Samoa for centuries, in straight lines to control workers and measure their work progress. Samoans usually planted coconut trees twenty to twenty-six feet from one another, while on foreign-owned plantations the average distance between trees was thirty-three feet.[51] Plantation owners also introduced clear divisions both among different plantation lands and within them, according to the different uses to which they were put. A map of the largest DHPG plantation in Mulifanua, twenty-five miles west of Apia, listed five distinct categories of plantation lands: ripe palm trees; not-yet-ripe palm trees; palm trees and cotton; cotton; and a mixed category including provisions, bananas, and sheds (see figure 6.3).[52] Plantation owners carefully classified every parcel of land into one of these categories and numbered them from 1 to 115. A system of pathways bounded and linked these different categories of land in straight and parallel lines. The resulting geometrical system of lands with different degrees of exploitation provided not only a grid of intelligibility for plantation managers but also allowed for better control of the workers who were tasked with making the land useful.

As the DHPG map indicated, copra plantations in colonial Samoa did include other crops (such as cotton, cocoa, and rubber), but grew increasingly monocultural as global demand for copra began to soar in the years around 1900.[53] And with crop monoculture came the imperative to protect the most valuable crop—the coconut—from threats by other plants and insects. A sturdy and fast-growing plant with a telltale name posed a particular threat to the efficient management of copra plantations: the touch-me-not (*Mimosa pudica*). Touch-me-not plants not only concealed fallen coconuts but also painfully stung human and animal feet. Concerned about the plant's spread, plantation managers weeded out older plants and allowed cattle, which only grazed on young touch-me-nots, to take care of the rest. In addition, plantation owners introduced Buffalo grass (*Bouteloua dactyloides*) from North America, which helped to push the touch-me-nots aside. Soon, however, the Buffalo grass threatened to replace the touch-me-nots it was meant to contain by depriving the soil of humidity. The Buffalo grass, in turn, then had to be checked by fires.[54] Alongside controlled fires and the free-roaming cattle, plantation workers had to bend down low to cut smaller bushes around the coconut trees

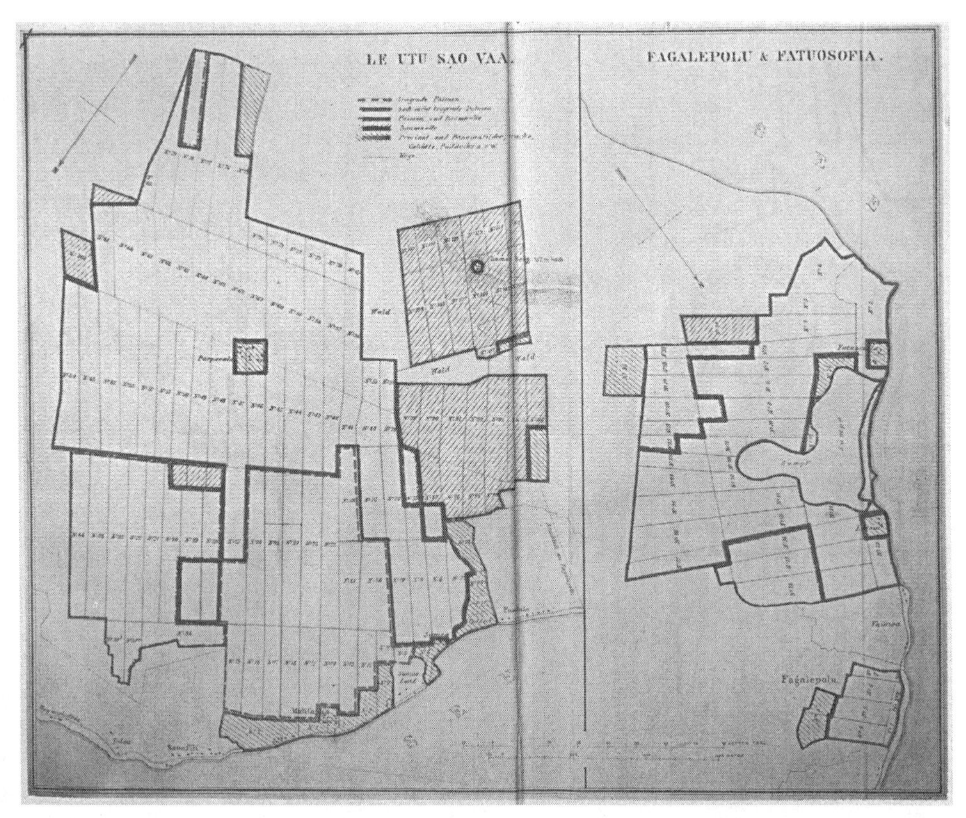

Figure 6.3. Map of DHPG plantation in Mulifanua, 1900. In *Ländereien und Pflanzungen der Deutschen Handels- und Plantagengesellschaft der Südseeinseln zu Hamburg in Samoa*, Hamburg Chamber of Commerce/Library of Commerce, 1964/546 Anhang, Table No. 5.

to help protect them from migrant plants and ensure their growth into cash crops.[55]

Another threat to the expanding plantation economy came in the shape of a rather large insect with an even larger appetite: the rhinoceros beetle (*Oryctes rhinoceros*). Native to South and Southeast Asia, the mighty beetle arrived in Upolu in 1909, probably hidden in rubber seedling pot plants from Sri Lanka, and quickly spread to the neighboring islands of Savai'i and Tutuila.[56] One of the largest and strongest beetles in the animal kingdom, the rhinoceros beetle is especially fond of the heart of coconut trees, and its arrival thus threatened the very heart of the Samoan plantation system.[57] In the following years and decades, the rhinoceros beetle proceeded to devastate Samoan copra plantations.[58] To combat the spread of the rhinoceros beetle, both German and American administrators passed regulations to force Samoans to collect the insects. At first, in American Samoa, the administration offered four cents for an adult rhinoceros beetle and two cents for larvae. Over the next year, the US naval administration paid cash to Samoans who delivered beetles and their larvae, but the threat to copra plantations remained.

As a consequence, in July 1911, Governor William Crose reformed the existing financial incentive scheme with a more draconian measure: all male Samoans were from then on required to devote every Wednesday morning to seeking and eradicating beetles and larvae—without payment.[59] Around the same time, the German colonial administration in Upolu passed a similar regulation to combat the rhinoceros beetle that empowered the police to enforce mandatory collecting of beetles on plantations. However, these more aggressive attempts to combat the voracious beetle had only limited effect. In the summer of 1915, the US consul in Apia, Mason Mitchell, reported that the rhinoceros beetle had destroyed more than half of the coconut trees in infected areas in Tutuila.[60] In German Samoa as well, the beetle continued to eat away at the copra plantation economy. Susceptible to newcomers such as Buffalo grass and the rhinoceros beetle, plantation monoculture fundamentally changed the Samoan environment.

Beyond plants and insects, the Samoa environment experienced the impact of another, much smaller, kind of organism: microbes. While Samoa was free of malaria, a virus more common in Europe and North America proved all the more dangerous to Samoans: the H1N1 influenza. Samoa had been hit by flu outbreaks throughout the 1890s.[61] After 1903, the newly arrived Chinese contract laborers were susceptible to the flu due to the new disease and labor environments in which they found themselves. But that was little preparation for the pandemic that would hit the islands fifteen years later. In the fall of 1918, trailing the destruction and malnutrition of World War I, the flu spread across the world like a bushfire. Sailing from Auckland on October 31 and arriving in Apia on November 7, the New Zealand steamer *Talune* brought good news about the armistice in Europe and, at the same time, several flu-infected passengers who would spell bad news for Western Samoa. As the pandemic quickly spread throughout Upolu and Savai'i, the New Zealand military administration failed to quarantine the islands. As a result, more than 8,500 Samoans, representing over a fifth of the total population, died of the flu in a matter of weeks.[62]

In fact, the New Zealand military administrator, Colonel Robert Logan, had deliberately cut the wireless link to American Samoa, which made it impossible to determine the severity of the pandemic and learn about potential ways to keep it from spreading further.[63] While Logan blamed "ill-disciplined" Samoan nurses for this human disaster and deliberately refused outside help, his counterpart in American Samoa, Commander Poyer, managed to prevent the epidemic from entering, and no casualties resulted.[64]

A strict quarantine enforced by Samoans who patrolled the shores to keep unwanted visitors from Upolu away and a good amount of luck kept the flu epidemic away from Tutuila. Partly in response to this drastic difference in colonial governance, Samoan leaders in Western Samoa presented a petition to the new Governor Tate in January 1919, demanding unification with American Samoa.[65] An Epidemic Commission, launched by the New Zealand authorities after the pandemic, blamed the high mortality rate on a general administrative failure by New Zealand and British colonial officials.[66] The dismal handling of the flu helped to alienate many Samoans from the New Zealand administration, paving the way for greater calls for self-determination in the 1920s.

Conclusion

Over the course of the nineteenth century, Samoans sold large tracts of their land to Euro-Americans to be turned into profitable cash crops. As a result, long-grown social and economic structures such as the *matai* system and household-centered farming came under pressure but survived through the colonial era. At the same time, the forceful introduction of plantation monoculture also altered the natural and social landscape of colonial Samoa in lasting ways. Migrating organisms—such as Buffalo grass, the rhinoceros beetle, and, most consequentially, the influenza virus—reshaped and sometimes devastated the islands and their inhabitants. Through it all, Samoans defended their sustainable food agriculture and adapted ecological changes on plantations to their own needs.

Notes

This chapter draws on sections of my book *Coconut Colonialism: Workers and the Globalization of Samoa* (Cambridge, MA: Harvard University Press, 2021). Many thanks to the editors James Beattie, Ryan T. Jones, Edward Melillo, and the anonymous reviewers for University of Hawai'i Press for their valuable comments.

1. In 1900, German Sāmoa had 32,815 and American Sāmoa 5,499 inhabitants. Twelve years later, these numbers increased only slightly to 33,554 and 6,659, respectively. Based on official censuses from the German colonial and US naval administrations in Felix M. Keesing, *Modern Samoa: Its Government and Changing Life* (London: G. Allen & Unwin, 1934), 33.

2. Mary L. Stover, "The Individualization of Land in American Samoa" (PhD diss., University of Hawai'i, 1990), 45; John W. Coulter, *Land Utilization in American Samoa* (Honolulu: Bernice P. Bishop Museum, 1941), 28.

3. Terry L. Hunt, and Patrick V. Kirch, "The Historical Ecology of Ofu Island, American Samoa, 3000 B.P. to the Present," in *Historical Ecology in the Pacific Islands:*

Prehistoric Environmental and Landscape Change, ed. Patrick V. Kirch and Terry L. Hunt (New Haven, CT: Yale University Press, 1997), 106.

4. Augustin Krämer, *Die Samoa-Inseln: Entwurf einer Monographie mit besonderer Berücksichtigung Deutsch-Samoas, Vol. II* (Stuttgart, Germany: E. Schweizerbart, 1902/03), 111f.

5. Eni F. H. Faleomavaega, *Navigating the Future: A Samoan Perspective on U.S.-Pacific Relations* (Carson, CA: KIN Publications, 1995), 24.

6. David C. Pitt, *Tradition and Economic Progress in Samoa: A Case Study of the Role of Traditional Social Institutions in Economic Development* (Oxford: Clarendon Press, 1970), 203.

7. Peter H. Buck (Te Rangi Hiroa), *Samoan Material Culture* (Honolulu: The Bishop Museum, 1930), 545.

8. Keesing, *Modern Samoa*, 269–270; Lowell D. Holmes, *Samoan Village* (New York: Holt, Rinehart & Winston, 1974), 41; Robert W. Franco, *Samoan Perceptions of Work: Moving Up and Moving Around* (New York: AMS Press, 1991), 24–25.

9. F. K. Lehman and David J. Herdrich, "On the Relevance of Point Field for Spatiality in Oceania," in *Representing Space in Oceania: Culture in Language and Mind*, ed. Giovanni Bennardo (Canberra: Australian National University Press, 2002), 187.

10. Lehman and Herdrich, "On the Relevance of Point Field," 185.

11. J. Tim O'Meara, *Samoan Planters: Tradition and Economic Development in Polynesia* (Orlando, FL: Holt, Rinehart & Winston, 1990), 60; Franco, *Samoan Perceptions of Work*, 42–43.

12. Franco, *Samoan Perceptions of Work*, 48–49.

13. Ibid., 53.

14. Holmes, *Samoan Village*, 43.

15. Cited in Pitt, *Tradition and Economic Progress*, 200.

16. Franco, *Samoan Perceptions of Work*, 63.

17. Cit. in Peter Mühlhäusler, "Samoan Plantation Pidgin English and the Origin of New Guinea Pidgin," *Papers in Pidgin and Creole Linguistics* (1978): 70; Darrell T. Tryon, and Jean-Michel Charpentier, *Pacific Pidgins and Creoles: Origins, Growth and Development* (Berlin: Mouton de Gruyter, 2004), 245.

18. For a detailed description of Samoan material culture, see Buck, *Samoan Material Culture*, 551.

19. Franco, *Samoan Perceptions of Work*, 63.

20. Ibid., 59.

21. Buck, *Samoan Material Culture*, 550.

22. William T. Pritchard, *Polynesian Reminiscences: Or, Life in the South Pacific Islands* (London: Chapman & Hall, 1866), 170.

23. Franco, *Samoan Perceptions of Work*, 61.

24. R. F. Watters, "Cultivation in Old Samoa," *Economic Geography* 34, no. 4 (1958): 348.

25. Krämer, *Samoa-Inseln*, 143.

26. On Sāmoan baskets, see Buck, *Samoan Material Culture*, 189–208.

27. Pitt, *Tradition and Economic Progress*, 196.

28. Krämer, *Samoa-Inseln*, 89.

29. Ibid., 129.

30. Kurt Schmack, *J. C. Godeffroy & Sohn, Kaufleute zu Hamburg: Leistung und Schicksal eines Welthandelshauses* (Hamburg: Broschek, 1938), 112.

31. Doug Munro and Stewart Firth, "Samoan Plantations: The Gilbertese Laborers' Experience, 1867–1896," in *Plantation Workers: Resistance and Accommodation*, ed. Brij V. Lal, Doug Munro, and Edward D. Beechert (Honolulu: University of Hawai'i Press, 1993), 103.

32. On the effects of the US Civil War on global cotton production, see Sven Beckert, "Emancipation and Empire: Reconstructing the Worldwide Web of Cotton Production in the Age of the American Civil War," *American Historical Review* 109, no. 5 (2004): 1405–1438.

33. Schmack, *Godeffroy & Sohn*, 145.

34. Gordon R. Lewthwaite, "Land, Life and Agriculture to Mid-Century," in *Western Samoa: Land, Life, and Agriculture in Tropical Polynesia*, ed. James W. Fox and Kenneth B. Cumberland (Christchurch, NZ: Whitcombe & Tombs, 1962), 143–144; Stewart Firth, "German Recruitment and Employment of Labourers in the Western Pacific before the First World War" (PhD diss., University of Oxford, 1973), 61; Franco, *Samoan Perceptions of Work*, 149.

35. Gregory T. Cushman, *Guano and the Opening of the Pacific World: A Global Ecological History* (Cambridge: Cambridge University Press, 2013), 101–102.

36. Buck, *Samoan Material Culture*, 546.

37. For an analysis of the Sāmoan tridominium, see Holger Droessler, "Colonialism by Deferral: Samoa under the Tridominium, 1889–1899," in *Rethinking the Colonial State*, ed. Søren Rud and Søren Ivarsson (Bingley, UK: Emerald, 2017), 203–224.

38. Biermann to Caprivi, April 30, 1894, BArch R 1001/2539.

39. Buck, *Samoan Material Culture*, 544.

40. Rose to Hohenlohe-Schillingsfürst, August 30, 1896, BArch R 1001/2539.

41. Solf, "Notizen zur Landwirtschaft," BArch N 1053/6, 292.

42. Schmack, *Godeffroy & Sohn*, 291.

43. Total copra output in German Sāmoa rose only slightly from 7,792 tons in 1899 to 9,634 tons in 1913. Keesing, *Modern Samoa*, 300, 303.

44. Peter J. Hempenstall, *Pacific Islanders under German Rule: A Study in the Meaning of Colonial Resistance* (Canberra: Australian National University Press, 1978), 70.

45. Cit. in Richard P. Gilson, *Samoa 1830 to 1900: The Politics of a Multi-Cultural Community* (Oxford: Oxford University Press, 1970), 406.

46. Keesing, *Modern Samoa*, 260.

47. Hempenstall, *Pacific Islanders under German Rule*, 53.

48. Firth, "German Recruitment," 242.

49. DHPG Business Report 1907, StAH, 621-1/14, 25.

50. Rose to Hohenlohe-Schillingsfürst, April 12, 1898, BArch R 1001/2540.

51. Krämer, *Samoa-Inseln*, 138.

52. "Mulifanua-Planzungen," table no. 5, Ländereien und Pflanzungen der Deutschen Handels- und Plantagengesellschaft der Südseeinseln zu Hamburg in Samoa, Commerzbibliothek Hamburg.

53. Increased demand for copra around 1900 was mainly due to the growth of the luxury soap industry in Europe and the United States.

54. Otto Riedel, *Der Kampf um Deutsch-Samoa: Erinnerungen eines Hamburger Kaufmanns* (Berlin: Deutscher Verlag, 1938), 185.

55. Franco, *Samoan Perceptions of Work*, 148; Keesing, *Modern Samoa*, 301.

56. Frank P. Jepson, "The Rhinoceros Beetle in Samoa," *Fiji Department of Agriculture Bulletin* no. 3 (1912).

57. A. Dexter Hinckley, "Ecology of the Coconut Rhinoceros Beetle," *Biotropica* 5, no. 2 (1973): 111.

58. Riedel, *Kampf um Deutsch-Samoa*, 220.

59. Mitchell to Vreeland Chemical Co., July 24, 1911, NARA-CP, RG 84, Vol. 58, 33.

60. General Report on Samoa by Mitchell, June 10, 1915, NARA-CP, RG 84, Vol. 84.

61. B. Funk to Municipal Council, August 10, 1898, NARA-CP, RG 84, Vol. 162.

62. More than 675,000 Americans succumbed to the virus, representing 0.5% of the total population. Worldwide, the Spanish flu claimed the lives of 50 to 100 million people, or 3–5 percent of the total world population. Sandra M. Tomkins, "The Influenza Epidemic of 1918–19 in Western Samoa," *Journal of Pacific History* 27, no. 2 (1992): 181–197; Alfred W. Crosby, *America's Forgotten Pandemic: The Influenza of 1918* (Cambridge: Cambridge University Press, [1976] 2003), 203–263.

63. Tomkins, "Influenza Epidemic in Western Samoa," 195.

64. "Epidemic Commission," *Samoan Times*, June 21, 1919, 4.

65. Hermann Hiery, *The Neglected War: The German South Pacific and the Influence of World War I* (Honolulu: University of Hawai'i Press, 1995), 177.

66. Tomkins, "Influenza Epidemic in Western Samoa," 195.

Seven

"One Extensive Garden"? Citrus Schemes and Land Use in the Cook Islands, 1900–1970

Hannah Cutting-Jones

> *Kai Kainga*, or land eating [is] getting unjust possession of each other's lands [and is] a species of oppression.
>
> —*John Williams, missionary to the Cook Islands, 1838*

In Cook Islands' mythology, Rongo, the god of cultivated crops, and his brother Tangaroa, god of the sea, were born of the earth mother Papa into a universe shaped like the hollow half shell of a coconut. Rongo went on to create taro irrigation and obtain the first kumara (sweet potato) from the heavens. Thereafter, small wooden carvings of Rongo's image placed at the edges of kumara plantations blessed the harvest, and parcels of cooked taro presented to the god signaled peace (see figure 7.1). Similar to other island societies in Oceania, Cook Islanders held a sacred and ancient connection to their environment, out of which grew not only creation myths and genealogical lineages, but life itself. Kumara, taro, breadfruit, coconut, bananas—these foods provided more than subsistence; they represented wealth, status, and cultural survival.

Cook Islands soil would also be the arena of a century-long struggle over intensive, commercialized agriculture. Missionaries sent by the London Missionary Society (LMS) were the first Europeans to settle in the Cook Islands in the 1820s, and although they altered land-use patterns, the most significant environmental changes took place after the Cooks became a Protectorate of Britain in 1888 and accelerated even further when New Zealand annexed the islands in 1900. Waves of European settlement and influence by missionaries, traders, and government officials affected land tenure patterns. The exploitation of natural resources pulled Cook Islanders into a global marketplace and permanently altered their ecology and landscape. New Zealand's direct political involvement from 1900 shaped and directed citrus production for export, in particular. As land and labor for oranges and

Figure 7.1. Rongo, the Māori god of kumara and cultivated plants and Mangaian god of agriculture and war. Source: Internet Archive Book Images, no restrictions, via Wikimedia Commons.

other cash crops displaced subsistence production, a greater reliance on imported labor and foodstuffs developed, with migration both feeding and responding to this cycle.

Unlike the experience of many other Pacific Islands, however, Cook Islanders retained ownership of their land throughout the political, economic, and cultural changes brought about by the establishment of an export market. Leaders never permitted land sales or allowed the implementation of "alien plantation economies," though hopeful settlers repeatedly sent letters asking for permission to plant lucrative crops in the islands.[1] The relationship between Cook Islanders and their land proved strong yet flexible enough to survive participation in a century-long agricultural export economy—a market that eventually collapsed, at least in part because of the clash between European and Indigenous land-use customs. By 1970, the rainforest had begun to re-envelop abandoned citrus plantations, and Cook Islanders, now running an independent nation, reconsidered land use—this time on their own terms.

Rarotonga, the largest, most populous, and capital island of the Cook group, is located 1,634 miles northeast of Auckland, New Zealand. With its fifteen islands occupying over 850,000 square miles of Pacific Ocean, the Cooks have a total landmass of 88 square miles, with just 16 percent suitable for agriculture.[2] The southern islands, including the capital Rarotonga, are primarily volcanic and support a variety of crops, while the northern group are atolls where few plants grow and people traditionally subsisted on a diet of coconuts and fish.

Polynesian settlement in Rarotonga goes back a thousand years, with two independent parties of immigrants arriving at the end of the twelfth century. Two men, Tangiia of Tahiti and Karika of Samoa—who would both become *ariki*, or chiefs, of Rarotonga—proceeded to ally with settlers already inhabiting the island. The leaders then allocated a slice of land reaching, as one local recounted, "from the sea to the mountain," or a *tapere*, to the head of each family group.[3] These plots fell across three zones, the coastal, lowland, and upland, with most plantings traditionally located in the lowland areas.[4] Kin groups bickered constantly, yet "every inch of land on the island was claimed by one party or another."[5] These quarrels did not prevent the planting of food crops.

Considered living descendants of the gods, *ariki* could institute *ra'ui*, or sacred prohibitions, over lands or lagoons to secure resources for feasts or important occasions. For generations the practice of *ra'ui* provided a way to avoid the overexploitation of coconut groves, taro plantations, and fishing grounds that otherwise might lead to scarcity and starvation. *Ariki* might also declare a piece of land or particular food permanently *tapu*, or forbidden, and therefore beyond the reach of lower-status individuals. Breaking *tapu* or *ra'ui* was punishable by death. Those who lived on and planted an *ariki*'s land lived by these guidelines and contributed crops as tribute, thus affirming the chief's authority as their leader and protector.[6]

While titles were significant, in pre-contact Rarotonga "there was no conception of the sale of land or its produce," no individual ownership of surplus resources, and no trade. This gave little incentive to produce food beyond what was necessary for subsistence, tribute, feasts, or gifts.[7] Islanders usually worked on their kin group's *tapere* and were expected to contribute willingly to the productivity of that community. Rows of coconut palms or chestnut trees marked boundaries between breadfruit, bananas, taro, plantains, and coconuts grown in the rich soil between Rarotonga's rugged mountainous interior and the coast.[8] Gardens established in coastal swamp areas risked greater devastation from floods and hurricanes.[9] Wild orange trees, introduced in 1789 by the mutineers of the infamous *Bounty*, grew wild in the valleys.[10] Missionaries later criticized Rarotongans' cultivation patterns and intermittent destruction of crops.[11] Initially, however, they were impressed.

Missionaries Influence Land Use

Shortly after his visit to Rarotonga in 1827, John Williams, the indefatigable LMS representative, painted the island as a picture of beauty, abundance, and successful agriculture. Williams planned to continue Christianizing the Cooks, a process begun by the two Tahitian missionaries he left on the island in 1823, Papeiha and Rio. As Williams wrote:

> The whole island was in a high state of cultivation [and] there are rows of superb chestnut-trees planted at equal distances, and stretching from the mountain's base to the sea, with a space between each row of about half a mile wide [and] divided into small taro beds, which are dug four feet deep, and can be irrigated at pleasure. . . . The pea-green leaves of the Taro, the extraordinary size of the Kape [gigantic taro] lining the sloping embankment, together with the stately bread-fruit trees on the top, present a contrast which produces the most pleasing effect.[12]

In 1830 Williams sailed again to Rarotonga, this time with Reverend Charles Barff and his family (see figure 7.2). The Barffs joined the Pitman and Buzacott families, who had arrived in 1827 and 1828, respectively. Stunned anew by the island's lush vegetation, Williams noted in his journal:

Figure 7.2. An engraving depicting the confiscation and destruction of idol gods by European missionaries in Rarotonga. John Williams, *A Narrative of Missionary Enterprises in the South Sea Islands*, 1837.

> June 3. Saw the fine island of Rarotonga. . . . It had a fine romantick [*sic*] appearance from the vessel, the lofty mountains separated by deep ravines and all covered with a beautiful foliage, formed a majestic [*sic*] landscape. The extent of Cultivation was to us a novel sight, almost every Individual having his . . . small farm cultivated with plantains . . . taro, yams, etc., so that the whole settlement appeared one extensive garden.[13]

These glowing descriptions suggest Rarotongans cultivated the land to its utmost, industriously mixing crops to utilize shade and soil. Leaders expected every able-bodied person to farm, with individuals publicly shamed if they shirked their duties.

Most Islanders resided in kin groups near their lowland plantations. Missionaries swiftly targeted non-nuclear living arrangements as a potential spiritual stumbling block. Thus, one of their first goals was to relocate Rarotongans from their ancestral lands and communal homes to individual houses for nuclear families situated near one of three missions constructed at Avarua, Aorangi, or Ngatangiia. But tensions developed as Islanders settled on others' lands or abandoned common-law spouses; some chose to return to their gardens.[14]

By the early 1830s, missionaries had introduced the first cash crops and encouraged locals to grow sweet potatoes, coffee, copra, and citrus to trade with passing merchant and whaling ships in exchange for a variety of goods. As Cook Islands historian Ron Crocombe asserted, in support of both the growing cash economy and church projects "settlement became nucleated and concentrated on the coastal plain" while cultivation continued in the valleys.[15] Market houses constructed next to each mission allowed the *ariki* of corresponding regions to oversee commercial exchanges by 1840.[16]

Rapid conversions and social restructuring impacted land use in unexpected ways. The "extra wives" of polygamous relationships returned to their families of origin and their children went on to inherit land through the maternal rather than paternal line.[17] Epidemics drastically reduced populations.[18] Lands abandoned through death or relocation reverted to the heads of various descent groups or were absorbed by neighbors.[19] All the while interactions with the outside world increased as Rarotonga became a well-known pit stop where ships could re-provision and recruit deckhands.

Missionaries worked to eradicate or modify cultural practices they considered ungodly, but they also advocated for Cook Islanders to retain the titles to their lands.[20] This approach was not altruistic, but part of a range of actions to shore up their own power and lessen the influence of the worldly whalers and beachcombers increasingly washing up on shore, and the guns and liquor they brought with them. Further, missionaries for decades reaped the rewards of cash-crop production in support of church projects. Similar to Tonga, Hawai'i, and Tahiti, *ariki* in the Cooks agreed to work within mission-proscribed laws to consolidate their authority and gain access to imported goods. Within just two decades of European settlement, in fact, the chiefs were proudly exporting produce. In 1852, E. H. Lamont, recently shipwrecked on Penrhyn, the northernmost island in the Cook Group, recounted with surprise that the three *ariki* of Aitutaki presided over regular markets and had already organized a sizable shipment of oranges to California. As a result, they "were in great glee, hoping it was the commencement of a new and successful trade."[21] The *ariki* also understood that increased settlement would lead to more commerce, and by 1855, as the whaling industry waned and market houses folded, they negotiated limited European settlement and allowed individual traders to open shops. *Ariki* continued to control crop production and distribution—primarily wild oranges at this time—although ideas of a free market and elective democracy remained anathema to Cook Islands custom for many years.[22]

Loads of fresh produce began arriving in New Zealand from Rarotonga in the 1860s, primarily through a renewal of Māori–Cook Islands connections. Paora Tūhaere, paramount chief of the Ngāti Whātua in Auckland, worked to establish trading and cultural relations with the Cook Islands in the early 1860s. On July 5, 1863, Tūhaere's ship the *Victoria*, which had sailed to Rarotonga on June 19, returned to Auckland carrying 43,000 oranges, 1,000 coconuts, 270 pounds of pears, 200 pounds of arrowroot, 8 tons of taro, and an important Rarotongan chief, Kainuku Tamako. Nineteen New Zealand Māori made the voyage, including Tūhaere himself.[23] The visit marked a Māori reconnection that would be strengthened through future commercial exchange, migration, and intermarriage. By 1865, the Cook Islands sent between ten and fifteen shipments of oranges annually from Rarotonga to New Zealand.[24]

In the 1870s, with New Zealand's demand for tropical produce increasing, and against the admonitions of missionaries, *ariki* began permitting larger numbers of prospective traders and planters free entry to Rarotonga. The chiefs also bought several schooners and organized most of the inter-island trade in the Cook group as well as commerce with Tahiti and New Zealand. By 1885, *ariki*-led trade was flourishing, "worth an estimated 60,000 pounds a year."[25] In addition, many chiefs formed partnerships to open teashops and trade stores.[26] *Ariki* took full advantage of new economic prospects.

As export revenue increased, however, European traders tried to wrest profits away from *ariki*. Avarua store owner J. H. Garnier complained in an 1890 letter of Tahitian traders bribing Rarotongan *ariki*, outbidding local traders such as himself and making away with "thousands of dollars' worth of produce which should have been entering the harbour of Auckland." Garnier's message was clear: "I am most anxious to see the entire trade of these beautiful and fertile islands secured to New Zealand," and to his own wallet.[27] Overall, the Church's influence lessened during this period as other Europeans and New Zealanders intensified their efforts to control the bourgeoning citrus trade.

Still, missionaries to the Cooks and other Pacific islands felt their work of civilizing as well as planting the seeds of faith had borne fruit. William Wyatt Gill, first a missionary to Mangaia and later to Rarotonga, wrote in 1876 that "the outward condition of these islanders has been marvellously improved since the introduction of Christianity. The soil is better cultivated, waste lands have been reclaimed [and] numerous places once sacred to the gods are now planted for the good of mortals."[28] They had successfully transplanted a Protestant vision of land management. Yet looking back in 1885, Reverend Gill also noted the unintended environmental effects of the mission era:

> The woods of Rarotonga, when I first knew the island some thirty-two years ago, were everywhere vocal with the song of birds . . . [But now] I have more than once ridden round the island without hearing the cry of any but sea birds. The stillness of the forest would be intolerable but for the pleasing hum of insects as the sun declines.[29]

Scholars have corroborated Gill's impressions; as native land bird stocks became depleted and could no longer serve as a food source, attention turned to agricultural expansion.[30]

As James Beattie and John Stenhouse have illustrated, Christian contemporaries of Reverend Gill who settled in New Zealand also worked to "make wild nature bountiful" by introducing European farming practices to Māori Christians. And like their Māori relatives, Cook Islanders soon pursued the production of produce for trade, as well.[31] The relationship that both Europeans and Islanders shared with the natural world was complex and varied, with some seeing extractive potential and others leaning toward conservation.

Ariki privilege superseded customary occupation rights of common planters during this period, for example, and the chiefs acquired great wealth in the process. In a juxtaposition of the traditional and the modern, chiefs turned again to ancient customs of land use, reinstituting restrictions on planting lands, or *ra'ui.* Now, however, rather than safeguarding resources, they adapted the practice to commercial trading by fixing the price of island-grown produce and fining those who broke *ra'ui* by selling below the set price. Unscrupulous chiefs could declare crops *ra'ui* in wait for the highest bidder.[32] European traders fought against the use of *ra'ui*, but government officials, following a *laissez-faire* policy and wanting to shore up Cook Islanders' trust, upheld the local planters' rights. Even so, Europeans influenced the market by controlling the sale of copra bags, fruit cases, and other manufactured items needed for the trade.

In their bid to further control the citrus market, by 1900 foreign traders had formed a "fruit ring" of several hundred members to set prices and trading terms. Cook Islanders organized protests and boycotts.[33] The rise of the Union Steamship Company—the New Zealand business founded in 1875 and known by the turn of the century as the "Southern Octopus" due to its far-reaching grip on Southern Hemisphere trade—further disrupted Cook Islander–run exports by transporting oranges to New Zealand from Tahiti and reducing demand for Cook Islands produce there.[34] Meanwhile, locally owned and expensive-to-run schooners struggled to compete with Union and other large steamers arriving from California, Australia, and

New Zealand. *Ariki*-led trade declined further following the formal annexation of the Cooks in 1900.

THE LAND COURT, 1903–1910

As New Zealand began the process of governing the Cook Islands, land reform sat at the top of the colonial agenda. One of the primary goals of administrators was to increase fruit production solely for New Zealand consumers, which would ostensibly benefit both economies, as Cook Islanders could spend cash-crop earnings on imported goods manufactured in New Zealand.[35] One agriculturalist noted that although "the total acreage is not large, the soils [in the Cook Group] are capable of producing a wide range of tropical crops, all of which are needed by this Dominion."[36] Cook Islands' growers at first seemed happy to comply with New Zealand administrators as long as they received timely and fair compensation. The main focus for all involved became the increased production of bananas, tomatoes, and particularly oranges at the expense of subsistence crops like kumara and taro, if necessary. As a result, export agriculture increasingly modified plant distribution and ecological diversity.

But establishing a successful fruit export industry during this political transition faced steep obstacles. The *ariki* balked at relinquishing control as growers experimented with various, and often disappointing, cash crops. Coffee trees, for example, were easily planted but, when prices faltered, neglected in the bush. By 1900, hundreds of acres of arable land sat unused.[37] Walter Gudgeon, the first resident commissioner appointed after annexation, regarded traditional land tenure and the control wielded by *ariki* as the primary obstacle to commercial development. Only land redistribution, Gudgeon believed, could overcome this problem.

And so, striking at the heart of Cook Islands land traditions by modifying people's relationship with their chiefs and resources, Gudgeon conducted a series of hearings from 1903 until his retirement in 1909. Collectively termed the "Land Court," the hearings attempted to erode *ariki* privilege by reallocating land ownership according to use and occupation, not customary status. Gudgeon argued that the *ariki* had co-opted common peoples' land rights during the missionary era. He hoped the chiefs might go along with ideas of modernization and use their influence to support the court's goals. Officials assumed a more equitable distribution of land rights would incentivize and thus increase cash-crop production for individual planters. In addition, smaller plots might increase the number of long-term leases offered to Europeans residing in the Cooks, whom administrators viewed as models of efficient farming methods.[38] However, loyalty to one's *ariki* and kin group—largely reflected through regular contributions and sharing of food resources—formed the basis of the Cook Islands' social structure. As Richard Gilson observed, with *ariki* in charge of the market, most Islanders could (literally) "not retain all the fruits of their labour," even with the promise of a paycheck. This, combined with the fact that exports through the 1920s mainly depended on simple and immediate returns on wild-growing oranges, meant many lost interest in planting cash crops. Gilson concluded that "one of the greatest mistakes of the European merchants [and officials] was to assume that commercial opportunities and credit would encourage the islanders to cultivate their land more intensively."[39]

The Land Court's decisions were ultimately problematic, and its achievements limited. New Zealand's belief in the superiority of individual land ownership clashed with Cook Islanders' tradition of quick earnings and communal distribution. New land titles ensured security of tenure for Indigenous planters but resulted in "excessive fragmentation of ownership" and ever-smaller, co-owned plantations that left little incentive to cultivate cash crops. The court (re)awarded to *ariki* over half of the lands in question as many Islanders were either unaware of the court proceedings or too intimidated to submit a claim. Steep tributes to *ariki* were formally abolished, however, as were the assumed rights of "parasitic relatives," while potential monetary gains for planters remained complicated.[40] Overall, the process eroded *ariki* support for an administration that ended their absolute control over the land, and with it, much of their wealth.[41]

Not only did the court's decisions disrupt cultural norms, but it was increasingly evident that intensive cash-crop production also threatened subsistence farming and caused environmental problems. In the early 1900s, planter Varopaua M. Mana Taiava in Aitutaki complained bitterly about the effects of the Land Court rulings:

> All our natural food supply we used to have in abundance before the investigation is no more. Each man is required to put his hands in the soil all the time now in order to get a living or else start stealing which is about the rule of the day. There is no more "tapu" and the sacredness of the "raui" now is a thing of the past. The water supply is bad and filthy.[42]

He went on to say that "the court chained us" when residents, required to "put [their] land into cultivation," had to then pay all expenses incurred. Otherwise, he faced the "seizure of his property." People had little recourse to challenge Gudgeon's rulings, but some resented what they perceived as forced participation in a market economy.[43]

For all of these reasons, the Land Court's reforms did not immediately lead to a profitable export market. After two years, a frustrated Gudgeon asked New Zealand officials to empower him to force Cook Islanders to plant their lands; they refused.[44] Wide-scale planting had not been undertaken by World War I, either, when New Zealand officials asked the Cooks to contribute not only soldiers but surplus foodstuffs to the war effort. Planting continued, but unsystematically. Another large push to efficiently export produce from the Islands would have to wait until the 1930s.

Orange Replanting Schemes and Environmental Effects

Wild oranges comprised the bulk of Cook Islands exports between 1900 and 1930, with the 1920s being the most profitable decade.[45] Taking cues from the thriving Southern Californian citrus market, New Zealand officials attempted to put the Cook Islands' orange industry—which then consisted primarily of fruit from scattered inland groves—on a more scientific and technological footing.[46] The first director of agriculture to the Cooks, Mr. Bouchier, died from injuries he sustained while attempting to save his botanical research during a 1935 hurricane.[47] Affected by the global economic depression, Cook Island growers petitioned New Zealand for assistance in 1936. The government responded, taking control of the marketing and export of fruit in 1937.[48] Maurice Baker, a citrus expert from Jamaica, arrived the same year to replace Bouchier and reinvigorate the industry. Baker worked quickly to implement a series of Citrus Replanting Schemes (CRS) and presented new strains "evolved by grafting exotic orange buds to lemon stock." He pressed growers to establish model groves of ninety trees each supplied by the government's nursery, and promised these would produce fruit in only six years.[49] Baker and others had high hopes for the one hundred plantations created under the scheme and predicted a huge increase in overall output by 1950.

But Baker's plans failed. The initial scheme of 1939 was "virtually stillborn," according to New Zealand geographer W. B. Johnston. Even after officials agreed to organize cultivation and give 50 percent of the profits to landowners once their debts were repaid, growers still rejected the plan due to concerns over land titles and planting methods. As Johnston wrote in frustration, "The most jealously guarded heritage of the native is his right to the land which, by law, he cannot sell. The Cook Islanders detested leasing their land, and the administration failed to gain their cooperation."[50] In response, officials passed an Occupation Rights law of 1946 allowing multiple landowners to grant one owner full planting rights on shared property, dependent on his continuous occupation of the land. Cook Islanders remained wary, but by 1960 724 plots had been established on 450 Rarotongan acres and produced almost 50 percent of the total agricultural exports from the Cooks.[51] Exports increased again in the 1950s, but shipping and storage problems continued. A canning plant constructed in 1961 on Rarotonga was soon using most of the island's excess oranges to make "Raro," a popular but low-value juice sold in New Zealand.

This partial success held mixed results for Cook Islanders. They desired the imported goods oranges could buy; entire families gathered the fruit between April and July from wild and cultivated trees scattered far from packing sheds. But, according to Ron Crocombe, for some growers "the planting of citrus" became "a strategy for protecting their land rights rather than a commitment to [commercial] citrus production."[52] In addition, intercultivation, or growing plants for local consumption, such as watermelons, manioc, kumara, and taro, among the wild orange trees, was common practice and encouraged casual growth of citrus. Interdependent crops provided shade, lessened erosion, and bolstered food security.

The benefits that more traditional agriculture provided were important because industrialized citriculture came with serious environmental consequences. Pacific Islands have highly dynamic ecosystems "particularly vulnerable to rapid and irreversible changes resulting from human activities."[53] Pests like fruit flies and European-introduced rats consumed as much as one-third of Rarotonga's subsistence crops in the first years of the twentieth century and attacked the wild orange groves that made up an "overwhelming proportion" of fruit trees until the 1950s.[54] Before officials initiated the replanting schemes, they reported that "most of the [orange] trees were old and were suffering from a variety of untreated diseases. As they were planted at random through bush and undergrowth, caring for them was arduous and time-consuming."[55] Under the citrus schemes, workers removed wild trees in suitably flat locations to make room for cultivated orange groves or tomatoes, at four months a fast-growing cash crop option. This "short-sighted exploitation of the islands' forest resources" later returned to haunt those involved in transforming the land.[56]

Environmental impacts were soon evident. As resident commissioner H. F. Ayson wrote in 1941, "The island is rapidly becoming unfertile due to erosion by rain and also due to a lack of humas [*sic*] in the soil aggravated by the lack of shelter trees."[57] Attempts to convince growers to spray, fertilize, and prune their new trees were initially unsuccessful, as Cook Islanders questioned both how these practices affected the land and whether (and when) they would result in higher returns.[58] Officials continued to invest in the new plans, even holding an Agricultural Field Day in 1941 advertising "the progress [already] made in the replanting scheme and for the purpose of interesting the leaders of the island in replanting shelter trees."[59]

An expanded Department of Agriculture transformed the land. They paid workers to plant thousands of trees, shelter belts, and cover crops, all the while managing spraying and pruning on a wide scale. By 1940, the department had constructed twenty-eight gassing rooms on Rarotonga, Aitutaki, Atiu, and Mauke. New chemical fertilizers reduced waste caused by disease and pests from 30 percent to 1 to 5 percent, but at significant environmental cost.[60] Pesticides and chemicals coated the trees and leached into the soil and streams; workers extracted centuries-old chestnut trees. In their place, Baker directed growers to plant "quick-growing pistache" trees (*Albizzia falcata*) brought from Samoa, whose wood was also used to construct fruit crates.

To the frustration of local administrators and resident Europeans, Cook Islands planters found creative ways to resist the new guidelines, such as leaving the impostor trees on their land "untrimmed and unchecked." They also carried out more violent opposition, such as uprooting the newly planted shelterbelts and leaving their gardens exposed. Through these actions Cook Islanders pushed back against altering the landscape in such extreme and permanent ways.[61] Today, some blame the replanting scheme pesticides for long-term health effects and lagoon pollution and cite the widespread culling of chestnut trees with critical loss of shade for a variety of crops.

The impact of intensive agriculture was also emotional and cultural. Geographer Kenneth Cumberland wrote that even with limited economic growth created by the Citrus Replanting Schemes, the post-war period was "accompanied by both economic and spiritual depression amongst the Cook Islanders," as shipping again lagged, fruit spoiled, and workers abandoned agricultural for wage jobs both in Rarotonga and in New Zealand.[62] Initially, growers had been happy to accept loans, but came to feel the administration lured them into debt and measly reimbursements. In addition, Cumberland estimated that due to the focus on cash crops over food production, between 1945 and 1960, people's traditional diet was "replaced by a diet composed largely of store products" consisting of tea and bread for breakfast and tinned meats for lunch.[63] The story of commercial citriculture can be directly linked to import dependency and its resultant health problems in the Pacific Islands.

Nor were the consequences of modern agriculture and the measures used to protect cash crops limited to the land. In the Pacific Theater of World War II (1939–45), fishing diminished in the lagoons of Rarotonga. Rules of *tapu* and *ra'ui* had been nearly abolished, and this combined with increased fishing with toxic plants and explosives led to a shortage of the former dietary staple.[64] Over-fishing and new agricultural techniques exacerbated and accelerated soil erosion and insect infestation. By the end of the war, "incipient erosion of the hill-country" due to excessive scrub burning and other factors carried large quantities of silt into the lagoons, which, along with runoff from pesticide use, poisoned inland streams on Rarotonga and further impacted fish populations and the health of the reef ecosystem. By the 1960s, fishing, according to some visitors, was virtually non-existent.[65] Severe leaching and erosion on Atiu and Mangaia also led to the implementation of reforestation projects in 1951 and 1959, respectively, primarily to produce more fruit-case timber.[66] In 1963, a Rarotongan elder, Tongareva, spoke out against plans to establish a joint Japanese-New Zealand tuna fish cannery on the island, arguing, "Next they will be asking for land." The cannery was never established.[67]

Conclusion

Cultivated orange trees take six years to produce fruit. As Cook Islands planters waited—and watched their young people move away or take up wage labor—they sank ever deeper into debt and disillusionment. Shipping and transportation problems, low prices, labor disputes, competition, migration—all of these played a part in the final collapse of the citrus industry in the early 1970s, but none were as significant as incompatible ideas of land use and the determination of Cook Islanders to control their own environment and resources.

In 1965, the Cook Islands gained political independence but remained in "free association" with New Zealand, and plans to construct an international airport in Rarotonga commenced. With its completion in 1974, tourism became the new economic mainstay of the Cook Islands, replacing agriculture. Citrus replanting efforts continued, albeit on a much smaller scale, and the islands continued to produce fruit for New Zealand, now mainly in the form of canned "Raro" juice. Opportunities in New Zealand and Australia pulled many away, some of whom never returned. Most adults balanced limited subsistence farming with other jobs, producing enough root crops and fruit for their families and selling any surplus at local markets.[68]

The migrant ecology envisioned and imported by missionaries, Europeans, and New Zealanders—one that attempted to implement intensive export agriculture and transform traditional land tenure—failed, due to a large extent to Cook Islanders' resistance and cultural resilience. As a result, land use in the Cook Islands today is probably quite similar to what it was two hundred years ago. The amount of land dedicated to cultivating subsistence crops has lessened significantly, but commercial cash crops are relegated to history. An industrialized system of agriculture proved unsustainable, as evidenced by the repeated attempts of the Land Court and replanting schemes to modify land use. Valuing land simply as a commodity clashed with the values of Cook Islanders, both *ariki* and commoner, and with the health of island ecosystems. Poisoning their beautiful lagoons, increasing erosion, and introducing pests, commercial agriculture never gained a secure place in these islands.

An example from my time in the Cooks helps illustrate the enduring power of connections there between people and place, environment and culture. A series of rainstorms pummeled Rarotonga the week before my research trip in 2015. While at the National Archives a few days later, the main archivist in hushed tones described two landslides that occurred on the island the previous day. I assumed flooding or erosion caused them, but she understood the events as signs foretelling the impending deaths of prominent Cook Islanders. The next day a beloved and well-known Catholic nun died, once again linking legacies of outside influence with ancient beliefs. I was reminded that land on Pacific Islands is "imbued with the spirits of the ancestors and binds together those who share rights in it."[69] On the way back to my apartment, I stopped by an outdoor market to pick up locally grown taro, breadfruit, bananas, and kumara. I then visited a small family run store to purchase tinned biscuits and coffee. Cook Islanders, in step with their traditions and environment, continue to cultivate an adaptive and ever-changing landscape.

Notes

Epigraph: John Williams, *Missionary Enterprises in the South Sea Islands* (London: J. Snow, 1867–1869), 116.

1. Richard Gilson, *The Cook Islands: 1820–1950* (Wellington: Victoria University Press, 1980), 140; Michael Bellam, *The Citrus Colony: New Zealand-Cook Islands Economic Relations* (Wellington: New Zealand Coalition for Trade and Development, 1981), 9.
2. Ron Crocombe, *Land Tenure in the Cook Islands* (Melbourne: Oxford University Press, 1964), 3.
3. Cook Islands Native Land Court Records, minute book Vol. 1 (Microfilm, originals in Rarotonga, University of Auckland), 173.
4. Caradoc Peters, "Human Settlement and Landscape Change on Rarotonga, Southern Cook Islands." PhD diss., University of Auckland, 1994, 105.
5. Crocombe, *Land Tenure*, 63–64.
6. Cook Islands Land Court Records, minute book Vol. 1, 13.
7. Crocombe, *Land Tenure*, 19.
8. Ibid., 16.
9. Peters, "Human Settlement," 142.
10. Ibid., 139.

11. Ibid., 74.

12. John Williams, *A Narrative of Missionary Enterprises to the South Sea Islands* (London: J. Snow, 1837), 204.

13. "A Journal of a Voyage Undertaken Chiefly for the Purpose of Introducing Christianity among the Fegees [Fiji] and Haamoas [Samoa] by Messrs Williams & Barff in 1830," p. 9, cited in Peters, "Human Settlement," 129.

14. Crocombe, *Land Tenure*, 66.

15. Peters, "Human Settlement," 315.

16. Rodney Hare, "Food Marketing and the Problem of Imported Food in Rarotonga, Cook Islands." PhD diss., University of Auckland, 1980, 35.

17. Crocombe, *Land Tenure*, 67–68.

18. By 1867 the population of Rarotonga had dropped from approximately 7,000 to 1,856, and in 1846 almost 20 percent of children on Mangaia were orphans.

19. Crocombe, *Land Tenure*, 68.

20. Crocombe, *Land Tenure*, 72; Cathy Banwell, "'Back Seat Drivers': Women and Development on Rarotonga." MA diss., University of Auckland, 1985, 138. Foreigners wanting to move to the Cooks initially met strong opposition from the missions, which enacted laws to bar them from settlement. Later, government officials would uphold this.

21. E. H. Lamont, *Wild Life among the Pacific Islanders* (London: Hurst and Blackett, 1867), 90.

22. Crocombe, *Land Tenure*, 193.

23. *Te Waka Māori o Ahuriri* 1, no. 4 (July 25, 1863): 4, accessed July 26, 2016, https://paperspast.natlib.govt.nz/newspapers/waka-Māori. This Māori newspaper was published from 1863 to 1871.

24. Crocombe, *Land Tenure*, 176.

25. G. F. Mills, *Islands in the Shade: 60 Years of NZ Rule in the Cook Islands* (Wellington: Socialist Forum, 1962), 1.

26. Gilson, *Cook Islands*, 82.

27. A. D. Couper, "Protest Movements and Proto-cooperatives in the Pacific Islands," *Journal of the Polynesian Society* 77, no. 3 (September 1968): 263–274.

28. William Wyatt Gill, *Life in the Southern Isles; or, Scenes and Incidents in the South Pacific and New Guinea* (London: The Religious Tract Society, 1876), 15, quoted in Peters, "Human Settlement," 130–131.

29. William Wyatt Gill, *From Darkness to Light in Polynesia* (London: The Religious Tract Society, 1894), 127.

30. As Peters noted, "finally, the population turned to the swamp lands, between the other two zones, to plant taro" and other crops. Peters, "Human Settlement," 69.

31. James Beattie and John Stenhouse, "Empire, Environment and Religion." *Environment and History* 13, no. 4 (November 2007): 419.

32. Crocombe, "Land Tenure," 187–189.

33. Cooper, "Protest Movements," 264.

34. Bellam, *Citrus Colony*, 18.

35. "Tropical Products for NZ," *The Press* 57, no. 10811 (November 12, 1900): 4.

36. Geoffrey Sylvester Peren, *Agriculture of Samoa, Cook Islands, and Fiji* (Palmerston North, NZ: Massey Agricultural College, 1947), 16.

37. Gilson, *The Cook Islands*, 136. Out of 8,000 acres, less than 1,000 were under cultivation, with about 300 being leased to Europeans. Cook Islanders owned and operated

their own ships and ran inter-island shipping in the last decades of the 1800s, and by all accounts did very well. It is possible that Cook Islanders' interest in cash crop production declined as the new government started to control shipping.

38. Richard Gilson, "The Background of New Zealand's Early Land Policy in Rarotonga," *Journal of the Polynesian Society* 64, No. 3 (September 1, 1955): 275; Gilson, *Cook Islands*, 148–149.

39. Gilson, *Cook Islands*, 82–83.

40. Crocombe, *Land Tenure*, 129, 145.

41. Bellam, *Citrus Colony*, 9; Gilson, *The Cook Islands*, 142.

42. Cook Islands Native Land Court Minute Book Vol. 2 (Microform), 1903–1908.

43. Crocombe, *Land Tenure*, 129.

44. Walter Gudgeon to the Hon. C. H. Mills, Minister administering the Islands, Wellington. "Cook and Other Islands," July 20, 1905, in *Appendices to the Journal of the House of Representatives* (Session II, A-3, No. 24, 1906), 10–11.

45. New Zealand Department of Island Territories, *Cook Islands: New Zealand's Tropical Province* (Wellington: New Zealand Department of Island Territories, 1950), 12.

46. Edward Dallam Melillo, *Strangers on a Familiar Soil: Rediscovering the Chile-California Connection* (New Haven, CT: Yale University Press, 2015), 104.

47. At least this was the official story. According to Jean Mason, who cited the former speaker of the Cook Islands Parliament, Raituti Taringa, locals said Bouchier died as punishment for his part in the destruction of the ancestral orange trees of Rarotonga.

48. Crocombe, *Land Tenure*, 143. Until that point various traders had established a monopoly "fruit ring" that controlled prices and shipping space.

49. It normally took ten years for Cook Islands wild orange trees to bear fruit, versus six for cultivated oranges, four for bananas, and four months for tomatoes.

50. W. B. Johnston, "The Citrus Industry of the Cook Islands," *New Zealand Geographer* 7, no. 2 (1951): 126.

51. Crocombe, *Land Tenure*, 144–145.

52. Kevin Barry Short, "Factors Contributing to the Decline of Production in the Rarotongan Citrus Industry." MA diss., University of Auckland, 1980, 40.

53. Jeremy Carew-Reid, "Conservation and Protected Areas on South-Pacific Islands: The Importance of Tradition," *Environmental Conservation* 17, no. 1 (1990): 34.

54. Gilson, *Administration in the Cook Islands*, 11; Johnston, "The Citrus Industry," 130.

55. Crocombe, *Land Tenure*, 143.

56. Ibid., 130.

57. Cook Islands "Annual Report, 1941/42" (file accessed at National Archives, May 2015), 5.

58. Crocombe, *Land Tenure*, 143.

59. Cook Islands "Annual Report, 1941/42," 5.

60. New Zealand Department of Island Territories, *Cook Islands: A New Zealand Province*, 21–22. (My italics.)

61. Johnston, "Citrus Industry," 131.

62. K. B. Cumberland, quoted in Mills, *Islands in the Shade*, 2.

63. Cumberland, quoted in Mills, *Islands in the Shade*, 3.

64. Peren, *Agriculture*, 23.

65. Quote from Jackson Webb, "Cook Islands Face a Dark Future?" *New Zealand Herald* (March 28, 1970). Information from both Webb and Peren, *Agriculture*, 24. Rarotongans are "burning off fern and scrub quite unnecessarily"—he says there's no real reason

for it, and it is "leading the erosion," which carries "large volumes of silt into the lagoons" and affects the fish. Also, problems with insects and fungi.

66. Crocombe, *Land Tenure*, 147, 140.

67. Dick Scott, *Years of the Pooh-Bah: A Cook Islands History* (Rarotonga: CITC; Auckland: Hodder and Stoughton, 1991), 287.

68. Pollock, *These Roots*, 161.

69. Ibid., 163.

EIGHT

Settler-Colonialism, Ecology, and Expropriation of Ainu Mosir
A Transnational Perspective

Katsuya Hirano

This chapter examines a cross-pollination of the concept *terra nullius* (unoccupied or uninhabited land) across the Pacific Ocean in the settler colonization of Ainu Mosir (now Hokkaido) and its violent effects on the Ainu's relationships with the natural world. It argues that recounting Ainu people's encounter with settler-colonial policies grounded in *terra nullius* confirms the truism that "settler colonialism is ecological domination"[1] as much as political and economic domination, and that any meaningful inquiry into settler-colonial history demands a transnational perspective that probes the ways in which colonial ideas and practices, including those that cause serious ecological damage, are transmitted across borders. In other words, this chapter offers a decolonial reading of the Ainu's encounter with a modern settler-colonial form of domination.

AINU-WAJIN RELATIONS IN THE EARLY MODERN ERA

Matsuura Takeshirō (1818–1888) traveled from 1844 to 1858 in Ainu Mosir or Ezo (in Japanese)—now Hokkaido, Sakhalin, and the Kuril Islands—and became closely involved with the Indigenous Ainu, learning their language and documenting his encounter with Ainu culture and the natural environment.[2] He filled a massive number of notebooks with minute observations and lively drawings of Ainu customs as well as flora and fauna he "discovered." With Ainu help, Matsuura made the region's most detailed maps, with names of rivers, mountains, bays, and settlements in the Ainu language. Contrary to his fellow Wajin (Japanese) merchants who discriminated and exploited Ainu for very lucrative trade in fishery products, Matsuura developed respect for the Indigenous people and grew increasingly critical of the

Matsumae clan, a Tokugawa vassal who ruled the southwestern part of Ezo and oversaw the trade between Wajin and Ainu. For example, Matsuura lamented in his 1858 report that one-third of 111 Ainu villagers, all of whom were men of prime age, were taken away from Nibutani, an Ainu village in south-central Ezo, by the Wajin merchants to perform forced labor for sardine fishing. Only the elderly, women, and children were left in the village.[3] Those men were confronted by harsh working conditions and were not allowed to leave the workplace without the permission of their supervisors.

Matsuura's vast knowledge of Ezo and special connections with the locals earned him fame as an expert on the region and led to a position in the new Meiji government (1868–1912). The government appointed him as an officer for the newly established department of *kaitaku* (settler-colonial development) in 1869 and had him rename Ezo. He named the land Hokkaido, which means a "northern land belonging to the Indigenous people."[4] However, merely six months after his participation in the new government, Matsuura resigned from the post in sharp disagreement with the government on its continuous exploitation and expropriation of Ainu labor. Matsuura never returned to Hokkaido.[5]

In 1869, despite Matsuura's protest, the Meiji government declared the Ainu lands to be *terra nullius* (*mushuchi*) and began aggressive settler colonization. The purpose was twofold: to establish Japan's sovereignty over the island against Russia and to start the cycle of capitalist accumulation by expropriating water, trees, minerals, fishes, and animals from the Ainu. The Ainu's relations with the natural world would forcibly become altered as a result of Wajins' complete domination over the lands.

As Matsuura recorded, Ainu people called their land Ainu Mosir—"peaceful land for humans"—and believed that every material form, including plants, insects, and animals, and natural phenomena such as sun, wind, thunder, water, and fire, were the manifestations of and precious gifts from spiritual beings (*Kamuy*). Among the many Kamuy, the most important were *Repun Kamuy*—the killer whale, who was the god of the sea, fishing, and marine animals; *Kim-un Kamuy*—the brown bear, who was the god of mountains; *Kotan Kor Kamuy*—Blakiston's fish owl, who protected Ainu communities; and *Kamuy Fuchi*—senior goddess of fire and hearth. The Ainu considered Kamuy essential to their material and spiritual existence and therefore expressed their gratitude through daily rituals and seasonal ceremonies. According to Matsuura's reports, this richly spiritual view of the natural world was directly intertwined and deeply entangled with the material conditions of Ainu communities. Ainu dwellings were sited near sources of drinking water and fishing and hunting grounds, as their livelihood was based on these activities. One of the most important factors was the spawning ground of salmon, a major food source that sustained Ainus' diet all year round. Thus, the houses were usually situated on or near the edge of a river terrace close to the spawning grounds.[6]

Although it was the Meiji government's settler-colonial policies that violently altered the Ainu's relations with the natural world, Wajins' aggressive intrusion into Ainu lands began much earlier during the Tokugawa period (1603–1868).[7] In the early seventeenth century, the Matsumae, a clan that was a direct vassal of the Tokugawa shogun, was authorized to control all the Wajin settlements in southwest Ezo. They built a small castle in the town of Matsumae and started setting up trading posts. While imposing restrictions on Ainus' extensive trading networks, the Matsumae gave their loyal vassals the exclusive rights to operate the posts, allowing them to facilitate the lucrative trade in rice, cotton, and artifacts for dried salmon, herring, and kelp. The shogunate granted the Matsumae the privilege to limit Ainus' trade only to these posts. As the vassals steadily turned over the business of trade to well-established merchants (mainly from Osaka, but especially to traders from Ōmi near Kyoto), the merchants quickly transformed their voluntary "trading" relationship with the Ainu into coerced corvée labor.[8]

From the mid-eighteenth century on, as the Matsumae put increasing pressure on the merchants to yield more profit from the trade, Ainus' working conditions in the fisheries worsened, and Wajins' unfair and deceptive trade practices became rampant. Dishonest merchants traded sake to the Ainu in barrels with false bottoms and diluted the sake with water. They encouraged Ainu to drink in order to make them intoxicated and accept unfavorable trading terms. Matsumae authorities also prohibited Ainu from raising crops and buying seeds or hoes. The prohibition of farming kept the Ainu dependent on trade for rice for their own consumption.[9]

The Ainu people's steady subjection to contractual relations with Tokugawa merchants in the late eighteenth century transformed many Ainu from trading partners to corvée laborers. As their anger and frustration grew, the Ainu from Furukamafu initiated the Menashi–Kunashir War in 1789. Ainu attacked Wajin at the Kunashir trading post and on a ship nearby, leaving at least seventy-one dead. In retaliation, the Matsumae authorities, with assistance from the shogunate, captured eighty-seven Ainu and executed thirty-seven.[10] After Wajin conquered Kunashir Ainu, they forcibly recruited Ainu in Shari and Abashiri as replacement labor and compelled them to move to Kunashir.[11] The fisheries system also allowed Wajin fishery managers to create a division of labor based on gender difference, splitting Ainu families by sending husbands and wives to distant fisheries across Ezo. Wajin managers took advantage of this gender-segregated arrangement of labor by instituting "a system of sexual colonization, forcing Ainu women to serve as mistresses or 'local wives' for mainland Japanese bosses, or subjecting them to sexual assault at the hands of Wajin laborers."[12]

Some historians consider this process of Wajins' encroachment in the southern part of Ainu Mosir and violent subjection of Ainu in an exploitative trading relationship as an exemplary case of settler colonialism.[13] I agree with this assessment to the extent that the process created Japanese settler communities in Ainu lands and destroyed the Ainu people's social relations of production, mode of livelihood, and connection to the places deeply integrated with spiritual practice. However, two decisive features of settler-colonial formation—namely, systematic dispossession of land (Indigenous peoples' means of livelihood) and demographic replacement or elimination—were not part of the Matsumae's colonial policies. As Patrick Wolfe, Lorenzo Veracini, and Glen Coulthard have incisively argued, the logic and structure that drive settler-colonial policies aim not for exploitation but elimination.[14] Wolfe argues:

> The primary object of settler-colonization is the land itself rather than the surplus value to be derived from mixing native labour with it. Though, in practice, Indigenous labour was indispensable to Europeans, settler-colonization is at base a winner-take-all project whose dominant feature is not exploitation but replacement.[15]

Neither the Tokugawa shogunate nor the Matsumae clan attempted to systematically replace Ainu with Wajin settlers or took the elimination of Ainu communities as the precondition for their economic gains. Indeed, as argued above, they depended on the overexploitation of Ainus' labor to increase their profit. Although Tokugawa Japan's aggressive control of southwestern Ezo undermined long-established Ainu communities in the region, it was not until the Meiji government adopted the concept of *terra nullius* to systematically displace Ainu and expropriate their land that the logic of dispossession and thus the eliminatory impulse became the primary feature of Wajins' domination of Ainu Mosir.

Furthermore, what I would like to argue in this chapter is that modern settler-colonial dispossession is simultaneously a violent reconstitution of Indigenous ecology. It entails the aggressive conversion of interdependent relationships, which long sustained Indigenous peoples' material and epistemological worlds, into a relationship that is conducive to the cycle of capitalist production and accumulation. This drastic remaking of Indigenous ecology constitutes an essential and distinct characteristic of modern settler-colonial dispossession, as manifested in the notion of *terra nullius*, a genocidal concept that negated Indigenous peoples' centuries-long modes of living as primitive, obsolete, and meaningless. Ainu Mosir too came to be re-articulated not only as a Japanese borderland that defined the northern limits of the nation's sovereign territory but also a rich reservoir of natural "resources" for capitalist industrialization. Meiji settler-colonial policies epitomized a distinct logic of national capitalist formation grounded in expropriation and monopolization of the productive forces of the natural world while simultaneously converting the Indigenous inhabitants into a dispensable population or a remnant of the prehistoric past—living fossils with no relevance in a world dominated by the requirements of capitalist production and accumulation. As I will soon articulate, the combined forces of the Meiji government and the US experts and technology carried out this process of Indigenous displacement and dispossession.

Hokkaido as *Terra Nullius*

Language like *terra nullius* and "virgin lands" worked as a colonial tool to justify the total expropriation of Ainus' means of sustenance and fundamental negation of their relationship with the natural environment. These strategic idioms drew from United States' policies for the American West.[16] A narrative of "wide open spaces" in the West just waiting to be filled by enterprising white settlers underwrote US homesteading policies and made westward expansion a kind of nationalist moral imperative. The Meiji government implemented a similar set of colonial strategies, starting with the same linguistic moves. Indeed, Horace Capron, who hailed from the United States, working as the chief adviser for the Meiji government's colonial polices in Hokkaido, called Hokkaido a virgin land and wilderness and proposed in his 1872 report a Japanese version of his country's allotment and land redistribution policies for Indian Territory set out in the US Homestead Act of 1862.[17] By urging the Meiji government to adopt "settlement on liberal terms offered by the government of the United States" as a means to spur the speedy occupation of Hokkaido, Capron proposed that a program of public land grants to small farmers should enable each settler in the new frontier "to become the bona fide owner of a tract of 160 acres of the public domain without cost, except the payment of $10 to the land officer of the district."[18]

The translation of the Ainu's heterogeneous world into familiar colonial idioms marked the decisive moment when the aggressive policies of expropriation came to signify the positive value of "opening" or *kaitaku* in the name of supposed civilizational development and progress.[19] Once this inscription gained legitimacy in the public discourse during the 1880s and 1890s, the colonial logic of the "civilizing mission" meant that the Ainu's practical and conceptual world soon came to signify backwardness to be effaced from the earth. By the 1900s, the social relations and values that had sustained Ainu communities were commonly rendered as the direct cause of their displacement and deprivation. The Japanese government argued that the Ainu's struggle and poverty were due to their own innate inability to understand the concept of private ownership and learn a way of life beyond primitive hunting and fishing. The self-serving idea came to dominate the governmental and popular discourses: the Ainu were a feeble race destined to die out according to the universal law of social Darwinism because they could not compete with the Japanese or develop the "frontier." Furthermore, Japanese intellectuals, educators, and policy makers all came to harbor the notion that the only means by which the Ainu people could ensure their own survival was through cultural assimilation.

Modernizing Japan, American Experts, and the Remaking of Ainu Mosir

In 1871, Kuroda Kiyotaka, concerned about Russia's push eastward, visited the United States looking for a leader to oversee the initial settler-colonial development of Hokkaido.[20] On President Ulysses Grant's recommendation, Kuroda met with Horace Capron, who served in Grant's administration as commissioner of agriculture, and successfully persuaded him to accept an appointment as special adviser to the Japanese government. Kuroda hired Capron for $10,000 per year and additional funds for expenses to undertake the mission.[21] It is quite likely that Capron's earlier experience managing the forced removal of Native Americans, including those affiliated with Delawares, Shawnees, Creeks, Comanches, Kickapoos, Wichitas, and others, from Texas to new territories after the Mexican–American War, appealed to Kuroda and his government.[22] Capron remained in Japan from 1871 to 1875 with a singular task: "find [the] best way to utilize the resources of Yesso [Ezo] for the material enrichment and elevation of imperial Japan."[23] During his first tour in Hokkaido in 1872, Capron became convinced that because of climate and vegetation that were similar to the northeastern parts of the United States, Hokkaido would be an ideal place to adopt American farming methods. He wrote:

> I perceive a marked similarity to the climate of the same parallels in the interior of the American Continent. The oak, the beech, the ash, in short all the trees of the forests of New York, Pennsylvania and Ohio, even including the sugar maple, grow in abundance and to perfection in Yesso. In Yesso on the same parallel these trees thrive even on the high mountain slopes,

while in America at the same altitude they are gnarled and stunted. . . . The great fall of snow in Yesso is a great advantage, serving, as it does, to protect grains and grasses from the frost and to prevent the freezing of the ground to any depth. Carefully weighing these facts and considering my own experiences in the more northern American States, I am forced to the conclusion that the obstacles to a profitable and permanent development of the recourses of the island of Yesso lies neither in the soil nor in the climates.[24]

In 1873, Capron and Kuroda established a demonstration farm in Tokyo to promote agriculture using American plants, crops, and livestock and soon invited more US experts in these fields to visit his facility.[25] W. S. Clark, Capron's successor who arrived in Japan in 1876 to take up a teaching position at Sapporo Agricultural College (today, Hokkaido University), later echoed Capron that American farming methods would be a perfect fit for the land, noting some local vegetation such as mistletoe and magnolia that were found in Virginia.[26]

Capron brought in American civil engineer A. G. Warfield and chemist and geologist Thomas Anticell to investigate the soil quality and geography of the island. Before they embarked on their first expedition to Hokkaido, Capron and his scientific advisers created an extensive preliminary proposal based on the reports of the Hokkaido Development Agency, which later served as the groundwork for the Japanese government's Hokkaido ten-year plan:

1. Hokkaido's weather is similar to the more northern American States and is thus suitable for extensive farming. It is desirable to bring in more settlers to fully utilize the wealth of natural resources.
2. A thorough land survey should be carried out and the law of private property should be introduced to settle matters related to land ownership.
3. A machine factory needs to be built. It will help open a road that connects Ishikari and Muroran, Hakodate, and Sapporo. The machinery shall [also] elevate and increase the value of human labor. This is crucial for the wealth of a country in proportion to its aggregate labor.
4. Silver and lead in the Yu-rappu area, sulfur in southern Hokkaido, and coal in Ishikari are all valuable as an ever-increasing source of wealth. A Bureau of Mines should be established. But private management of the enterprise is more desirable for securing the highest productiveness, as proven in England and America. The government's role is to be limited to enacting and enforcing the necessary laws for the regulation of mining interests and rights of property.[27]

On the heels of these recommendations, Capron and his team set off on a tour of Hokkaido in 1872. During the expedition, Capron wrote, "[Yesso] is a wonderful island. But its true value has not been sufficiently appreciated. It is rich in mineral resources, its fishery is unlimited, lumber has excellent quality, and agriculture has unlimited possibilities. . . . It could easily provide for several million people."[28] Then he added, "The resources and climate of this island have been misunderstood, or misrepresented, at least to me. . . . If the natural products of a soil are any indication of its fertility or climate, this Island will compare favorably in these respects with some of the wealthiest and most populous portions of the United States."[29]

After returning to Tokyo, he advised Meiji leaders to invest in the development of coal mines, railroads, fishing, capital-intensive agriculture (including livestock), orchards, and irrigation systems, and to build a public school.[30] Capron began introducing capital-intensive farming, with American methods and implements, imported seeds for Western crops, and European breeds of livestock, including his favorite Devon and Durham cattle. He established experimental farms on Hokkaido, had the land surveyed for mineral deposits and farming opportunities, and recommended water, mills, and road improvements.

He also urged Meiji leaders to hire more foreign experts and bring in foreign (American) farmers and capital to initiate and speed up the process. Kuroda and his government followed most of Capron's advice, but while Kuroda was enthusiastic about inviting American capital and settlers to Hokkaido, the latter chose not to pursue foreign labor and investment as it was feared that they could give the United States a toehold for the virtual colonization and control of the island.[31] In 1876, based on Capron's advice, the Meiji government established Sapporo Agricultural College to foster new leaders and the skills needed for development.[32]

The Japanese settler colonization of Hokkaido was thus outlined and facilitated by the joint forces of the Japanese state and US experts and technology. Japanese leaders focused on the occupation of Hokkaido through the systematic migration of former samurai lords, samurai retainers, and ordinary citizens—in particular, displaced farmers and peasants—from the 1870s to the 1880s by supplying them with "free" land and financial support. Such change was also facilitated by depending on American experts, who offered various technologies of colonization, to reshape Ainu Mosir into a land suitable for Japan's capitalist farming.

The reality of settlers' lives in the colony was far from rosy. Settlers received subsidies and funds to start their new lives. However, shortages leading to 300 percent inflation and very slow progress in building infrastructure such as paved roads and railways inhibited economic production and aggravated people's lives. The obstacles eventually pushed many settlers to return to their home prefectures.

The prospect of the colonial policy's failure posed a major threat to the governmental vision of accelerated cultivation and privatization of Hokkaido's land. In 1872, the Meiji government tried to prompt wealthy Japanese landlords to invest their capital in purchasing land at a bargain price.[33] The policy did increase land sales but was most effective as a strong impetus for migration, generating a new wave of peasants and workers looking for employment on the newly purchased lands. In 1873, Japanese settlers numbered about 160,000; that population nearly quadrupled over the next four years.[34] Many of the settlers, however, did not stay in the colony very long due to the harsh winter weather and difficult living conditions caused by the lack of modern infrastructure such as roads, railways, hospitals, and schools.[35] Capital also stopped flowing in, as the real estate brokers often appropriated investors' money.

The Meiji government responded with a new strategy: while accelerating the construction of infrastructure such as telegraph lines in 1874, regular sea routes in 1880, railways in 1880 and 1883, a Mitsui bank branch in 1880, a lighthouse in 1883, mail steamer service in 1885, and the modern ports of Kushiro, Muroran, and Otaru in 1899, it introduced in 1886 the Regulation for the Sale of Hokkaido Land. The law shifted the focus from common settlers and wealthy landlords to emergent capitalists and large-scale modern corporations such as Mitsubishi and Mitsui for the development of trading networks as well as light and heavy industries. The government sold state-run factories and enterprises on Hokkaido such as breweries, farms, fishery processing factories, shipyards, soy sauce/soybean-paste factories, and sugar mills to private investors in 1881–1882;[36] it offered capitalists and corporations a free ten-year land lease with the additional incentive that if their enterprises succeeded, they could purchase the land at a steep discount. If they failed, they could simply return the land to the government. The new policy did not directly benefit working-class settlers, but the development of infrastructure and commerce and the establishment of banks in 1896 and 1899 stimulated rapid urbanization in the southern part of Hokkaido and brought a new wave of migration to major settlements such as Hakodate, Sapporo, Otaru, and Kushiro. In 1901, a new ten-year development plan announced the construction of more bridges, roads, and railway stations. The number of incoming immigrants began to surge after 1900 from 50,000 to 80,000 annually, leading to accelerated colonization of Ainu land. In 1909, the Japanese population of Hokkaido reached 1.5 million.[37]

Japanese settler communities continued to expand into not only Ainus' long-established settlements along rivers and coastal areas but also inland, where the Meiji government had forcibly relocated the Ainu. Meiji officials also took steps to prohibit traditional Ainu means of sustenance. In 1876, Japanese authorities outlawed the traditional Ainu bow and poison arrows, claiming that they represented barbarism.[38] According to Yamada Shinichi, Capron's view that "[the use of poison] is not only a wasteful but a barbarous practice" influenced the Hokkaido Development Agency's decision to implement the law.[39] Anticipating that the ban would drive the Ainu into grave difficulties, the law stipulated that the government would lend rifles to the Ainu with the condition that it would receive 20 percent of the deer hide they hunted.[40]

Ainu people made several appeals to the Hokkaido authorities asking them to lift the ban because "the use of rifles is not familiar, and when only one or two out of 10 people have learned to use them, they cannot supply for the family."[41] Their appeals were denied on the grounds that "staying consistent in policies is the best way to cleanse old customs."[42] At the same time, overhunting by Japanese settlers—armed with rifles—contributed to such a sharp decline of wild animal populations that the agency enacted other laws to regulate hunting in 1878. These laws were intended to protect not the Ainu, but Japanese hunters, whose reckless overhunting of animals such as deer, bears, and raccoon dogs—all major game animals central to the Ainu diet—began as part of the Meiji government's initial policy of extracting capital through predatory colonial development of Hokkaido. The Japanese hunters systematically slaughtered more than half a million Ezo deer between 1873 and 1878 alone and sold the resulting venison and deerskin to meet the high demand in China, France, and the United States.[43]

In addition, Benjamin Smith Lyman, an American mining engineer working for Capron and the Meiji government, advised the Hokkaido Development Agency in 1874 to encourage the extermination of bears, wolves, and wild dogs by "offering bounties, as is done in other countries," because their presence "in the mountains will perhaps be some hindrance to the introduction of sheep and even larger cattle."[44] This advice was put into law in 1877.[45] Lyman was not alone in advocating the mass killing of native wildlife species for the sweeping reconstitution of Ainu Mosir's ecosystem. Seeking to build livestock industries of the new colony, the Hokkaido Development Agency put together a team of American specialists who had gained expertise in the methods of settler-colonial reconstitution of the American West. Among them was Edwin Dun, who was a rancher from Ohio and lived in Hokkaido as an adviser for the development of farms from 1876 to 1883 (see figure 8.1). Soon after his arrival in this new Japanese colony, Dun started horse-breeding programs to raise high-quality ranch horses and

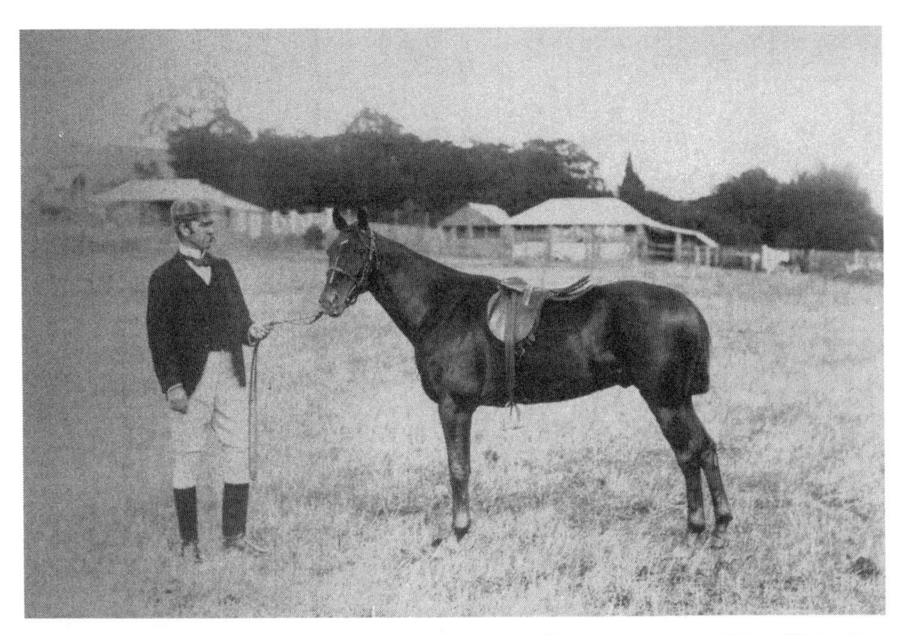

Figure 8.1. Edwin Dun and imported horse from the United States in 1878 in Hokkaido. Hokkaido University Library.

proposed the purchase of strychnine to poison wolves, bears, and wild dogs at new ranches such as the 35,000-acre Niikappu Ranch.[46] The overhunting of Ezo deer aided this quest to eliminate wolves, as hungry wolves lost a major source of prey and turned on horse foals, only to consume strychnine-laced flesh. In March 1880, the Hokkaido Development Agency reported to Tokyo that "the strychnine campaign had been a success and should be continued."[47] Dun even supported the Hokkaido Development Agency's decision to extend these policies of extermination to domestic dogs kept by Ainu people for the reason that they posed a threat to livestock.[48]

Deforestation for land development and lumber production further aggravated the ecological damage and displacement of Ainu communities. As "reclamation" progressed steadily based on the American model of "frontier" policies, and the settlers formed extensive communities in and around the Sapporo area, deforestation became a serious concern. The logging industry also contributed to the rapid deforestation. To formulate forest protection policies, the Meiji government sought the American advisers' opinions. In 1877, the Hokkaido Development Agency presented the following questions to Clark: 1) Which country should serve as the standard for implementing forest protection law? 2) What would be the proper measures by which to understand the need of forest protection? 3) What was the job of forest watchers? How many are needed? 4) What steps could prevent mountain fires? 5) What trees should be protected in the Sapporo area?[49]

Clark's responses were not of much help. Stating that infant nations such as the United States did not have forest protection policies, while older nations such as Great Britain had developed them as part of animal conservation for hunting as well as aristocratic estates, Clark discouraged the Development Agency from taking an active role in controlling forestry and recommended that settlers be in charge of clearing woodlands.[50] He reasoned that since Hokkaido was like the American frontier with rich natural resources, deforestation would not harm settler communities. Partial protection around river areas would be sufficient.[51] He made no mention of the Ainu and their reliance on natural woodlands for hunting and gathering.

Clark's noninterventionist approach and complete disregard for Indigenous communities directly reflected policies implemented in the United States. The 1862 Homestead Act prioritized public land grants to small farmers. Forests were considered an obstacle to this process of settler colonization and even became an incentive for "cut-and-run" logging.[52] Decades passed before the so-called frontier line disappeared and reckless deforestation became a serious concern. The US government finally implemented national forest management with the Forest Reserve Act of 1891.[53] In short, Clark's advice was consistent with the United States' general lack of environmental consciousness and indifference to the devastating damages of deforestation for its Indigenous communities.

Although Capron and Clark influenced the Meiji government's forest policies, the Hokkaido Development Agency nevertheless decided to introduce forest protection laws that restricted deforestation and sale of timber in 1878. They also appointed guards to prevent illegal logging. Logging in government and private forests was strictly forbidden, subject to heavy fines. Between 1910 and 1920, forestry produced up to 12 percent of the Hokkaido economy's total revenue.[54] Paper manufacturing companies such as Ōji and Tomakomai dominated the forestry and logging industries (see figure 8.2). Places such as Teshio in the northern part of Hokkaido, where vibrant Ainu fishing communities once thrived, became a boomtown for logging and paper manufacturing. Furthermore, from the 1880s on, the high volume of timber harvesting caused by the rapidly increasing number of Japanese settlers seeking to expand the agricultural land dried up rivers in Ainu communities.

During this time, Meiji authorities also moved to outlaw Ainu fishing practices, including nighttime fishing and traditional fishing nets (see figure 8.3). While the Japanese state during the early modern era had recognized nocturnal trout and salmon fishing in Ainu rivers and their tribu-

Figure 8.2. Ōji paper manufacturing company in Tomakomai in the 1900s. Hokkaido University Library.

taries as a legitimate Ainu activity, by the beginning of the Meiji period, this formal endorsement was considered a "failure to address a long-standing abuse" against natural resources.[55] The Hokkaido Development Agency summarily prohibited nocturnal fishing in 1878.[56] The Meiji government justified this action as a means of transferring the right to fish from the Ainu to the Japanese, while promoting commercial fisheries on a massive scale under the rubric of policies to "increase production in industrial enterprise" (*shokusan kōgyō*). From the 1870s to 1890s, as Japanese settler communities grew in size and number, fisheries made up more than 60 percent of Hokkaido's overall revenue, compared to agriculture (20 percent) and industries such as iron manufacturing, brewing, and paper making (15 percent). By 1910, however, fishery outputs fell to 15 percent of the revenue, while the agricultural economy grew to nearly 50 percent and industry to nearly 20 percent.[57]

Robbed of their crucial means of sustenance, the Ainu appealed to governmental agencies to delay the nighttime fishing ban, which covered not only major rivers, but smaller tributaries as well. The Development Agency denied the appeal on the basis of the "former natives' illiteracy and ignorance of law."[58] Wajin pioneers or "openers of the frontier," with the backing of the Meiji government, legally expropriated and monopolized the fishing industry in Hokkaido in the same way they had acquired the land. These displacement and assimilation policies had a devastating effect on Ainu health

Figure 8.3. This early modern illustration depicts Ainus' use of traditional fishing nets in the 1850s. Hokkaido University Library.

and well-being. The traditional staple foods, salmon and deer, were replaced by cultivated crops as the Ainu were denied their right to hunt and fish and required to farm for subsistence. Foraged wild plants, which had been an important part of their food consumption, also steadily lost importance to the Ainu, since the settlers' farming and livestock devastated patches of edible plants.

After the implementation of the Former Native Protection Law in 1899, some Ainu people managed to adapt fairly successfully to life based on farming. But, displaced from their eco-communal way of life, many Ainu failed, fell ill, and suffered starvation and impoverishment. According to the survey conducted twelve years after the inauguration of the law, many Ainu men returned to fishing while leaving farming to women and children. This virtually meant the abandonment of farming.[59] Another survey, which was carried out in 1916, reported that out of 4,007 Ainu households, about 57 percent of them lived on farming. The result appears to point to the new law's success in transforming Ainu people into farmers, but the Ainu's harvest was only one-fourth of that of an average Wajin farming family.[60] The land granted to Ainu families under the law between 1899 and 1910 was about 16,810 acres, but most of it was wasteland unsuitable for farming. Therefore, Ainu families were expected to labor extraordinarily hard to cultivate the land, and many eventually gave up. According to the law, land left uncultivated for fifteen years had to be returned to the government—and that was exactly what happened to many Ainu families. Thirty percent of the 57 percent of Ainu farming families actually became low-wage laborers in the fishing and farming industries.[61]

Conclusion

What the story of Hokkaido's settler-colonial experiences tells us is twofold: Japan's ideas and practices of settler colonialism were transpacific from the outset, shaped by frontier politics developed in the United States; and the cross-pollination of the settler-colonial project resulted not only in the massive displacement and dispossession of the Ainu people but also in the violent reconstitution of Indigenous ecology. It meant a genocidal transformation of the Ainu's "relationships with plants, animals, physical entities, and ecosystems of those places" from which they derived "economic vitality, cultural flourishing and political self-determination."[62] As Kyle Whyte succinctly puts it:

> Settler colonialism is deeply harmful and risk-laden for Indigenous peoples because settlers are literally seeking to erase Indigenous economies, cultures, and political organizations for the sake of establishing their own. Settler colonialism, then, is a type of injustice driven by settlers' desire, conscious and tacit, to erase Indigenous peoples and to erase or legitimize settlers' causation of such domination.[63]

It comes as no surprise that Ainu people came to be called by the 1910s a "vanishing race," similar to Native American and Australian Aboriginal peoples.[64] The adjective "vanishing" erased or legitimized the *historical* process of their dispossession, impoverishment, and near extinction carried out through colonial governments' migration and developmental policies—all in the name of the law of progress and capitalist modernization. And what accompanied this process of erasure was ecological violence. Ainus' encounter with Japan's systematic settler-colonial domination evinces the truism that modern settler colonialism, especially in its capitalist form, is not simply displacement of Indigenous peoples but also a total reconstitution of the natural world that sustained their livelihoods and cultures. I would like to close this chapter with the words of Emiko Chikappu, a late Ainu weaver, poet, and activist, who fought for Ainu sovereignty and human rights as well as environmental justice throughout her life. Reflecting on the history of Japan's settler colonization of Ainu Mosir, Chikappu contended in 1988 that the "opening" of Hokkaido by the Meiji government was the beginning of the destruction of the great chain of life, the deeply entangled relationships between nature and humans, rooted in Ainu livelihood as well as their conception of the world:

> In Ainu language, the phrase "thank you" literally means "I kill myself." Why does "thank you" mean "killing oneself"? You cannot understand it unless you know the backgrounds of Ainu life. We believe that wild animals offer their lives to us. "Thank you" expresses our understanding of life as a great chain that connects wild lives and humans. It means an endless circle of life. But this circle of life in Ainu Mosir was rapidly destroyed when the development of Hokkaido (under the Japanese government) began in 1869. Don't they know that destroying the circle of life leads humans to death?[65]

Chikappu asserts:

> Have people forgotten that we live on this earth? People have been cutting down trees. Human hands are threatening this beautiful earth. People, just like trees, make their livings by being firmly rooted in this earth, but they are taking away trees' lives. Ainu people call tree roots *shinritsu* which means ancestor. Tree roots are our ancestors. People have been destroying the great chain of life. Life that connects the natural world and the humans is being threatened. How much longer is such a (destructive way of) life sustainable?[66]

Notes

This chapter is an expanded version of my "Terra Nullius and the Modern Settler-Colonization of Ainu Mosir," in "Hokkaido 150: Settler Colonialism and Indigeneity in Modern Japan and Beyond," *Critical Asian Studies* 51, no. 4 (2019): 4–10.

1. Kyle Powys Whyte, "Settler Colonialism, Ecology, and Environmental Justice," *Environment and Society* 9, no. 1 (2018): 1.

2. Ezo or Ezochi was a derogatory name used by Tokugawa Japan (1603–1868) to refer to the northern islands. It meant "lands of barbarians."

3. Kaizawa Kōichi, "Nibutani ni Umarete," in *Ainu Minzoku no Fukken*, ed. Kaizawa Kōichi, Moriyama Hiroshi, Matsuna Takashi, and Okuno Tsunehisa (Kyoto: Horitsu Bunkasha, 2011), 4–5.

4. The Ainu did not have a writing system. Matsuura's diary and reports are now indispensable archives for our understanding of the Ainu culture, communal practices, and

conceptions of natural environment that existed prior to Meiji Japan's settler-colonial projects. The most representative work by Matsuura in a modern Japanese version is *Ainu Jinbutsushi* (Tokyo: Heibonsha, 2002).

5. Aida Kazumichi, *Matsu'ura Takeshirō, Kitano Daichi ni Tatsu* (Sapporo, Japan: Hokkaidō Shuppan Kikaku Center, 2017), 21.

6. For a more detailed account of Ainus' ecosystem, see Hitoshi Watanabe, *The Ainu Ecosystem* (Tokyo: University of Tokyo Press, 1972), 69–82.

7. Ann-elise Lewallen, in "Hokkaidō 150: Settler Colonialism and Indigeneity in Modern Japan and Beyond," *Critical Asian Studies* 51, no. 4 (2019): 10–17; Okuyama Ryō, *Ainu Suibōshi* (Sapporo, Japan: Mamiya Shobō, 1965), 86–162; Brett L. Walker, *The Conquest of Ainu Lands* (Berkeley: University of California Press, 2006). Also see Enomori Susumu, *Hokkaidō Kinseishi no Kenkyū* (Sapporo, Japan: Hokkaidō Shuppan Kikaku Sentā, 1997).

8. Kobayashi Masato, "Seiritsuki Bashoukeoi Sei no Seidoteki Kōsatsu," in *Basho ukeoi sei to Ainu*, ed. Hokkaidō-Tohoku Shi Kenkyūkai (Sapporo, Japan: Hokkaidō Shuppan Kikaku Center, 1998), pp. 42–111.

9. Okuyama, *Ainu Suibōshi*, 93–98.

10. Walker, *The Conquest of Ainu Lands*, 172–176; Okuyama Ryō, *Ainu Suibōshi,* 103–111.

11. Ann-elise Lewallen, "Signifying Ainu Space: Reimagining Shiretoko's Landscapes through Indigenous Ecotourism," *Humanities* 5, no. 3 (2016): 59.

12. Ibid.

13. Ibid., 5.

14. Patrick Wolfe, *Settler Colonialism and the Transformation of Anthropology* (London: Cassel, 1999); Lorenzo Veracini, *Settler Colonialism: A Theoretical Overview* (London: Palgrave Macmillan, 2010); Glen S. Coulthard, *Red Skin, White Masks* (Minneapolis: University of Minnesota Press, 2014).

15. Wolfe, *Settler Colonialism and the Transformation of Anthropology*, 163.

16. For the use of the concept of "empty land" in the United States, British Empire, and Imperial Germany, see Benjamin Madley, "Patterns of Frontier Genocide 1803–1910: The Aboriginal Tasmanians, the Yuki of California, and the Herero of Namibia," *Journal of Genocide Research* 6, no. 2 (2004): 167–192.

17. The argument presented in this section draws mostly on my earlier publication "Thanatopolitics in the Making of Japan's Hokkaidō: Settler Colonialism and Primitive Accumulation," *Critical Historical Studies* 2, no. 2 (2015): 191–218.

18. Horace Capron, *Reports and Official Letters to the Kaitakushi* (Lexington, KY: ULAN Press, 2015), 48.

19. Komori Yōichi, "'Rule in the Name of 'Protection': Vocabulary of Colonialism," in *Reading Colonial Japan*, ed. Michele M. Mason and Helen J. S. Lee (Redwood City, CA: Stanford University Press, 2012), 65.

20. Iguro Yataro, *Kuroda Kiyotaka* (Tokyo: Miyama Shobo, 1965), 50–51.

21. Ōsaka Shingo, *Kuroda Kiyotaka to Horace Capron* (Sapporo, Japan: Hakkai Taimusu, 1962), 123–135.

22. Capron often compared the Ainu with Native Americans in Texas in a more favorable light. He saw both as uncivilized but praised the former as polite and refined while complaining about the latter as savages. Horace Capron, *Journal of Horace Capron: Expedition to Japan, 1871–1875* (Sapporo, Japan: Hokkaidō Shinbun, 1985).

23. Ibid., 43.

24. Capron, *Reports and Official Letters to the Kaitakushi*, 41–43; Tawara Kōzō, *Hokkaidō Midori no Kankyōshi* (Sapporo, Japan: Hokkaidō Daigaku Shuppankai, 2008), 63.

25. Fujita Kinsuke, *Capron no Oshie to Genjutsu Seito* (Sapporo, Japan: Hokkaido Shuppan Kikau, 2006), 14–21.

26. Tawara, *Hokkaidō Midori no Kankyōshi*, 68.

27. Capron, *Reports and Official Letters to the Kaitakushi*, 41–52.

28. Capron, *Journal*, 151.

29. Capron, *Reports and Official Letters to the Kaitakushi*, 59.

30. Capron, *Journal*, 151.

31. Iguro, *Kuroda Kiyotaka*, 60–62.

32. Capron, *Reports and Official Letters to the Kaitakushi*, 50–51.

33. Sixty kilograms of rice cost two yen, and 3,300 square meters of fertile land was sold for 1.5 yen (for poorer land, the price was one yen or even half a yen).

34. Shinya Gyō, *Ainu Minzoku Teikōshi* (Tokyo: Kadokawa, 1974), 183.

35. Hideki Asada, ed., *Hokkaidō kaihatsu seisaku no rekishi, Meiji-hen* (Sapporo, Japan: Ishikarigawa Samitto Jikko Iinkai, 2004), 30–31.

36. Tatsuo Kikuchi, "Hokkaidō ni okeru shokuhinkōgyō no seiritsu yōin," *Hokkaidō Chiri* 70 (Sapporo, Japan: Hokkaido Chirigakkai, 1996): 13–19.

37. Hideki Asada, ed., *Hokkaidō kaihatsu seisaku no rekishi, Meiji-hen*, 55, 145.

38. Shinya, *Ainu Minzoku Teikōshi*, 189.

39. Shinichi Yamada, *Kindai Hokkaidō to Ainu Minzoku* (Sapporo, Japan: Hokkaidō University Press, 2011), 27.

40. Ibid., 37.

41. Ibid., 42–45.

42. Ibid., 45.

43. Kōichi Kaji, Masami Miyaki, and Hiroyuki Uno, *Ezoshika no Hozen to Kanri* (Sapporo, Japan: Hokkaidō daigaku shuppan-kai, 2006), 6–7.

44. Yamada, *Kindai Hokkaidō to Ainu Minzoku*, 117.

45. Ibid., 123–124.

46. Brett L. Walker, "Meiji Modernization, Scientific Agriculture, and the Destruction of Japan's Hokkaidō Wolf," *Environmental History* 9, no. 2 (April 2004): 263.

47. Ibid., 266.

48. Ibid., 260–261.

49. Ōkura Shō (Ministry of Finance), *Kaitakushi Jigyō Hōkoku Dai-ippen* (Hokkaidō University Library, 1885), 436–438.

50. Ibid., 438–439.

51. Ibid., 439–441.

52. Ibid. John Wright's *Rocky Mountain Divide* offers an interesting story of the opposing forces of development and conversation that have shaped the American West. John B. Wright, *Rocky Mountain Divide: Selling and Saving the West* (Houston: University of Texas Press, 2010).

53. Tawara, *Hokkaidō, Midorino Kankyōshi*, 67.

54. Ibid., 100.

55. Iwasaki Naoko, "Rekishi to Ainu," in *Nihon wa Doko e Ikunoka* (Tokyo: Kodansha, 2003), 209.

56. Ibid.

57. Tawara, *Hokkaidō, Midorino Kankyōshi*, 100.

58. Iwasaki, "Rekishi to Ainu," 211.

59. Sekiguchi Akira and Kuwahara Masato, eds., *Ainu Minzoku no Rekishi* (Tokyo: Yamakawa Shuppan, 2015), 197–198.

60. Ibid., 199–200.

61. Ibid., 201.

62. Whyte, "Settler Colonialism," 134–135.

63. Ibid., 135.

64. See Brian Dippie, *The Vanishing American* (Lawrence: University Press of Kansas, 1982); Patrick Brantlinger, *Dark Vanishings: Discourse on the Extinction of Primitive Races, 1800–1930* (Ithaca, NY: Cornell University Press, 2003).

65. Emiko Chikappu, *Kazeno Megumi: Ainu Minzoku no bunka to jinken* (Tokyo: Ochanomizu Shobo, 1991), 154.

66. Ibid., 119–120.

Nine

Pearl of the Empire
Conservation, Commerce, and Science in the Tuamotu Archipelago

William Cavert

In the 1880s, the French colonial government in Pape'ete faced a dilemma—how could it simultaneously conserve and exploit the undersea banks of pearl-bearing oysters in the lagoons of the Tuamotu archipelago. French presence in the lagoons was light and largely invisible despite the pearl-shelled oysters being one of the colony's chief exports. In order to address concerns that the oyster banks faced imminent exhaustion, the colonial administration solicited metropolitan authorities to fund a series of scientific missions. These missions, they argued, were necessary to accurately establish the nature of the lagoons and formulate a rational, cost-effective method to regulate the *nacre*, or pearl-shell, industry.

The local administration assumed that government-sponsored naturalists would support and strengthen their regulatory measures. Furthermore, they would establish a framework for a sustainable pearl oyster industry based on lagoon-farmed oysters that enriched the colony through export duties and licensing while providing employment for Islanders in the Tuamotu Archipelago and merchants in Pape'ete. The solicitation of Germain Bouchon-Brandely in late 1883 was the first occasion a naturalist had been employed to advise and undertake a comprehensive study of the lagoons. While his work would be feted in France, his recommendations were found untenable by those tasked with implementing them and irrational by local merchants and divers. The story of Bouchon-Brandely and the subsequent and aborted merchant rush to establish oyster farms in the lagoons constitutes a compelling episode in the interconnected environmental, cultural, and colonial histories of France and Oceania—an episode in which science served as a tool of state to assemble the lagoon as a manageable object.

The colonial state employed naturalists in the Tuamotu Archipelago to render the marine environment, the aquatic, unseen, and rarely visited space, into something governable and manageable from afar. Arun Agrawal observes a similar usage of science for the development of new technologies of governance in forestry practices in British colonial India. Through the generation of statistics and categories, the condition of the forest could be abstractly represented as data that allowed the state to expand its claim to be carrying out a modern rational regulation of forestry resources.[1] Scientific knowledge production was an important technology for environmental governance and one that promised valuable resources could be protected and exploited effectively and simultaneously by realizing a more rational, data-driven, and systematic model for resource management.[2]

In his study of scientists, scallops, and fishermen off the coast of France, Michel Callon highlights this as an issue of mistranslation. In this instance, naturalists from France brought with them their own conceptions of the environment and society. They translated their observations in Oceania through the lens of their preexisting ontology, which brought them into conflict with competing views of the social and natural world. Callon points to the role played by the identity of the storyteller, or actor making a claim to eco-authority, as an important consideration as well as the identity of the scallops, or oysters, studied by science and taken to stand as ambassadors for an entire species. Translation involved disruption as expert claims influenced a network of decision makers.[3]

The relocation of lagoon management away from traditional local decision-making toward distant administrators enabled an over-harvesting of the lagoons so that by 1950 the oysters were extinct in all but a few. The lagoons were subject to what Rob Nixon calls slow violence—a delayed destruction that is "dispersed across time and space, an attritional violence that is typically not viewed as violence at all." It is a violence that is often imperceptible and invisible due to its gradual unfolding in areas peripheral to power centers.[4] The lagoons, so distant from the colonial center in Pape'ete, itself distant from Paris, were out of sight to those with administrative power. Though naturalists were intended to bridge this gap and make the invisible visible, the status of the marine environment was obscured by statistics such as the ever-increasing export in pearl shell. The transitory nature of government agents and the steadily rising tonnage of exported shell masked any larger awareness that the oysters were declining in size and population in the distant lagoons.

Life on the Lagoons

European arrival brought swift change to the Oceanic lagoons of the black-lip pearl oyster, *Pinctada margaritifera*. For a millennium, Islanders had harvested the oysters for their shells to fashion fishhooks, razor-sharp edges, harpoon heads, and decorative embellishments. European ships brought a revaluation of the oysters as a commercial resource, a raw material for the manufacture of luxury goods, prized for its lustrous dark mother-of-pearl shell and ability to produce small black pearls. In 1798, in search of the oyster's smokey mother-of-pearl, the Australian Pearl Fishing Company outfitted visits to the atolls, beginning the practice of exchanging western-manufactured products for pearls and pearl shell. European manufactured goods, such as metal fishhooks, knives, and sharp edges, replaced the local usage of the oyster shell. By 1827, the burgeoning pearl-shell industry had transformed the oyster-rich lagoons of the archipelago into the center of a great commercial web as merchants arrived from both sides of the Pacific, recruiting divers regionally, and bringing with them cargoes of in-demand goods and foods.[5]

As the century progressed, divers harvested oysters in greater numbers, cutting them one at a time from the rocky coralline lagoon depths. On the surface Islanders sorted the shells, discarding those that were pitted or deemed too thin, too small, or blemished in some way. Merchants stowed the commercially viable oysters to be weighed before making the long journey to the auction markets of London and Hamburg. Most of these made their way to the hands of jewelers in France and Austria, their mother-of-pearl shell destined to be used as inlay on a multitude of luxurious goods such as buttons, cabinets, utensils, and combs. By the time France raised the flag of its protectorate over the Tahitian Kingdom in 1842, the potential for the pearl-shell oysters and their lagoons to be an important commercial resource was readily apparent.[6]

The profitability of the pearl-diving industry attracted foreign investment such as the transnational Hamburg-based trading firm Godeffroy & Co. The firm established agents in the Tuamotu Archipelago to purchase shells and sell European goods during the 1860s as part of a larger expansion into copra production in Samoa and transpacific shipping. However, the value of the pearl oysters proved unpredictable, subject to a boom-and-bust cycle determined by the vagaries of European fashion. Within a decade, Godeffroy & Co. divested its assets in the archipelago after pearl-shell prices hit an unusually low trough in 1867. The first official warning that the lagoons might not be able to sustain a rapacious extractive industry accompanied this greater commercial interest. In 1863, Edmund de Bovis, a naval officer overseeing the lagoons, recommended state regulation to stave off resource exhaustion.[7] In 1868, this took the form of a *rahui* placed over the lagoon at Anaa for three years so the oyster population could recover. In announcing a *rahui*, the government borrowed a local cultural practice in which access to a resource or place could be restricted for a time.[8] In 1873, the *rahui* was extended to five other lagoons and ten specific additional oyster banks, closing them off to oyster harvesting. That same year the first oyster farms were established by German merchant Christian Schmidt, at Kaukura, and a local chief, Mapuhi a Tekuravehe, at Takaroa. Their establishment proved fortuitous as the 1873 Vienna *Weltausstellung*, or World Exposition, prominently featured beautiful displays of nacre and nacre inlaying, which led to an increase in demand and prices. However, 1878 brought twin disasters to the industry; a destructive typhoon wiped out cultivation sites and the Bank of New Caledonia collapsed, leaving the industry without easy means to raise new capital.[9]

Despite the implementation of the *rahui,* or open/closed seasons for diving, the state engaged in little serious surveillance or enforcement of its regulations. Lieutenant Mariot, the naval *Résident* of the Tuamotus in 1875, reported that several formerly rich lagoons had become nearly emptied due to overharvesting. Mariot blamed the overexploitation on the "avarice of merchants" who encouraged indiscriminate harvesting of oysters, even those too small to have any commercial value. He believed the development of oyster farming was the best hope for the industry and described the favorable results of his own trials.[10] That year, another radical new development also arrived, the diving machine or *scaphandre*. Captain Clark of the American schooner *Florence Bayley* received the first license to use a *scaphandre* for oyster diving in the lagoon at Aratika. The *scaphandre* promised to increase production and efficiency by opening up the depths to anyone who could use the diving machine. It challenged the position of the islander divers as the sole intermediary between the world of the depths and the surface.[11]

The ability of the administration to regulate commerce in the lagoons was limited by the dispersed nature of the archipelago's vast collection of low-lying atolls. Though Tahiti was not more than a day's sail from the closest island, a tour of the major diving centers of the Tuamotu archipelago took well over two weeks to complete. The administrative presence was light and naval duties kept most officials assigned to the Tuamotu archipelago in Pape'ete instead.[12] The prominence of the pearl-shell industry further curtailed government action to restrain overharvesting. Paul Deschanel, future president of France and proponent of colonial expansion, judged, "that which is certain, is that the Tuamotu archipelago is the principal source of Tahiti's wealth." Exports of pearl shell were three times as valuable as the second-largest colonial export, copra.[13]

There was never any question of completely closing down the lagoons or the industry. The wealth of the lagoons contributed significantly to funding the local administration. Administrators believed the pearl-shell industry was a key pillar in the civilizing mission as it encouraged atoll residents toward wage and contract labor as divers, guides, sailors, and oyster cleaners. Yet, if the resource became exhausted the colonial budget and civilizing mission would both be in jeopardy.[14] It was in this context of uncertainty that Governor Morau appealed to France for a scientific mission to study the lagoons. The naturalist would make the distant lagoons visible through careful observation; he would take the unseen undersea habitat of the oysters and render it visible and therefore manageable. Quickly, government correspondence settled on one name for the mission: Germain Bouchon-Brandely.[15]

The Missionary of Science

Bouchon-Brandely was already conducting a study of the pearl oysters from Ceylon with the goal of acclimatizing them to the coastal waters of France when he was offered the mission. Under-secretary for the Ministry of the Marine, Félix Faure, wrote that Bouchon-Brandely was his first choice for the mission. Faure described Bouchon-Brandely as uniquely capable of undertaking a study of the oysters for their scientific value, as well as their commercial and economic potential.[16] The Ministry of the Marine presented Bouchon-Brandely a series of questions to answer regarding measures that could be taken to stop the exhaustion of, and efficaciously restock, the oyster banks. Despite concerns within the Ministry that the mission would be "long, difficult, and costly," arrangements came together quickly in France, and by the end of the spring he was on his way.[17]

When Bouchon-Brandely disembarked at Pape'ete on May 31, 1884, the paper *L'Océanie Française* heralded the arrival of a "Missionary of Science."[18] Despite the fanfare, the mission got off to a slow start. Bouchon-Brandely had trouble finding suitable boarding in Pape'ete and discovered the local administration had never secured him a vessel he could use to survey the fisheries. It fell on Governor Morau to step in and offer an apartment in his own residence for the naturalist and open up the local budget to hire a private vessel for Bouchon-Brandely.[19] Despite the work of the governor to get the mission back on track, Bouchon-Brandely spent his first four weeks in relative comfort as he made observations off the coast of Pape'ete. He experimented with different methods of raising young oysters brought over from the Tuamotu lagoon at Anaa.[20] The encircling lagoon of Tahiti was never renowned for oyster production, and it is a remarkably different environment than the Tuamotu atolls, which made it a peculiar place for Bouchon-Brandely to center his research. He made an attempt on June 24 to visit the Tuamotu archipelago, but the weather was poor and he landed on Tahiti's neighboring island of Moorea instead. For the next week he stayed on Moorea before departing on July 1 for the Tuamotu archipelago, cruising among the atolls for a little less than two weeks before returning again to Tahiti on July 13.[21]

The governor reported back to France that Bouchon-Brandely had quickly verified the lamentable state of the oyster banks and believed nothing short of urgent action could save them from imminent ruin. His initial research into artificial reproduction and oyster farming proved promising and guaranteed a bright future for the colony. To that end, the governor hoped Bouchon-Brandely would spend the rest of the year instructing and training students to carry on his important work.[22] The popularization of oyster research was an important goal of the project so that the findings could be instructive for individuals interested in oyster farming. Yet, Bouchon-Brandely offered no practical advice for lagoon oyster farming after he abruptly concluded his mission nearly five months early, announcing his departure, preliminary findings, and policy recommendations in a letter to the governor published in *L'Océanie Français*.[23]

In his final report to the Ministry of the Marine, Bouchon-Brandely identified four factors he believed responsible for the decline of the lagoons. Foremost he blamed "unscrupulous foreign merchants" whose greed drove abusive fishing practices. Second, he argued the lack of state supervision over the lagoons permitted divers and merchants to harvest oysters without any commercial value in order to fill out their cargo holds. Third, he reasoned that administrative measures intended to regulate the lagoons had been insufficient and scientifically unsound. Bouchon-Brandely interpreted the *rahui*, a ban on diving in a lagoon for two to five years, as analogous to the regulatory measures used in the Indian Ocean and Persian Gulf fisheries. He believed this prohibitory measure was based on a sound principle, but he failed to account for the special geography of the lagoons and differences in species between pearl-shell oysters in the Tuamotu archipelago and those of the East Indies. Bouchon-Brandely translated traditional local authority, the *rahui*, as a global practice, only so he could dismiss it as not applicable in the islands. Finally, he concluded that the absence of any "efficacious provisions for restocking the lagoons" prevented a quicker recovery of oyster populations.[24] His final report was critical of the local administration, which he believed had mismanaged the resource. He argued the transitory nature of the colonial officials prevented any stable set of government regulations from being implemented. Each new administration seemed caught unaware by issues in the nacre industry and struggled to formulate its own policies.[25]

Bouchon-Brandely wrote that after his preliminary report had been published, he had been contacted by French oyster farmers expressing a desire to immigrate to Tahiti. Bouchon-Brandely argued that the state should provide financial support to these "hard working, active, economical" oyster farmers to develop an industry of oyster cultivation among the islands.[26] Though the suggested immigration measures were never implemented, they did prompt accusations in the *Conseil Colonial,* the highest locally elected body in the colony, that Bouchon-Brandely desired to expropriate the communal property of the Tuamotu Islanders.[27] In his final report, it was not local knowledge or practices that could save the lagoons, but science that would reveal the reproductive nature of the oyster. Science would describe the principles upon which a sound regulatory system could be based—a system that appeared remarkably like that which was practiced by oyster farmers in the coastal waters of France.[28]

Contested Conclusions

The publication of Bouchon-Brandely's findings was well received in France and widely distributed by the Service des Colonies, which dispatched one thousand copies of the report to various chambers of commerce in France's coastal ports and the prefects of Brest, Lorient, Rochefort, Cherbourg, and Toulon.[29] Bouchon-Brandely's status as France's foremost expert on the pearl fisheries of the Tuamotus inspired Félix Faure to privately fund an attempt by Simon Grand, an expert oyster cultivator from France, to verify Bouchon-Brandely's theories in the colony. However, on Tahiti, Bouchon-Brandely's reputation as a highly qualified scholar of marine resources and fisheries was challenged after his departure. His findings were disputed by Edmond Liais and Clary Wilmot, members of the Conseil Général, who faulted Bouchon-Brandely as "a great savant, without a doubt, in the art of the reproduction of carps and trout," but one who had learnt nothing about the particular environment of the Tuamotu lagoons during his abbreviated stay.[30] The local government had requested his mission with the goal of improving the local oyster industry, and they were not satisfied with his suggestion that French oyster farmers relocate to the lagoons and establish an industry analogous to that of France. Wilmot, Liais, and others who asserted an intimate knowledge from working in the Tuamotu archipelago, as well as Grand who professed a practical knowledge from decades raising oysters in France, made competing claims to authoritative knowledge.

Simon Grand arrived with letters of introduction supplied by Félix Faure, silver medals in oyster cultivation from the 1878 Exposition Universelle, and a promise that he could establish practical applications for Bouchon-Brandely's work. The director of the interior in Pape'ete provided Grand with an initial grant of 2,000 francs and marine concessions off Tahiti at Motu-Uta and Fareutu to establish oyster farms. When Grand reported he was unable to reproduce Bouchon-Brandely's results during his initial experiments at Motu-Uta, he faced attacks that, like Bouchon-Brandely, he assumed too much similarity between France, Tahiti, and the Tuamotu lagoons. Subsequent funding requests from Grand were rejected when he could offer little in the way of results and an investigation sent by the Conseil Général found Grand had done no more than attempt to transplant oyster cultivation practices from France. His methodology discredited and accused of spending most of his grant money in the local saloons, Grand followed the recommendation of those government officials who still defended him and relocated to the colony's more distant Gambier Islands.[31]

After dealing with two experts from France who accomplished little, Victor Raoulx, an elected member of the Conseil Général, proclaimed that "since the passage of M. Bouchon-Brandely, we have no need of professors to teach us the great difficulties for the artificial reproduction of the pintadine."[32] From that point the Conseil Général privileged the production of knowledge by local experts in order to protect the nacre industry. Neither the work of Bouchon-Brandely nor that of Grand had produced effective technologies of governance or conservation. In 1887, the Conseil Général dispatched two of its own members, Liais and Wilmot, to study the Tuamotu archipelago and create a counter-authority to that of Bouchon-Brandely. The goal of their mission was to more accurately inform the metropolitan officials crafting legislation of the reality of the life in the lagoons.

In the first instance, Edmond Liais departed for the Tuamotu archipelago to investigate petitions received from island communities requesting government protection of their lagoons. Liais was directed to study the situation and report back on ways to improve basic governance over the islands.[33] His report painted a bleak picture for enacting or enforcing any meaningful regulation. The Tuamotu archipelago, he asserted, was as big as Europe and largely governed from distant Pape'ete by one naval officer. Liais reported that the illegal harvesting of young oysters proved the impotence of the state. Open defiance of state regulations engendered further exploitation by Islanders and merchants who understood they had little to fear. The report concluded that it was beyond the current resources of the state to regulate the industry, protect the nacre, and prevent the tragic exhaustion of the oysters.[34]

A second report was compiled by Clary Wilmot, who led the advisory Commission on Nacre in its response to Bouchon-Brandely. The Commission argued Bouchon-Brandely's findings were faulty, attributable to his failure to consult local experts and his inability to bridge the language barrier and consult Islanders.[35] The Commission argued that all size requirements for harvesting nacre were impractical and illogical. Size and weight were difficult, even impossible, for the diver to judge while ten to thirty meters below the lagoon surface; furthermore, the two shells of the oyster were often unevenly sized. How, they asked, could half an oyster be legal and the other half criminal?[36] Finally, the local experts on the commission argued that the lagoons had been threatened with exhaustion before the nacre industry. They claimed that rather than rapacious and unregulated commerce, the primary threat to the oysters was natural geological evolution. As the encircling reef belt gradually uplifted, the undersea canals that fed water into the lagoons became blocked, causing a progressive filling in of the lagoon. Happily, they reported, this could be averted with a few well-placed kilos of dynamite to open new passages between ocean and lagoon.[37]

Role of Technology

Critiques of overexploitation centered on the role played by the *scaphandre*, or diving suit, in the late 1880s. Thanks to weights and air pumped down from the surface, divers equipped with the *scaphandre* went deeper and stayed down longer as they harvested previously unreachable banks of oysters. The accompanying rise in exports gave the impression that the lagoons were stable and not on the verge of collapse. The diving suit masked the slow violence and degeneration taking place in the lagoons through the increasing tonnage of oyster shell exported during the 1880s and 1890s. For distant observers, market demand and price drove production, and the dire warnings of exhaustion from scientists or naturalists appeared to have been greatly exaggerated.

In 1885, Bouchon-Brandely celebrated the *scaphandre* as safer than the unassisted diving of the Islanders. It was also more productive as European divers in the *scaphandre* stayed down longer, sending up buckets of oysters, while the island diver had to be satisfied with only one or two oysters each dive.[38] The *scaphandre* was not cheap and required a team to operate. The naturalist Albert Seurat described a typical operation of ten men, two to don the diving gear and harvest in the depths, one to man the lifeline and communicate with the divers, four to turn the crank handle air pump, and two more to clean the oyster shells on the boat.[39] In 1888, Wilmot reported hiring a *scaphandre* crew to carry out observations on the oysters and their habitat; he detailed expenses of 6,000 francs for the *scaphandre*, two Islanders working the surface at 5 francs a day, one European diving at 10 francs a day, and supplies for the four of them at 10 francs a day.[40] The expense only increased in the 1890s after the Conseil Général, in response to petitions from Islanders in the Tuamotu archipelago demanding a total ban on the *scaphandre*, levied an annual 1,000 franc licensing fee and decreed diving suits could only be used at a depth of eighteen meters or below. They hoped these measures would limit the number of *scaphandre* in the lagoons, but the number only increased. The license became just one more expense for merchants to pay off, and they did so by extracting even more pearl shell than before.[41]

In 1892, the chamber of commerce in Pape'ete sent a letter to the Conseil Général in conjunction with the administrator of the Tuamotu, Aubert, requesting that the issuance of licenses be suspended and the use of the diving suit be restricted to only a few of the deepest lagoons. In the letter they argued that there was no enforcement of the eighteen-meter rule and that pearl oysters were so rare at that depth that merchants had to exploit shallower waters to cover their costs. They claimed that forcing diving suits to exploit only the deepest banks of oysters was scientifically unsound and impaired the ability of a protected oyster population to replenish the lagoon. François Cardella, longtime president of the Conseil Général, gave testimony in support for the ban from administrators and ship captains, all of whom agreed it was better to allow for diminished production than see the lagoons ruined. Cardella stated that island chiefs had witnessed diving suits eluding the depth restriction by descending in the deeper parts of the lagoon and then traveling underwater to the shallower areas to harvest oysters, communicating with the lifeline if authorities showed up. The undersea world of the lagoons, he declared, was the domain of poachers and absent any state surveillance.[42]

The ban on diving machines received a largely negative response from metropolitan officials who contended the ban was based on the clamoring of unscientific partisans and resulted in the lagoons being underutilized. Export statistics stood as proof that the use of diving machines had had no noticeable deleterious effect on the industry.[43] They argued that diving-suit operators encouraged oyster repopulation by exploiting deeper banks and spreading reproductive material from the oysters in shallower waters when they processed and cleaned them at the end of the day. Otherwise, they concluded, the deep banks of oysters sat inert, their reproductive material never rising to the shallower waters, and they became nothing more than "unhealthy conglomerations."[44] The diving suits, rather than exhausting the lagoons, cleansed them for repopulation; by this logic it was unhealthy to not use the diving suit. Little thought seems to have been given to how the oysters had naturally spread across the lagoon in the first place. What was certain was that technology could solve the problem.

The declining condition of the lagoons was difficult to discern for people who never saw them. Cries of exhaustion and scientific findings hit a flat note in the late 1890s for those who noted that the commerce had been successfully going on for a half century and, despite all the proclamations of doom, seemed to be doing just fine.[45] Arguments that the diving machine ban was irrational won out and it was lifted in 1902. Governor Gallet explained that while the interdiction of the diving suit was done with the laudable goal of ending the impoverishment of the lagoons, it had had in effect, "the contrary result of leaving the wealth of the lagoons unproductive and leaving the depths of the lagoons smothered and unable to regrow." The diving suit was necessary to excise abnormal oysters that had not grown into commercially viable commodities, the only goal of the oyster.[46] The end of the ban led to a round of protest from the islands and this time physical confrontation between Islanders and merchants in 1903. That year a terrible typhoon and tsunami decimated the islands and the diving industry; the surviving divers on Hikueru seized the equipment of newly arrived foreign diving machine crews and requested the government again ban the machines so they could use the industry to finance the rebuilding of their communities. In 1906, the local administration prohibited diving machines after yet another destructive typhoon hit the islands. The commercial success of the pearl-shell industry during World War I forestalled any question on the necessity of again lifting the ban. However, in 1925 when a new administrator and naturalist arrived, they decided the ban was unnecessary and lifted it yet again.[47]

Conclusion

By the 1950s, commercial production had entered a steep decline as pearl oyster populations were extinct or nearly extinct in every Tuamotu lagoon. With the diving industry apparently doomed, the administration funded a new research program to develop pearl oyster farming. In 1961, the Territorial Assembly passed legislation regulating oyster cultivation in the lagoons to help establish the industry. French nationals with a land claim in the Tuamotu archipelago could now apply for concessions for state property to begin farming oysters in their neighboring lagoon. The industry did not prove to be as sustainable as Bouchon-Brandely had hoped. Intensive oyster cultivation placed competitive pressure on the rest of the lagoon ecosystem for resources and space, which ultimately led to a new sort of slow violence as the declining lagoon environment negatively impacted the ability of the oysters to be farmed.[48] This reincarnation of an extractive low-level violence is driven by the same sort of distance and invisibility as the first; the Tuamotu lagoons remain the hidden production center for pearls and nacre marketed as "Tahitian."

The recommendations of Bouchon-Brandely took eighty years to be promulgated into law. The distance of the lagoons from Pape'ete and the distance from Pape'ete to Paris left the archipelago on the periphery of political power within the French colonial empire. The expedition of Bouchon-Brandely had been summoned to settle questions over environment and industry, and his findings provided a scientific pretext for the expansion of state power and authority over the lagoons, but the distance was too great. The presence of a few gendarmes or police and one administrator meant that most of the archipelago and commerce were largely free from state surveillance. The lagoons were subjected to a slow violence in spite of scientific and official recognition that overharvesting oysters was damaging the lagoon environment. Answers to the dilemma of how to conserve and exploit a resource simultaneously were circumscribed by the spacial and jurisdictional position of the Tuamotu lagoons and constantly contested by different eco-interpretations.

Notes

1. Arun Agrawal, *Environmentality: Technologies of Government and the Making of Subjects* (Durham, NC: Duke University Press, 2005), xiii, 4.

2. Ibid., 8–13.

3. Michel Callon, "Some Elements of a Sociology of Translation: Domestication of the Scallops and the Fishermen of St. Brieuc Bay," in *Power, Action and Belief: A New Sociology of Knowledge?* ed. J. Law (London: Routledge, 1986), 198–199, 211–212.

4. Rob Nixon, *Slow Violence and the Environmentalism of the Poor* (Cambridge, MA: Harvard University Press, 2011), 2.

5. FR ANOM 3800 COL 40, *Rapport à M. Le Sous-Secrétaire d'Etat aux colonies, par M. Bouchon-Brandely, Secrétaire du Collège de France, chargé par le gouvernement d'une mission en Océanie.*

6. Bengt Danielsson, *Work and Life on Raroia; An Acculturation Study from the Tuamotu Group, French Oceania* (Stockholm: Saxon & Lindströms Förla, 1955), 88–89; Moshe Rapaport, "Oysterlust: Islanders, Entrepreneurs, and Colonial Policy over Tuamotu Lagoons," *Journal of Pacific History* 30, no. 1 (1995): 40.

7. Germain Bouchon-Brandely, *Rapport au Ministre de la Marine et des Colonies par M. Bouchon-Brandely sur la pêche et la culture des huitres perlières a Tahiti* (Paris: Imprimerie de Journal Officiel, 1885), 29.

8. Tamatoa Bambridge, "Introduction, the *Rahui*, a Tool for Environmental Protection or for Political Assertion," in *The* Rahui*: Legal Pluralism in Polynesian Traditional Management of Resources and Territories*, ed. Tamatoa Bambridge (Canberra: Australia National University Press, 2016), 1–3.

9. Clary Wilmot, *Notice sur l'archipel des Tuamotu: Ses lacs, ses habitants, la formation de la propriété, la culture de la nacre* (Pape'ete, Tahiti: Imprimerie Civile Léonce Brault, 1888), 36; Procès-Verbaux des Séances du Conseil Général (hereinafter CG), Sessions Ordinaire, 1888, 130. [Gallica Online], FR ANOM 3800 COL 40, Letter 12 Septembre 1883, Le Gouverneur en Papeete à Monsieur le Ministre de la Marine et des Colonies.

10. A. Mariot, "Note sur Taïti et les Tuamotu," *Revue maritime et coloniale* 45 (1875): 82–83.

11. CG (1888): 128.

12. Ibid., 290.

13. Paul Deschanel, *Les intérêts Français dans L'Océan Pacifique* (Nancy, France: Imprimerie Berger-Levrault et Co., 1888), 92.

14. Pierre-Yves Toullelan, *Tahiti Colonial (1860–1914)* (Paris: Publication de la Sorbonne, 1984), 157–158.

15. FR ANOM 3800 COL 40, Rapport au Sous-Secrétaire d'Etat, 12 Février 1884.

16. FR ANOM 3800 COL 40, Letter 9 Janvier 1884, Le Sous Secrétaire d'Etat de la Marine et des Colonies, F. Faure à Monsieur le Ministre de l'Instruction Publique.

17. FR ANOM 3800 COL 40, Rapport au Sous-Secrétaire d'Etat, 12 Février 1884.

18. *L'Océanie Française*, 3 Juin 1884.

19. FR ANOM 3800 COL 40, Letter 14 Juin 1884, Monsieur le Gouverner à Monsieur le Ministre.

20. Bouchon-Brandely, 47.

21. *L'Océanie Française,* 24 Juin, 8 Juillet, 15 Juillet 1885.

22. FR ANOM 3800 COL 40, Letter 17 Juillet 1884, Monsieur le Gouverner à Monsieur le Ministre.

23. *L'Océanie Française*, 26 Août 1884. *Le Messager de Tahiti: Te Vai no Tahiti*, 1 Juin 1885.

24. Bouchon-Brandely, 37–38.

25. Ibid., 40.

26. Ibid., 67–68.

27. CG, Sessions Ordinaire (1884), 142–143; CG (1888): 82.

28. Bouchon-Brandely, 41–43.

29. FR ANOM 3800 COL 40, Letter 31 Juillet, 1885, Le Ministre de la Marine et des Colonies à Monsieur le Directeur des Journaux Officiels; Letter 30 Septembre 1885, Le Ministre à M. le Vice-Amiral Commandant en Chef, Prèfet à Cherbourg, Brest, Lorient, Rochefort, Toulon.

30. CG (1888): 123–125.

31. CG (1886): 436–438; Simon Grand, "Méthode de culture de l'huitre perlière dans les lagons de Tahiti." *Revue maritime et coloniale* 125 (1895): 579–590.

32. CG (1886): 441.

33. CG (1887): 40–43.

34. Ibid., 68–70.

35. Ibid., 123–125.

36. Ibid., 132–135.

37. Ibid., 130–131. Wilmot, 13.

38. Bouchon-Brandely, 35.

39. L. G. Seurat, "La Nacre et la Perle en Océanie—Pêche—Origine et mode de formation des perles," *Bulletin du Musée Océanographique de Monaco* 75 (1906): 8.

40. Wilmot, 15.

41. CG, Sessions Ordinaire (1892): 253.

42. CG, Sessions Ordinaire (1891): 379–385.

43. G. Darboux, J. Cotte, P. Stephan, and F. Van Gaver. *Exposition coloniale de Marseille; nos richesses coloniales 1900–1905: L'industrie des pêches aux colonies, Vol. 1.* (Marseille: Barlatier, 1906): 378–380.

44. Edwin Pallander, *The Log of an Island Wanderer: Notes of Travel in the Eastern Pacific* (London: C. Arthur Pearson, Ltd., 1901), 233.

45. GC (1897): 417.

46. GC, Sessions Ordinaire (1901): 98–99.

47. Rapaport, 45–46. Colin Newbury, *Tahiti Nui; Change and Survival in French Polynesia 1767–1945* (Honolulu: University of Hawai'i Press, 1980), 286.

48. Rapaport, 49.

TEN

From Boki's Beans to Kona Coffee
The 'Ōiwi (Native) Roots of an Exotic Species

Edward Dallam Melillo

He ali'i nō ka 'āina, he kauā wale ke kanaka.
[The land is a chief, the people merely the servants.]
—*Mary Kawena Pukui,* 'Ōlelo No'eau: Hawaiian Proverbs & Poetical Sayings

In 1866, a young correspondent for California's *Sacramento Daily Union* visited the Kingdom of Hawai'i and declared, "I think that Kona coffee has a richer flavor than any other, be it grown where it may and call it by what name you please."[1] The reporter's name was Samuel Langhorn Clemens, better known to his readers as Mark Twain. The beverage that Twain so effusively praised was brewed from the fruits of an exotic species, Arabian coffee (*Coffea arabica*), which had taken root on Ka Pae 'Āina 'o Hawai'i (the Hawaiian Archipelago) only a few decades earlier.[2]

Today, coffee connoisseurs concur with Twain's appraisal. Coffee grown in the Kona District on the western side of the Island of Hawai'i—known to many as the Big Island—ranks among the world's most desirable (and expensive) varieties.[3] Despite such international renown, Kona coffee has received little attention from historians. Its development has been discussed in fragmented and episodic accounts.[4] This chapter plunges beneath the surface of Hawai'i's fabled brew to uncover a history that complicates our basic assumptions about environmental change in the Pacific World.

Three arguments frame this discussion. First, although coffee cultivation evokes the terrestrial imagery of verdant hillsides and loamy soils, its expansion depended upon a watery world of ocean crossings. Ships often served as vectors for the long-distance dispersal of coffee plants, causing botanical relocations that did not always follow established trajectories. Second, most historians have assumed that major Polynesian influences on the composition of the Pacific's flora and fauna receded in the eighteenth century, giving way to the dominant environmental impacts of European explorers. The story of Kona coffee upends this conventional narrative. It was a Kanaka Maoli (Native Hawaiian)—namely O'ahu's governor Boki Kamā'ule'ule (c. 1785–c. 1830)—who brought thirty coffee plants from Brazil to Hawai'i in 1825. The cuttings from Boki's bushes became the stock for the first successful Kona coffee crop. Third, expressions of a unique *'ōiwi* (native) Kona coffee identity, and its associated notions of the value added by the region's volcanic landscapes and local cultivation techniques, developed far earlier than scholars and industry analysts have previously thought. This extended chronology emerges only through extensive research among nineteenth-century Hawaiian-language newspapers. Taken together, these three arguments frame a new interpretation of coffee culture and environmental change in Hawai'i and beyond.

Before focusing on the unique attributes of Hawai'i's coffee history, this chapter reviews the celebrated bean's thousand-year transcontinental journey from Ethiopia to Brazil. This overview situates coffee as an Islamic beverage that captivated European taste buds and social institutions, and it illuminates why Boki's visits to London and Brazil were crucial maritime stopovers in coffee's journey to Hawai'i.

How an Ethiopian Beverage Shaped Latin America's Geography

The Horn of Africa was the cradle of coffee domestication.[5] According to popular legend, an Ethiopian goat herder named Kaldi discovered the stimulating effects of the coffee plant sometime around 850 CE. After noticing that his goats became frisky when they nibbled on the bright red berries of a fragrant bush, Kaldi chewed on the fruits and experienced their exhilarating effects. He enthusiastically shared his discovery with an Islamic monk at a nearby Sufi monastery, but the holy man disapproved of the mood-altering substance and threw the coffee cherries (as the seed-containing ripe fruits are called) in the fire. The enticing aroma that wafted forth from the charred pile persuaded the monk to rake the beans from the embers, grind their roasted hulls, and dissolve them in hot water. This auspicious incident yielded the world's first cup of coffee.[6]

Regardless of the veracity of this time-honored fable, we know for certain that by the ninth century, coffee culture and its associated flora had crossed the Red Sea, reaching the Arabian Peninsula aboard lateen-sailed ships known as dhows. The Persian polymath Rhazes penned the first written testimonial to the novel beverage. In his medical text, *Al-Haiwi* (The Continent), Rhazes chronicled the rousing effects of *bunchum* (coffee) made from the seeds of a plant called *bunn*. As he explained, "*Bunchum* is hot and dry and very good for the stomach."[7]

Sufi mendicants embraced coffee's nourishing properties. The beverage heightened their mental focus during lengthy devotional ceremonies known as *dhikr* (literally, "remembrance"). Coffee reached Mecca by the end of the fifteenth century, and public coffeehouses became a prominent fixture of Islam's holiest city shortly thereafter.[8] Following the sixteenth-century Ottoman conquest of Yemen, Turkish pashas from Istanbul asserted control over the increasingly lucrative coffee trade. For the next three centuries, the Yemeni port of Mocha retained its status as the world's leading coffee market. It is no coincidence that "Mocha" is still a metonym for high-quality beans. What is less readily apparent to today's consumers of "the world's favorite beverage" is that they are drinking one of Islamic civilization's many contributions to global culture.[9]

Countless new ecocultural networks emerged during the Age of Sail (1571–1862).[10] In earlier times, one might have expected that Europe's first coffeehouses would have sprung up along the Mediterranean in close proximity to their Middle Eastern forebears. Indeed, coffee had become a popular Venetian beverage by the 1640s. However, Britain's sprawling nautical linkages in the Early Modern Era allowed the Near Eastern brew to find a new home in the West. Oxford (1650) and London (1652) hosted Europe's first coffeehouses.[11]

Coffee's maritime expansion continued in the early 1720s when French naval officer Gabriel-Mathieu Francois D'ceus de Clieu transported several *Coffea arabica* plants from Paris' *Jardin royal des plantes* to the Caribbean colony of Martinique. Reflecting on the journey, de Clieu made much of his heroic sacrifices to safeguard the fragile seedlings en route. He described his vessel's narrow escapes from a wily Tunisian corsair and a furious tempest, both of which had threatened to annihilate passengers and crew. When the ship began to run out of supplies, de Clieu surrendered his own "scanty ration of water to moisten [the plants]" in their time of dire need. De Clieu's delicate wards were among the first *Coffea arabica* bushes to arrive in the Americas.[12]

No less embellished are the tales of coffee's arrival in South America. In 1727, Sergeant Major Francisco de Melho Palheta planted Brazil's first coffee bushes in the northern state of Pará. Prior to this inaugural event, Palheta had been sent to Cayenne, French Guiana, by the governor of Belém under the pretense of resolving a diplomatic spat between French and Dutch Guiana. Palheta's underlying motivation was to abscond with coffee, a crop over which the French and Dutch exerted a duopoly. According to legend, Palheta became entangled in a passionate affair with the wife of French governor Claude d'Orvilliers. As her lover was preparing to depart, d'Orvilliers' wife presented the Brazilian with a bouquet in which she had concealed a branch laden with ripe coffee berries. Coffee chronicler Anthony Wild dismissed this tale as a prime example of "coffee mythology," but other scholars have been more hesitant to reject it completely.[13]

Regardless of the anecdote's authenticity, we do know that coffee fever quickly consumed Brazilians. *Coffea arabica* had become a mainstay of Rio de Janeiro's gardens and orchards by the 1750s; a century later, Brazil was producing more than half of the world's coffee.[14] The southeastern states of Minas Gerais, São Paulo, and Paraná offered ample slave labor and a tropical climate ripe for plantation crops. Brazil's nineteenth-century coffee boom was so far-reaching that it threatened national food security. As a British observer noted in 1886, "Brazil is suffering severely for having overdone Coffee cultivation and neglected the raising of food products needed by her people."[15] Although it came at enormous cost to other crops, the entrenchment of *Coffea arabica* in Brazilian soils would prove vital to the maritime diffusion of coffee elsewhere in the world.

Coffee Arrives in Hawai'i in the Wake of a Tragedy

Historians credit a flamboyant Spaniard named Don Francisco de Paula Marín (1774–1837) with the first attempt to cultivate coffee in Hawai'i. Marín was born in Jerez de la Frontera, a city in southwestern Spain that had once been a thriving center of Islamic culture—and, thus, of coffee consumption—under four centuries of Moorish rule (711–1231). On July 30, 1789, the fifteen-year-old Marín departed the entrepôt of Cadíz as an apprentice pilot with Alessandro Malaspina's scientific tour of the Pacific. Shipboard life was not to Marín's liking. As soon as Malaspina's vessels reached Alta California's Monterey Bay, the young sailor deserted. Soon thereafter, he found his way to the Kingdom of Hawai'i aboard the US fur-trading brig *Lady Washington*.[16] Marín never looked back.

By all accounts, Don Francisco was a raffish jack-of-all-trades. Despite a lack of formal training, he served as a military adviser, bookkeeper, physician, and interpreter for King Kamehameha I and his favorite wife, Ka'ahumanu. Within a few years, Marín had acquired at least three intimate partners of his own, with whom he fathered no fewer than twenty-three children. His fertile tendencies also extended to the soil. Hawaiian historians frequently credit Marín with introducing at least forty plant species to Hawai'i, although visitors who called on the well-known Spaniard likely brought most of the seeds and cuttings as gifts for him.[17]

Marín's estate near Wai Momi (sometimes known as Pu'uloa or Pearl Harbor) was home to an astounding array of food crops. These included onions, pineapples, horseradish, cabbages, asparagus, corn, chili peppers, limes, lemons, oranges, coffee, carrots, plums, figs, mangos, lettuce, olives, avocados, parsley, peas, guava, apricots, peaches, pears, apples, papayas, eggplants, potatoes, tea, cotton, and cocoa. Marín was also Hawai'i's first vintner. Today, Honolulu's Vineyard Boulevard offers an enduring reminder of Don Francisco's grape-growing legacy.[18]

Despite these horticultural triumphs, the Spanish émigré's *Coffea arabica* plants were not destined for greatness. In 1857, twenty years after Marín's death, a visitor wrote to the *Pacific Commercial Advertiser* that they had toured the site of the Spaniard's estate and found the tangled vestiges of coffee bushes among the remnants of other fruitless botanical experiments. Most of Don Francisco's vegetal aspirations had been reduced to gnarled limbs and weed-choked garden plots.[19]

It took a catastrophe of global proportions to secure coffee's central place in Hawaiian history. In 1823, King Kamehameha II (born Liholiho), his beloved queen Kamāmalu, and their retinue traveled to Great Britain aboard *L'Aigle* (*The Eagle*), a rickety, 114-foot whaling ship. The repurposed vessel was quite a sight to behold. On the day of its November 27th departure, "the decks were crowded with queens and chiefs, pigs and poultry. Of pigs there were about 300; goats, 36; sheep, 6; and bullocks, 4; with 8 dozen of fowls, and 4 dozen of ducks—all adrift together; and potatoes and powey [poi, or pounded taro root] from stem to stern."[20] At the helm of the 476-ton ship was Nantucket-born Valentine Starbuck, a salty (and often inebriated) veteran of the Pacific whaling fleet whose surname would achieve immortality in Melville's 1851 novel *Moby-Dick* and later as the trademark for the world's foremost coffee retailer. The *ali'i* (royalty) aboard Starbuck's ship sought an audience with King George IV to discuss a diplomatic alliance against the escalating imperial incursions of the United States and Russia. *L'Aigle*'s intrepid voyage across the world's two largest oceans came at the highest possible price; tragically, Liholiho and Kamāmalu died of measles—a disease for which they lacked immunity—two months after arriving in England. Although they toured London, visiting Westminster Abbey and the Royal Military Asylum, and attended the theater, the opera, and an assembly hosted by Countess Bathurst (wife of the secretary of state for the colonies), their much-anticipated meeting with the British monarch never took place.[21]

Following the untimely deaths of Kamāmalu and Liholiho at Osborn's Caledonian Hotel, George IV invited the survivors of the Hawaiian company to his court. The group, now led by Boki and his wife, Liliha, received the king's condolences, heard his assurances of British interest in the protection of the Hawaiian Kingdom's sovereignty, and accepted his offer to send the bodies of the *mō'ī* (king) and *mō'ī wahine* (queen) back to their homeland aboard the 46-gun frigate HMS *Blonde*. The funerary expedition departed Portsmouth on September 8, 1824, under the command of Admiral George Anson Byron, cousin of the famous poet George Gordon Byron. The voyage was as much a scientific exploring mission as it was a ceremonial duty. Among the distinguished crew were botanist James Macrae, naturalist Andrew Bloxam, and horticulturalist John Wilkinson.[22]

After an uneventful Atlantic crossing, the ship arrived at the port of Rio de Janeiro on November 27, 1825. Bloxam was distressed by the barbarities of the transatlantic slave trade that he witnessed along the wharves. As the twenty-three-year-old Oxford graduate remarked in his diary: "During our short stay, not less than three full shiploads arrived with their wretched beings. There is a regular slave market where they are kept and sold like so many sheep and oxen."[23] The *Blonde* soon sailed southward along the coast to Ilha de Santa Catarina (Santa Caterina Island) where it anchored on December 24 to acquire provisions and take on fresh water. Boki and Macrae disembarked and obtained thirty coffee plants, which they took aboard the ship and deposited in soil-filled crates on deck.[24] This seemingly inconsequential act would have a lasting impact on Hawaiian and world history.

The *Blonde*'s procurement of these Brazilian cultivars was Boki's brainchild. He had first encountered coffee while exploring London's vibrant café scene. The city's celebrated "penny universities" had aroused the political ambitions of an emerging middle class from the mid-1600s onward.[25] Their legacies also stirred the imaginations of many political theorists. In *The Structural Transformation of the Public Sphere* (1962), German sociologist and philosopher Jürgen Habermas portrayed London's coffeehouses as central institutions in the creation of eighteenth- and early nineteenth-century bourgeois public life (*bürgerliche Öffentlichkeit*), arguing that these sobering venues for conversation kindled democratic social exchanges. According to Habermas, the caffeinated milieu of the coffeehouse fostered the rapid expansion of public opinion in the political realm and laid the foundations for an age of democratic revolutions.[26] Despite his tremendous ingenuity, Habermas was unaware of how these venues also shaped the aspirations of a perceptive Hawaiian whose newfound respect for coffee would alter the environmental history of an archipelago in the north Pacific.

Before departing from England, Boki had convinced John Wilkinson, a grower with experience managing sugarcane and coffee plantations in the West Indies, to join the voyage. The governor of Oʻahu, ever the entrepreneurial opportunist, hoped that Wilkinson would remain in Hawaiʻi to supervise the cultivation of sugarcane and coffee there. Coffee, which entered the Hawaiian vocabulary in the 1820s via the loanword *kope*, was generally unknown in Hawai'i at the time. Until European sailors introduced brandy and rum in the 1790s, the social beverage of choice for most of Hawaiian history had been *'awa* (*Piper methysticum*). Known as kava (and by a multitude of other names) elsewhere in the Pacific, the dried and ground roots of this heart-leafed shrub can be chewed or mixed with water and drunk to relieve tensions, encourage camaraderie, consecrate ceremonial transactions, and induce a host of health benefits. The nineteenth-century Kanaka Maoli historian Samuel Mānaiakalani Kamakau remarked that "*'Awa* was a refuge and an absolution. Over the *'awa* cup were handed down the tabus and laws of the chiefs, the tabus of the gods, and the laws of the gods governing solemn vows and here the wrongdoer received absolution of his wrongdoing."[27]

Unbeknownst to Islanders in the 1820s, coffee was also destined for cultural prominence in Hawai'i. Boki had many obligations to meet before he could begin experimenting with his exotic plants. After a brief stop at Lahaina on the island of Maui, the *Blonde* anchored at Honolulu Harbor on May 6, 1825. Thousands of Hawaiians, including the powerful queen regent Ka'ahumanu, her co-regent Kalanimoku, and the twelve-year-old Kamehameha III (born Kauikeaouli), greeted the ship with displays of profound sorrow and *kūmākena* (mourning).

Once the funeral ceremonies and diplomatic exchanges were over, Wilkinson supervised the planting of Boki's Brazilian coffee bushes, alongside a thirty-acre crop of sugarcane, in a lush valley near the present-day University of Hawai'i, Mānoa campus. Boki and Wilkinson paid Hawaiian laborers twenty-five cents per day to prepare the land with *'ō'ō*, or traditional digging sticks.[28] Unfortunately, on September 17, 1826, just as the coffee bushes were bearing their first fruits, Wilkinson died of a mysterious ailment. Missionary Elisha Loomis chronicled the event in his journal: "Boki called after dinner for some medicine to take to Mr. Wilkinson the English planter lately established at Manoa, he being near his end. . . . But the poor man was insensible when Boki reached his house and expired shortly after."[29]

The "expired" English planter left behind a robust botanical legacy. In the late 1820s, the tender cuttings—known to horticulturalists as "slips"—of Boki's and Wilkinson's Mānoa bushes served as stock for various attempts to grow coffee elsewhere in Hawai'i. A Scot named Captain Alexander Adams was among the first to plant some of Boki's cuttings; his bushes in Kalihi and Niu Valleys on O'ahu "produced excellent coffee."[30] In 1828, Samuel Ruggles, a thirty-three-year-old missionary from Brookfield, Connecticut, cultivated several of these slips in the soils *mauka* (inland and up the mountain) from Kailua Kona on Hawai'i Island's west coast.[31] Ruggles could hardly have chosen a more suitable location for the plants. The western slopes of Kona's volcanoes—Hualālai and Mauna Loa—provide one of the world's most suitable microclimates for coffee cultivation. Kona's *Coffea arabica* plants thrive between the altitudes of 800 and 2,500 feet. The region's sunny mornings, cloudy afternoons with light to moderate precipitation, and mineral-rich volcanic soils have long fostered a thirty-mile stretch of small farms. This combination of favorable conditions has allowed the Kona region to achieve the world's highest yield per acre of *Coffea arabica* coffee.[32]

From the 1840s onward, Hawaiian coffee cultivators benefited from extensive state support. In 1842, the Hawaiian government enacted legislation to accept payment of land taxes in either coffee or pigs.[33] Six years later, the Kingdom of Hawai'i's minister of finance, Gerrit Parmele Judd—an American physician and missionary who renounced his US citizenship and became a trusted adviser and translator to Kamehameha III—wrote to the French consul in Hawai'i: "It is also desirable, as I have already mentioned in conversation, to impose a Duty of one or two cents per pound upon the Coffee and Sugar of Manilla [*sic*], in order to sustain for a few years our infant agricultural establishments, which are now just rising from great discouragements and hopes to supply the new markets in our neighborhoods."[34] During the mid-1800s, unexpected commercial outlets frequently appeared overnight. Skyrocketing demand from California's gold rush boomtowns caused a sevenfold increase in Hawai'i's coffee production between 1849 and 1850.[35]

While such commercial stimuli proved short-lived, they lasted long enough to give Hawai'i's coffee production a stable foundation. Addressing members of the Royal Hawaiian Agricultural Society in 1853, the organization's president, Judge William Little Lee, emphasized, "The Society has taken no measures to introduce new seed from other coffee growing countries; for it is generally admitted by the best judges, that the Kona coffee of Hawai'i is not surpassed by any in the world."[36] Among the early successes were the luxuriant bushes raised by English merchant Henry Nicholas Greenwell. His Kona coffee, cultivated at an elevation of 1,500 feet, was considered among the best being grown in Hawai'i at the time and even won international accolades.[37] Attempts to produce coffee elsewhere in the archipelago were less successful. By the 1860s, droughts, labor shortages, *kakani* (a Hawaiian term for tropical plant blights), and competition from the sugar industry had shuttered all of Hawai'i's coffee farms, with the notable exceptions of those in Kona and Hāmākua on Hawai'i Island.[38]

The extent of Kānaka Maoli (Native Hawaiian) involvement in shaping the dimensions of the burgeoning Kona Coffee business is uniquely documented

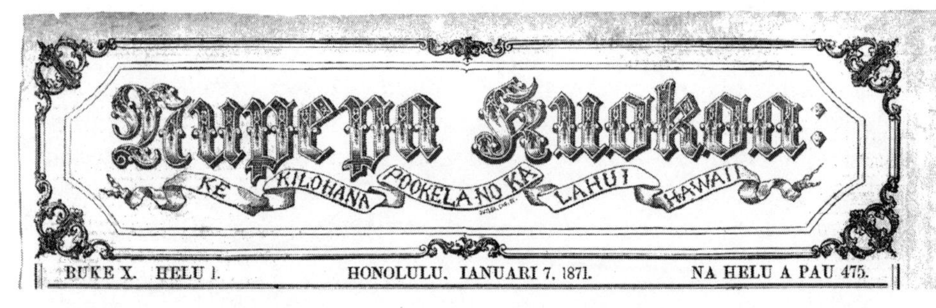

Nupepa Kuokoa:

KE KILOHANA POOKELA NO KA LAHUI HAWAII

BUKE X. HELU 1. HONOLULU, IANUARI 7, 1871. NA HELU A PAU 475.

Figure 10.1. Masthead from the January 7, 1871, edition of *Ka Nupepa Kuokoa*. Among the most popular of the Hawaiian-language newspapers, this broadsheet circulated from 1861 to 1927.

in *nā nūpepa ʻŌlelo Hawaiʻi* (the Hawaiian-language newspapers) (see figure 10.1). The magnitude of this literary output is stunning. In the words of Hawaiian language scholar Puakea Nogelmeier, "In just over a century, from 1834 to 1948, Hawaiian writers filled 125,000 pages in nearly 100 different newspapers with their writings."[39] As well as providing clearinghouses for the latest news from home and abroad, *nūpepa* featured *mele* (chants and songs), *uē* (laments), *kaua paio* (intellectual debates), *moʻolelo* (histories and legends), and *kaʻao* (folklore). An exceptionally erudite population of Kānaka Maoli craved this material. The rapidity and extent to which Hawaiians mastered an orthography developed in the 1820s by missionaries have few precedents in world history; by the mid-1800s, Hawaiʻi's remaining population of 69,800 had attained nearly universal literacy.[40]

Early Hawaiian-language articles on Kona coffee often adopted promotional or didactic approaches. An 1857 editorial in *Ka Hae Hawaii*, a government-run paper, suggested that coffee was among the crops "by which one is wealthy" with "almost no end to consumers' desire." The author added that the farmer's "responsibility is minimal. The yield of coffee is high on a small piece of land when done correctly."[41] Other stories offered instruction to aspiring Kānaka Maoli growers. As a writer for *Ka Nupepa Kuokoa* noted in 1862, "Gather cherries that have properly matured and ripened sufficiently; do not gather cherries that are only half-ripe or perhaps pale, do not gather withered cherries that have dried up in the sun and the like. Aerate until completely dry and firm, unless said coffee is left someplace for a few years, it will not become moldy, rancid, or bumpy."[42]

Haole (foreign) planters dominated coffee-cultivation efforts in the first half of the 1800s, but Kānaka Maoli were prominent players in the development of Kona's coffee business during the quarter century 1860–1885. When coffee trader John Gaspar Machado arrived on Hawaiʻi Island in 1872,

"*Kanackas* [*sic*] were the only coffee planters. They lived down on the beach and went up to the coffee patches only to pick coffee. Coffee trees grew wild without being hoed or pruned."[43] Kona's prominent *ali'i* also planted coffee. In 1878, a correspondent for the widely circulated Hawaiian-language newspaper, *Ka Nupepa Kuokoa*, described a prolific grove of coffee bushes in South Kona that were "laden with fruit." Princess Miriam Likelike Kekāuluohi Keahilapalapa Kapili—sister of the last two ruling monarchs of the Kingdom of Hawai'i—had planted them "with her own hands."[44]

In the decades following the turn of the century, *nūpepa* focused on the global reach of Kona Coffee's product identity. At Seattle's Alaska-Yukon-Pacific Exposition, held in the summer of 1909, coffee was a focal point of events at the Hawai'i Building. During the eighteen-week fair, which hosted 3,700,000 visitors, a correspondent reported to his Hawaiian readers, "Kona coffee is also being prepared to pour for those who will gather during those days." He added, "[We are] bringing some fine young girls from Hawai'i to pour coffee for the public."[45] Hawai'i was on display for the world to see, and a crop that had found a receptive home on its mountainsides less than a century earlier was among the archipelago's most distinctive features.

Appeals to a sense of *aloha 'āina*—or patriotic duty to the Hawaiian homeland—were also a leitmotif of early twentieth-century coffee advertising. The Honolulu-based McChesney Coffee Company asked its Hawaiian-language readers:

> When it is your time to buy coffee do you ask for Kona coffee? Do you look at the outer seal so you can know if what you are getting is Hawaiian coffee or adulterated coffee? We ask you to do so the next time you go buy. Why take the coffee that was adulterated? Why not invest in the land by buying Hawaiian coffee? Help Kona's fine people by buying their coffee.[46]

The existence of such entreaties as early as 1913 shows that the much more recent efforts of "buy local movements" have deep roots.

By the beginning of the twentieth century, the dominant variety of coffee grown on Kona's farms was no longer descended from Boki's bushes. In 1892, a German expatriate named Herman Widemann introduced a Guatemalan coffee variety that attained increasing popularity with cultivators for its higher yields and reduced fertilizer needs. By 1900, Widemann's variety had become known as "Kona typica."[47] Despite the fact that the descendants of Boki's Brazilian bushes had jumpstarted the Hawaiian coffee industry and provided it with nearly seventy years of growth, these progeny are now found only growing wild in gullies and isolated patches of forest.

During the twentieth century, Kānaka Maoli growers continued to play key roles in the ongoing growth of the Kona coffee industry. In 1938, a group of these farmers orchestrated a lobbying campaign "in our own language" to persuade the Hawaiian Territory's United States congressional delegate Samuel Wilder King to secure a contract with the US Department of Defense to purchase 300,000 pounds of Kona coffee annually for US soldiers stationed in Hawaiʻi. Ironically, Brazilian coffee had previously occupied this market niche.[48]

Simultaneously, other ethnic groups discovered their own roles in an enterprise that fostered unusual opportunities for newcomers. Kona's coffee farms offered a refuge from the harsh labor conditions of Hawaiʻi's sugarcane plantations. It is no coincidence that immigrants, many from the western Pacific, fled the harsh circumstances of Hawaiʻi's low-elevation sugarcane fields for the more hospitable conditions of upland coffee cultivation. As ninety-one-year-old Japanese-born immigrant Torahichi Tsukahara recounted in 1980:

> There were lots of people who had run away from sugarcane plantations before their contracts had expired. They came to Kona because it was a big place. There were some people who changed their last names. I knew this because some of them told me that their real name was such-and-such. There were lots of them who ran away from the plantations, breaking their contracts. And most of them started in coffee farming.[49]

Some of these refugees ended up as the owners of small coffee-farming operations. Anthropologist Carol A. MacLennan has noted that, as of 1932, Kona was home to 1,077 coffee farms (see figure 10.2). Japanese managed 959 of these, Filipinos owned 58, and the rest were in the hands of Hawaiians, Portuguese, Puerto Ricans, and Koreans.[50]

From a cursory read of the literature on Kona coffee, one gets the impression that the worldwide success of this specialty beverage was the result of forces from beyond Hawaiian shores. For example, in their chapter on Kona coffee for a book titled *Guide to Geographical Indications: Linking Products and Their Origins*, scholars Danielle Giovanucci and Virginia Easton argue that "Kona's reputation was built through alliances with larger industry players whose marketing and distribution networks brought Kona to a wide audience."[51] Unfortunately, such conclusions mimic a colonial mindset and further mystify the links between this product and its origins.

As the tale of Boki's maritime transplantation of Brazilian coffee bushes to Hawaiʻi and the documentary evidence on display in *nūpepa* stories from

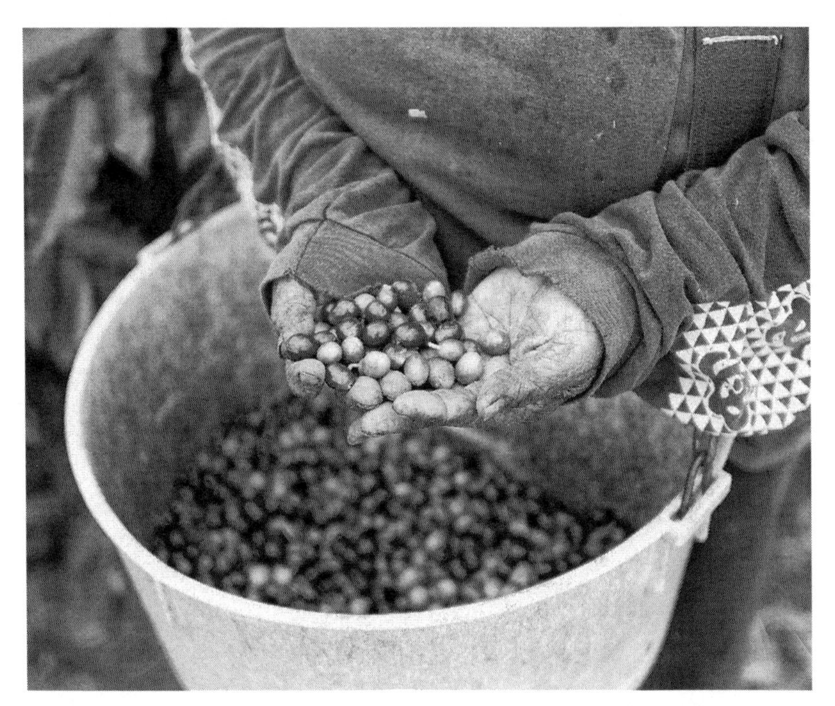

Figure 10.2. A woman harvests Kona coffee cherries on Hawai'i Island.

the nineteenth and early twentieth centuries so vividly attest, Kānaka Maoli built Kona coffee's reputation from the ground up, quite literally.

Despite repeated attempts by settlers and continental corporations to establish industrial production regimes, Kona coffee cultivation remains in the hands of family farmers whose holdings rarely exceed five acres. Rocky, volcanic soils on steep grades have limited the ability of production up-scaling. Meanwhile, an array of idiosyncratic traditions, such as Japanese sun-drying techniques introduced in the late 1800s, have frustrated the efforts of larger operations to enter this lucrative market.[52] The persistence of such small-scale, locally owned operations in the face of countervailing trends in global coffee production suggests strategies for mobilizing the discourse of terroir—the taste of place—to support a more ambitious agenda than previously realized.

The unique, local history of Kona's most important crop is celebrated annually at the weeklong Kona Coffee Cultural Festival. Each November, since 1970, coffee farmers on Hawai'i Island have welcomed their harvest of the bright red "cherries" that contain their prized beans.[53] Although the event showcases cultural activities—ranging from traditional Hawaiian quilt-making to the stringing of flower *lei*—organizers have not highlighted the *ʻōiwi* roots of a plant brought to Hawaiʻi by one of its most prominent nineteenth-century *aliʻi*, nor have they emphasized that coffee cultivation is a tradition that has been nurtured and promoted for decades by Kānaka Maoli and Hawaiian-language *nūpepa*. Such explorations would more deeply historicize and localize the "richer flavor" of Kona's celebrated beverage.

Notes

Epigraph: Mary Kawena Pukui, *ʻŌlelo Noʻeau: Hawaiian Proverbs & Poetical Sayings* (Honolulu: Bishop Museum Press, 1983), 62 (ʻōlelo #531). Translation edited by author. I am not a Kanaka Māoli of Native Hawaiian heritage. As an outsider, I have attempted to write this chapter with *kuleana*, or profound care and responsibility. I would like to thank my *kumu ʻōlelo Hawaiʻi*, N. Haʻalilio Solomon, for his wisdom and guidance.

1. Mark Twain, *Mark Twain's Letters from Hawaii*, ed. A. Grove Day (New York: Appleton-Century, 1966), 206. That year, Twain spent four months in Hawaiʻi, writing dispatches on assignment for the leading Pacific Coast newspaper. Twain was, in many ways, an astute observer and a profound admirer of Hawaiian culture (he even tried surfing), but the limitations of a nineteenth-century *haole* perspective suffuse his writing.

2. Two species of coffee dominate today's global consumer market. *Coffea arabica* beans impart superior flavors and aromas, making them more expensive than their more bitter-tasting cousin, *Coffea canephora* (known more commonly as *Coffea robusta*).

3. Kona coffees regularly earn scores in the nineties (of a possible 100) from *Coffee Review: The World's Leading Coffee-Buying Guide*, http://www.coffeereview.com/ (accessed May 5, 2021). This prized status is reflected in its price. As of 2021, a pound of roasted beans retailed for as much as US$100. For example, Lee and Karen Peterson's award-winning Hula Daddy Kona Coffee Farm sells a bean called "Kona Sweet," which (as of May 2021) retailed at $94.95 per pound. See https://www.huladaddy.com/catalog-coffee.

4. The two historical monographs on Kona coffee are Gerald Kinro, *A Cup of Aloha: The Kona Coffee Epic* (Honolulu: University of Hawaiʻi Press, 2003); and Don Woodrum, *Kona Coffee from Cherry to Cup* (Honolulu: Palapala Press, 1975). Neither book uses Hawaiian-language sources, nor does either author consider most of the developments discussed here.

5. The Oromo People of southeastern Ethiopia affirm that coffee sprouted from the tears of Waaqa, the supreme sky god. For centuries, coffee has been a centerpiece of Oromo religious rituals, land-tenure customs, and culinary practices. See Bula Sirika Wayessa, "Buna Qalaa: A Quest for Traditional Uses of Coffee among Oromo People with Special Emphasis on Wallaga, Ethiopia," *African Diaspora Archaeology Newsletter* 14, no. 3 (2011): Article 3.

6. Bennett Alan Weinberg and Bonnie K. Bealer, *The World of Caffeine: The Science and Culture of the World's Most Popular Drug* (New York: Routledge, 2001), 3–4.

7. Rhazes, as quoted in William Harrison Ukers, *All about Coffee* (New York: The Tea and Coffee Trade Journal Co., 1922), 12. Rhazes was the Latinized name of Abū Bakr Muhammad ibn Zakariyyā al-Rāziī (854–925 CE).

8. Ralph S. Hattox, *Coffee and Coffeehouses: The Origins of a Social Beverage in the Medieval Near East* (Seattle: University of Washington Press, 1985), 26–27.

9. Giancarlo Casale, *The Ottoman Age of Exploration* (New York: Oxford University Press, 2010), 183.

10. For more on these networks, see James Beattie, Edward Melillo, and Emily O'Gorman, *Eco-cultural Networks in the British Empire: New Views on Environmental History* (New York: Bloomsbury Press, 2015).

11. W. Scott Haine, "Introduction," in *The Thinking Space: The Café as a Cultural Institution in Paris, Italy and Vienna*, ed. Leona Rittner, W. Scott Haine, and Jeffrey H. Jackson (Burlington, VT: Ashgate, 2013), 4. Venice's famous *Caffè Florian* did not open its doors until 1720. On the first British coffeehouses, see Brian Cowan, *The Social Life of Coffee: The Emergence of the British Coffeehouse* (New Haven, CT: Yale University Press, 2005), 25, 90.

12. De Clieu's account is described at length in William Harrison Ukers, *All about Coffee* (New York: The Tea and Coffee Trade Journal Company, 1922), 6–9. Numerous accounts repeat the story of de Clieu's pioneering role in the spread of coffee to the Americas. For an example, see Henri Welter, *Essai sur l'histoire du café* (Paris: C. Reinwald, 1868), 18–20. However, other sources note that coffee had been planted in Dutch Suriname several years prior to the Frenchman's arrival. See Jean Benoît Désiré Cochet, *Galerie Dieppoise: Notices biographiques sur les homes célèbres ou utiles de Dieppe e de l'arrondissement* (Dieppe, France: E. Delevoye, 1862), 178.

13. Anthony Wild, *Coffee: A Dark History* (New York: W. W. Norton & Co., 2004), 173. Other scholars have pointed out that evidence abounds to confirm Palheta's introduction of coffee into Brazil following his visit to the French colony. For example, see Fernando E. Vega, "Review of *Coffee: A Dark History*, by Anthony Wild," *Quarterly Review of Biology* 82, no. 3 (2007): 283–284.

14. Boris Fausto, *A Concise History of Brazil* (New York: Cambridge University Press, 1999), 103. Steven Topik and Mario Samper, "The Latin American Coffee Commodity Chain: Brazil and Costa Rica," in *From Silver to Cocaine: Latin American Commodity Chains and the Building of the World Economy, 1500–2000*, ed. Steven Topik, Carlos Marichal, and Zephyr Frank (Durham, NC: Duke University Press, 2006), 124.

15. Edwin Lester Linden Arnold, *Coffee: Its Cultivation and Profit* (London: W. B. Whittingham & Co., 1886), 253. During ensuing decades, coffee remained at the heart of Brazilian agriculture and economics. The "café com leite" period is how Brazilian history textbooks still refer to the half century between the country's abolition of slavery in 1888 and the Great Depression of the 1930s. Indeed, this "coffee and milk" era was dominated by a powerful alliance between coffee and dairy farmers, which exerted unprecedented control over national policy by manipulating exchange rates, redirecting public investments, and shaping trade and labor policies to their benefit. See Daniel R. Reichman, "Big Coffee in Brazil: Historical Origins and Implications for Anthropological Political Economy," *Journal of Latin American and Caribbean Anthropology* 23, no. 2 (2018): 5.

16. Several historians have suggested that the *Lady Washington*'s Captain John Kendrick was the first foreigner to harvest *'iliahi* (Hawaiian sandalwood, or *Santalum paniculatum*) in 1791, leaving two sailors on Kaua'i to gather sandalwood and pearls for the lucrative China trade. However, in 1790, Captain William Douglas of the schooner *Grace* had dropped off two men on the same island for identical purposes. Historian Ralph Simpson Kuykendall suggested that the two captains knew each other, and that Kendrick may have been following Douglas' lead. See Kuykendall, *The Hawaiian Kingdom*, 3 vols. (Honolulu: University of Hawai'i Press, 1938–1967), 1:434 (Appendix C).

17. For the most complete accounting of these, see Kenneth M. Nagata, "Early Plant Introductions in Hawai'i," *Hawaiian Journal of History* vol. 19 (1985): 35–61.

18. "Aligator Pear," *Pacific Commercial Advertiser* (July 22, 1871), 3.

19. Ross H. Gast, *Don Francisco de Paula Marin: A Biography by Ross H. Gast & the Letters and Journal of Francisco de Paul Marin, edited by Agnes C. Conrad* (Honolulu: University Press of Hawai'i for the Hawaiian Historical Society, 1973), 55.

20. "Sketches of Society: Greenwich Hospital," *The Literary Gazette, and Journal of Belles Lettres, Arts, Sciences, &c.* (London) 389 (July 3, 1824): 430.

21. A few, short accounts of their visit have been published. See Kuykendall, *The Hawaiian Kingdom*, 1:77–81; and J. Susan Corley, "Kamehameha II's Ill-Starred Journey to England aboard *L'Aigle*, 1823–1824," *Hawaiian Journal of History* 44 (2010): 1–35.

22. Lord George Anson Byron, *Voyage of H.M.S.* Blonde *to the Sandwich Islands, in the Years 1824–1825* (London: John Murray, 1827).

23. Andrew Bloxam, *The Diary of Andrew Bloxam, Naturalist of the "Blonde" on Her Trip from England to the Hawaiian Islands, 1824–25*, ed. Stella M. Jones (Honolulu: Bishop Museum Press, 1925), 11.

24. James Macrae, *With Lord Byron at the Sandwich Islands in 1825: Being Extracts from the MS Diary of James Macrae, Scottish Botanist*, ed. William Frederick Wilson (Honolulu: n.p., 1922), 1, 4, and 34n33.

25. On English coffeehouses as "penny universities," see William Harrison Ukers, *All about Coffee* (New York: The Tea and Coffee Trade Journal Co., 1935), 69–70.

26. Jürgen Habermas, *The Structural Transformation of the Public Sphere: An Inquiry into a Category of Bourgeois Society*, trans. Thomas Burger (Cambridge, MA: MIT Press, 1989), esp. 3037. For a longer-term examination of the coffeehouse as a cornerstone of metropolitan life, see Markman Ellis, *The Coffee House: A Cultural History* (London: Weidenfeld & Nicolson, 2011).

27. Samuel Mānaiakalani Kamakau, *Nā hana a ka po'e kahiko* [The works of the people of old], trans. Mary Kawena Pukui, ed. Dorothy B. Barrère (Honolulu: Bishop Museum Press, 1976), 43.

28. Stephen Reynolds, "Reminiscences of Hawaiian Agriculture," *The Transactions of the Royal Hawaiian Agricultural Society* 1, no. 1 (1850): 50.

29. Elisha Loomis Journal, 1824–1826 (entry for September 17, 1826). Hawaiian Mission Houses Archives, Honolulu, Hawai'i.

30. Reynolds, "Reminiscences of Hawaiian Agriculture," 50–51.

31. *Kona* is a spatial designation meaning the leeward or dry side of an island. This term is counterpoised to *ko'olau*, which refers to the windward or the wet side of an island. Ruggles and his family are briefly profiled in Hawaiian Mission Children's Society, *Portraits of American Protestant Missionaries to Hawaii* (Honolulu: Hawaiian Gazette Co., 1901), 6.

32. Joel Schapira, Karl Schapira, and David Schapira, *The Book of Coffee and Tea*, 2nd rev. ed. (New York: St. Martin's Griffin, 1996), 75.

33. Lorrin A. Thurston, ed., *The Fundamental Law of Hawai'i* (Honolulu: Hawaiian Gazette Co., 1904), 122.

34. Judd Collection (G. P. Judd Papers), Manuscript Group 70, Box 9, Folder 4, Item 7 (letter dated May 31, 1848). Bernice Pauahi Bishop Museum Archives, Honolulu, Hawai'i.

35. Malcolm Rohrbough, "'We Will Make Our Fortunes—No Doubt of It': The Worldwide Rush to California," in *Riches for All: The California Gold Rush and the World*, ed. Kenneth N. Owens (Lincoln: University of Nebraska Press, 2002), 58.

36. William L. Lee, *Transactions of the Royal Hawaiian Agricultural Society at Its Third Annual Meeting in June, 1853* vol. 1, no. 4 (Honolulu: Government Press, 1853), 5.

37. Greenwell Farms continues to produce award-winning Kona coffee. See https://www.greenwellfarms.com/ (accessed May 5, 2021).

38. On blights in the late 1850s and early 1860s, see "The Coffee Blight," *The Polynesian* (August 13, 1859), 2; and J. W. Keopaa, "Kakani ke Kope o Kona," *Ka Hoku o ka Pakipika* (April 17, 1862), 1. Hawai'i's coffee exports plummeted from 311,300 pounds in 1857 to 58,100 pounds the following year. Conversely, sugar exports rose from 701,000 pounds to 1,205,000 pounds during the same period. Robert C. Schmitt, *Historical Statistics of Hawaii* (Honolulu: University of Hawai'i Press, 1977), 551.

39. M. Puakea Nogelmeier, *Mai Pa'a I Ka Leo: Historical Voice in Hawaiian Primary Materials, Looking Forward and Listening Back* (Honolulu: Bishop Museum Press & Awaiaulu, 2010), ii. As Nogelmeier notes, to date, well over a million letter-sized pages of this corpus remain untranslated.

40. Noenoe K. Silva, *Aloha Betrayed: Native Hawaiian Resistance to American Colonialism* (Durham, NC: Duke University Press, 2004), 55. The population figure is from Robert C. Schmitt, *Demographic Statistics of Hawai'i: 1778–1965* (Honolulu: University of Hawai'i Press, 1968), 10.

41. "No Ka Mahiai—Helu 8," *Ka Hae Hawaii* (August 5, 1857), 1.

42. J. W. Kalaiokona, "Kope o Kona," *Ka Nupepa Kuokoa* (August 30, 1862), 3.

43. Gaspar, as quoted in Baron Goto, "Ethnic Groups and the Coffee Industry in Hawai'i," *Hawaiian Journal of History* vol. 16 (1982), 115. In 1880, Gaspar built Hawai'i's first coffee mill near Kealakekua Bay. Popular histories of Hawai'i often refer to him as "Machado," but Gaspar's surname is Spanish, so English speakers should defer to his paternal name (Gaspar), not his material name (Machado).

44. W. I. Y., "Ke Kope o Kona," *Ka Nupepa Kuokoa* (November 16, 1878), 3.

45. *"Hawai'i 'ia nō ka hō'ike'ike ma Seattle"* ["The Exhibition in Seattle Is Made Hawaiian"], *Ka Nupepa Kuokoa* (April 2, 1909), 1.

46. "McChesney Coffee Co.," *Ka Nupepa Kuokoa* (April 25, 1913), 2.

47. Charles Johnson, "Coffee and Coffee Tourism in Kona, Hawai'i—Surviving in the Niche," in *Coffee Culture, Destinations and Tourism*, ed. Lee Jolliffe (Buffalo, NY: Channel View Publications, 2010), 118.

48. "WAKINEKONA," *Ka Hoku o Hawaii* (March 2, 1938), 3. At the time the article appeared, King was representing Hawai'i in the US House of Representatives. He later served as the state's 11th governor from 1953 to 1957. Incidentally, King was the first governor of the state who had Native Hawaiian lineage. His mother, Charlotte Holmes Davis, was descended from Oliver Holmes, governor of O'ahu under Kamehameha I.

49. Torahichi Tsukahara, interviewed by Michiko Kodama (Captain Cook, Kona, Hawai'i: November 14, 1980, trans. Mako Mantzel) in *A Social History of Kona*, 2 vols. (Honolulu: University of Hawai'i at Mānoa Ethnic Studies Program, 1981), 2:1270.

50. Carol A. MacLennan, *Sovereign Sugar: Industry and Environment in Hawai'i* (Honolulu: University of Hawai'i Press, 2014), 165–66. The most comprehensive work on the history of Kona's Japanese community is David K. Abe, *Rural Isolation and Dual Cultural Existence: The Japanese-American Kona Coffee Community* (Cham, Switzerland: Palgrave Macmillan, 2017).

51. Daniele Giovannucci and Virginia Easton Smith, "The Case of Kona Coffee, Hawaii," in *Guide to Geographical Indications: Linking Products and Their Origins* (Geneva: International Trade Centre, 2009), 177.

52. Capital has always struggled to navigate the treacherous straits between the profitability and pitfalls of ecosystems. Plants (even genetically engineered ones) only grow

so fast, monocultures simplify harvests but magnify disease susceptibility, fossil fuels offer kinetic solutions for rapid economic development, but these energy sources come with very high long-term costs. For explorations of such themes that build upon rigorous empirical evidence, see W. Scott Prudham, *Knock on Wood: Nature as Commodity in Douglas-Fir Country* (New York: Routledge, 2005); and John Soluri, *Banana Cultures: Agriculture, Consumption, and Environmental Change in Honduras and the United States* (Austin: University of Texas Press, 2005).

53. The festival's website is http://konacoffeefest.com/ (accessed May 5, 2021).

Eleven

Maunalua

Shifting Nomenclatures and Spatial Reconfiguration in Hawaii Kai

N. Ha'alilio Solomon

> In surveys of marine algae, reef fish, and invertebrates, a higher percentage of introduced species was found in Maunalua Bay (18%), than in Waikīkī (6.9%). Inside Koko Marina, the percentage of introduced species reaches 40%, the highest percentage recorded in Hawai'i.[1]

The above passage appeared in a 2016 newsletter published by the Hawaiian Islands' Humpback Whale National Marine Sanctuary under the Office of National Marine Sanctuaries, a division within the National Oceanic and Atmospheric Administration. These statistics constitute a microcosm of the greater, statewide threat facing Hawai'i's marine environments. The colonized hard bottom of mushy brown silt on Maunalua Bay's floor is the result of five decades of sustained terrestrial drainage through nearby urban channels.[2] The ecological invasion of Maunalua Bay is analogous to the changes happening on "land."[3] Similar to the upheaval beneath the water, the transformation sweeping across terrestrial Maunalua is equally foreign and invasive (see figure 11.1).

On the southeast end of Honolulu—Maunalua—O'ahu has undergone massive transformation since the 1960s. When Hawai'i became the fiftieth state of the Union in 1959, new legislation spurred the urban channelization of the region's natural streams. This coalesced with residential development, generating changes that rippled across Maunalua, creating a new sense of place bearing a new name, Hawaii Kai.[4] Other environmental historiographies detail the dynamics in spaces of transcultural contact. In his book *Islands of Truth*, scholar Daniel W. Clayton describes the "imperial refashioning" of the Pacific Northwest as a "geopolitical shell" under which the colonial enterprise advanced upon the formation of Western ideologies,

Figure 11.1. View of Maunalua from atop Kohelepelepe, Pu'uma'i, taken in 1915 by L. E. Edgeworth. Courtesy of Bishop Museum Archives.

including "imaginative geographies" that contribute to the creation of imperial space and underpin geopolitical control.[5] Historian Debjani Bhattacharyya details the "technologies of property" that have transformed the Bengal Delta into what it is today through processes of "forgetting," predicated on tensions between geographic naming systems, land ownership, and territorial appraisal.[6]

In Hawai'i, the colonial overlay not only erased the traditional Maunalua toponym (herein referred to as an endonym) but effaced several other place names as well. The area's shifting nomenclature and its spatial reconfiguration are the parameters that motivate, premise, and inform this chapter. Its linguistic transformation is symbolic of Maunalua's ongoing physical, ecological, economic, and sociopolitical transformation. Such change is most visibly marked by the dredging of Ke-Ahupua-o-Maunalua, a traditionally built and managed fishpond whose prolific yield once sustained a large portion of the local population.[7] Cement tycoon Henry Kaiser dredged parts of the fishpond as early as the 1960s.[8] This quite literally paved the way for Hawaii Kai's ongoing urban sprawl. The fishpond has since been repurposed as a small-boat harbor and marina for aquatic recreation.

Previously dissuaded by territorial law that excluded aliens from private property ownership, Hawai'i's new "state" legislation eased the anxieties prospective transplants to Hawai'i like Henry Kaiser likely felt before the state's legal transformation took effect. To meet the demands of Hawai'i's growing population, retailers and restaurateurs set their sights on Hawaii Kai's shores. Corporate outcroppings enjoyed the burgeoning consumerism that accompanied this influx. In 1970, the most symbolic maneuver of Hawaii Kai's metamorphosis occurred when several trustees of the Bishop Estate—the largest private property holder in Hawai'i—evicted some 150 multigenerational families from their homes in Kalama Valley.

In this chapter, I use endonyms, traditional place names, and Indigenous narratives to reconstruct Maunalua before its transformation into Hawaii Kai. Against colonial and English names that currently map the space in its contemporary context, this recasting remembers the area's Hawaiian sense of place before it succumbed to urbanization and residential development. By tracing the changes throughout the last five decades, I argue that Maunalua's place names, when rescued from obscurity, stitch together a holistic islandscape imbued with historical information and cultural value. Further, they reconstitute the land-sea continuum typical to spatial understandings of islands and access to natural resources. I contend that this continuum has been severed by the colonial logic that favors a binary land-sea boundary, separating them not only in the cognitive imagination, but also, by implementation, as two distinct physical spaces. Importantly, here, I analyze the abstractions of the kai-uka directionality significant in a Hawaiian cosmology, often misconstrued by English translations as *mountain and sea*, or effectively, *mountain or sea*. I also compare English landscape terms (marina, land, harbor) against Hawaiian landscape terms (loko i'a, 'āina, awa), whose spatial and utilitarian delineations contrast, reflecting larger fundamental differences that emerged during the area's transcultural contact.

These distinctions have shifted the ideologies of generating, procuring, and supplying natural resources. Hawaii Kai is but one example of this among hundreds across the island chain. Such disruption of islandscape prefigures other colonial ideas like urban channelization, which, by the close of the 1970s, led to the conversion of fifteen natural streams into drainage canals that empty into Maunalua Bay. The impact of watershed urbanization in Hawaii Kai constitutes the most severe, most visible, and most lasting ecological transformation that undermines traditional practices of resource access, sustainability, and food production. The local ecology has been permanently impacted by the incessant development of real estate and retail space presumed by the colonial logics of urbanization and subdivision.[9] The area's marina houses yachts and fishing boats, indicative of Hawaii Kai's coastal geography, though no different in appearance than boat harbors in San Diego, California or Tampa, Florida. Residential neighborhoods now obstruct most views of the nearby shore and lake, as well as access to them. Yet, the consequences of privatized access are mitigated in light of the defunct marine environment, where fish and seafood populations have significantly declined. Together, all of these factors have degraded Maunalua in its Hawaiian sense of place, grossly repurposing it to suit the demands of Americanism.

The recovery of traditional place names, then, challenges the current land-use practices in Hawaii Kai by re-envisioning traditional land and water use as typified in a Hawaiian islandscape, worldview, and cosmology. However, if my goal is to reconstruct Maunalua in its pre-colonial vision, then it is necessary to address certain tensions that surface during such an endeavor. For reasons obscured by the marginalization of Hawaiian indigeneity vis-à-vis the colonial legacy, the matter of reviving endonyms as a way to re-center indigeneity and destabilize colonial hegemony poses ethical consequences.[10] The difficulty is rooted in the Hawaiian concept of kuleana. Mary Kawena Pukui, one of Hawai'i's most celebrated Native intellectuals, defines kuleana as: "Right, privilege, concern, responsibility, title, business, property, estate, portion, jurisdiction, authority, liability, interest, claim, ownership, tenure, affair, province; reason, cause, function, justification."[11]

To understand kuleana, then, is to understand how and where we pledge our loyalty, responsibility, and stewardship, such that we may gain the privileges afforded by such kuleana.[12] To use (recover) an endonym requires kuleana on the user's behalf. While the constitution of kuleana may have evolved over the last century for reasons beyond the scope of this chapter, at the very least, those who wish to employ endonyms must fulfill their kuleana to that place that bears the name. The result of fulfilling this kuleana generates more kuleana, as in the form of privilege, maintained by ongoing accountability. Essentially, using the name Maunalua is one step toward fulfilling kuleana as a resident in this place, but mere usage (recovery) of the place name should not be the extent of that kuleana.

If fulfilled, kuleana affords one privilege by accountability. The same can be said to the contrary: Does one who neglects kuleana to a place deserve to know its traditions? Its lore? Its names? To this arrives the matter of spatial multiplicity, or layers of differently constructed space existing simultaneously in the same area. As an example, I reference Kahikina de Silva, professor at the University of Hawai'i at Mānoa and a kama'āina (native-born) of Ka'ōhao. Today, after radical subdivision, soaring real estate value, and the commercialization of a world-famous beach destination, the imposition of the area's exonym "Lanikai" atop Ka'ōhao prevails. Regarding her hometown, de Silva asserts that there was community discussion about the appropriateness of reviving its endonym. Her reasoning was that even in spite of the contemporary expansion and recovery of its endonym, the question still remains: "Who really lives where?" To answer, de Silva asserts, "I may live in Ka'ōhao, but some of my neighbors certainly live in Lanikai."[13] Her words not only acknowledge spatial multiplicity but also begin to address the struggle between the privilege of using an endonym and fulfilling the kuleana by which one deserves to use it.

De Silva's statement also justifies the anxieties surrounding one's decision between the endonym and the exonym, between Ka'ōhao and Lanikai, between Maunalua and Hawaii Kai, in a way that recognizes two distinct spaces, two histories, two communities, and two names imposed unto one another at the same time and in the same space. Given these considerations, it is perhaps easier to spare the tension and acknowledge current space by current names, as they seem more suited to the way those spaces exist, that is, the means for which they are purposed today. Ka'ōhao and Maunalua name Hawaiian wahi pana (traditionally storied places imbued with lore), whereas Lanikai and Hawaii Kai name spaces where there is a marked transformation in the way of urbanization, real estate, and retail space, all of which favor the transplantation of non-local homeowners and beachfront vacation rental occupants at the visible displacement of Hawaiian kama'āina and other locals. This reconfiguration of place has obscured the Indigenous lens on the land insomuch as the endonyms for these places often feel foreign and out of place.

Before the Hawaiian Kingdom's overthrow in 1893 and subsequent illegal annexation to the United States as a territory in 1898, the moku (district) to which Maunalua belonged was ambiguous. Even recently, it seems to have vacillated between two delineations. As of 1859, the Hawaiian Kingdom declared it as an ahupua'a (smaller district within a moku) in the moku of Kona.[14] During other periods, such as the Māhele (1848), it was designated as an 'ili (smaller district within an ahupua'a) of the Waimānalo ahupua'a, in the moku of Ko'olaupoko. In a 2014 report by the Office of Hawaiian Affairs, cultural specialist Holly Coleman notes:

> Traditionally, land divisions in Hawaiian society reflected shifting resource use and availability; it is likely that the areas which were considered to be part of Maunalua changed over time. Historical records suggest that Maunalua was alternately considered an ahupua'a (land division) and an 'ili (small land parcel) of Waimānalo or Honolulu (Maly & Wong, 1998). In the late 1700s, Maunalua was considered to be an 'ili of the ahupua'a of Waimānalo in the moku (district) of Ko'olaupoko.[15]

Coleman's own research suggests why Maunalua moved historically between two neighboring moku, Kona and Koʻolaupoko. While land divisions depended on yield and availability of resources, it may be that aliʻi ʻaimoku (district-ruling chiefs) passed Maunalua between both chiefs' respective purviews based on the seasonal productivity of Ke-Ahupua-o-Maunalua fishpond, a traditional Hawaiian aquaculture technique; if the deficit in one moku was the abundance in the other, Maunalua's moku shifted to meet production demands.

In her 1983 publication *ʻŌlelo Noʻeau: Hawaiian Proverbs and Poetical Sayings*, Pukui displays an *ʻōlelo noʻeau* (proverbial saying) of cartographical nature: "*Kona, mai ka puʻu ʻo Kapūkaki a ka puʻu ʻo Kawaihoa*," meaning "Kona, from Kapūkaki Hill (the petulant land shell) to Kawaihoa Hill (the setting aside)."[16]

Kapūkaki is the endonym for the widely used exonym Red Hill, an area near Makalapa Crater. At the opposite end of the Kona district, Kawaihoa is the Southeasternmost point of Kuamoʻokāne, which divides Maunalua and Hanauma.[17] The spatial relationship between Kapūkaki Hill and Makalapa Crater is similar to that of Kuamoʻokāne Hill and Puʻumaʻi Crater in terms of distance (visible proximity) and direction (uka-kai). Kapūkaki and Kawaihoa are in the land sections of Moanalua (two encampments) and Maunalua (two mountains), respectively.[18] As typified in Hawaiian memory culture, mnemonic function can be seen in the alliteration and assonance of both names, as well as the suffix—lua (two), encoding numerical information.[19] Beyond these relations, Kapūkaki and Kuamoʻoakāne are similar in shape and size, and their resemblance facilitates geographical epistemology in Hawaiian oral tradition and memory culture. This type of cartography is exemplified in oral histories, as noted by anthropologist Julie Cruikshank, who comments on conclusions made by fellow anthropologist Frances Harwood:[20]

> Harwood contends that mnemonic function of place names may reflect universal cognitive processes. She suggests two possible axes for ordering events, a temporal one for literate societies and a spatial one for non-literate societies. Place names, then, are not decorative embellishments but structural markers, dividing the corpus into cognitive units and spatially anchoring stories so that they can be recalled by remembering the land.[21]

Since 1859, the district's eastern border has moved, possibly more than once. According to archaeologists Elspeth P. Sterling and Catherine C. Summers, "the many previous acts referring to Oahu districts never did make this sufficiently clear . . . the descriptions of Honolulu and Koolaupoko districts clarified this point." In 1932, the territorial government moved the border from Kawaihoa to Makapu'u (Makapu'u Head), although Maunalua still remained within the Kona district, as either an ahupua'a or an 'ili, or both. In recent years, the boundary between the Kona/Ko'olaupoko moku was re-situated east of the Kuli'ou'ou 'ili in the Waikīkī ahupua'a, excluding Maunalua. This relocation now placed it as an 'ili in the Waimānalo ahupua'a in the moku of Ko'olaupoko.[22]

Maunalua's fluctuation between two moku may be seen as the prescient foretaste of its current condition. On the eastern boundary of Kona and the southern boundary of Ko'olaupoko, it has historically occupied a liminal and peripheral position, both on the land and in the imagination. Coleman demonstrates such liminality, asserting, "Historical records indicate that the occupation of certain villages in Maunalua was not always sustained or permanent, and that Maunalua likely had a shifting population."[23]

In a place marked by such egregious outmigration and extinction, it is easy to imagine the ease and expedience with which spatial reconfiguration happened in the area. The reconfiguration of Maunalua began via the imposition of exonyms at the expense of endonyms. Researcher John Clark writes, "In 1936, when Bishop Estate converted some pastureland at the base of Koko Head into a subdivision, Albert F. Judd, Bishop Estate trustee and Hawaiian historian, named the subdivision after Captain Portlock. Kawaihoa Point was then also named Portlock Point."[24]

The deliberate replacement of Indigenous toponyms by colonial designations—names of the father—is no foreign practice to those of any region reeling in the aftermath of imperial conquest. By the colonialist agenda, replacing Indigenous place names with new and foreign exonyms seems obligatory, even ceremonial, and the practice of stamping a name of the father upon the new shore prevails as celebratory and commemorative of the visiting party's "discovery."

The erasure of traditional Hawaiian place names and markers of wahi pana indicates more than a nominal transformation. In many instances, the replacement of the endonym, both discursively and cartographically, is a superficial indication revealing deeper motivations and intentions to transform the space, to reimagine it, and to reconstruct it for new purposes.[25] R. K. Herman comments on the practice of imposing names of the father that "mark a second phase of naming, a second overlay of toponyms that correspond with the 'modern' period of the Islands. . . . This layer of Western family names was largely imposed during the Hawaiian Kingdom, and as one additional means by which the resident foreigners asserted their control over a territory not their own, but over which they desperately sought to gain control."[26]

Portlock was likely the first exonym to replace the Kawaihoa endonym and eventually the entire (Puʻu ʻo) Kuamoʻokāne (Hill called Kuamoʻokāne). This endonym also has a variant, Moʻokua o Kāne ʻĀpua. Both kuamoʻo and moʻokua refer to the backbone of Kāne. Kāne is the deity and provider of fresh water, said to have visited Maunalua alongside Kanaloa, a fellow deity, with his ʻōʻō digging stick to spade the earth and create freshwater springs.[27] The reason for naming the hill after a backbone is evident in its spine-like appearance. According to Hawaiian historian and Native informant J. K. Mokumaiʻa, traditional land use in this area is likely suggested in one of many of Kāne's bodily forms as Kāneʻāpua, where ʻāpua is a type of fish trap.[28] Etymologically, the name Moʻokua o Kāneʻāpua marks an area named for the "spine of fish-trapping Kāne." This endonym connects to one adjacent, Ke-Ahupua-o-Maunalua, or "Maunalua's Shrine of Young Mullet." Both endonyms attest to the ecological value of the area and map the natural resources found there.

Given the cultural, ecological, and topological value encoded in endonyms, replacing them with exonyms is the first step toward spatial reconfiguration. Below, I present a list of endonyms in the Maunalua area that have been replaced by colonial exonyms. I utilize the Papakilo Database, a digital archive of Hawaiian-language newspapers, to track this shift discursively (see table 11.1). The column labeled "Last" indicates the last year the respective endonym appears in the Hawaiian-language newspapers. The column labeled "First" indicates the first year the corresponding exonym began circulating in Hawaiʻi's Hawaiian and/or English newspapers. Further, I note the traditional land use of the particular space in contrast to its contemporary use, while providing the meanings of the endonyms to provide an idea of Hawaiian place-naming traditions. It should be noted that the Hawaiian-language newspaper archive represents a national dialogue in which most Hawaiian and non-Hawaiian Kingdom and Territory subjects were participating. I use this archive as a means to reveal public discourse by way of print media.[29]

There is a scarcity of moʻolelo (Hawaiian history and folklore) for which Maunalua is the prominent setting around which the story is centered. In traditional Hawaiian literature, there are a handful of references to Maunalua or places within, cited as a stopover more than a destination. Hiʻiaka, a patron deity of hula, is said to have passed through on her way to another archipelagic destination during her journey across the island chain.[30] Another story tells of the shape-shifting trickster Kamapuaʻa, who chases Hawaiʻi's volcano deity, Pele, so Kapo, Pele's sister, flies her kohe (vagina) to distract Kamapuaʻa, giving Pele time to rest.[31] Kapo's kohe left an indentation on the rim of Puʻumaʻi, only visible from the Wāwamalu (Awāwamalu) side. Other Hawaiian folklore relates to ancestral stones of resident deities and

Table 11.1. Endonyms changing to exonyms in Maunalua (Hawaii Kai), their respective uses, and years when renaming happened, as attested to in historical local newspapers

Endonym (meaning)	Last	Traditional Use	Exonym	First	Current Use
Awāwamalu Wāwamalu (shaded valley)	1930 1924	Beach	Sandy Beach	1950*	Beach park
Hālona (peering place)	1922	Cove, blowhole, lookout	Cockroach Cove	n/a	Beach
Kawaihoa Pu'uokawaihoa Keawahili (setting aside) (smiting harbor)	1980 1877 1865	Easternmost point of Maunalua Bay Former house site of Kamehahema I	Spitting Caves	n/a	Cliff-jumping site
Kahauloa (long hau tree)	n/a	Plain	Koko Head	1877	Shooting range
Ka'ili'ili Kaloko (pebble) (pond)	1865 n/a	Fishing shrine Pond	Alan Davis	1980*	Recreational area
Ke-Ahupua-o-Maunalua (shrine of young mullet)	n/a	Inland lake, aquacultural farm	Koko Marina	1972*	Marina, small boat harbor
Kuamo'okāne (backbone of Kāne)	1885	Hill	Portlock	1988*	Subdivision
Maunalua Mauna Lua (two mountains)	1936 1926	'Ili/Ahupua'a land division	Hawaii Kai	1979*	Municipality, subdivision
Kamilonui (large milo tree)	1985	Valley	Mariner's Ridge	n/a	Subdivision
Pai'olu'olu (lift gently)	n/a	Point	Witches' Brew	n/a	Point
Palea (brushed aside)	1866	Point	Toilet Bowl	n/a	Point

(*Continued*)

Table 11.1. (Continued)

Endonym (meaning)	Last	Traditional Use	Exonym	First	Current Use
Pu'uokīpahulu (fetch from over-farmed soil)	n/a	Historic feature	Pele's Chair	1996*	Landmark, hiking spot
Pu'uma'i Kohelepelepe (genital hill) (vaginal fringe)	1877 n/a	Peak, crater, historic feature	Koko Crater/ Koko Head (Stairs)	1988*	Hiking destination

Source: Papakilo Database, https://www.papakilodatabase.com/ (accessed March 31, 2021).

* Denotes an exonym found in English-language media only and does not seem to appear in the Hawaiian-language newspapers.

kama'āina who were solidified as rock formations across the coasts and valleys, whose names became the endonyms, most of which have slipped away.

Literature hardly identifies specific, long-standing stewardship of the area, for reasons likely relevant to Maunalua's shifting territory. Like its local population, stewardship, land use, and resource distribution changed over time. However, Missionary Levi Chamberlain identified a settlement of eighteen houses belonging to Kalola, a chiefess.[32] Further, a 1921 map by the Bishop Estate names Kuhina Nui Victoria Kamāmalu as the owner of the 'ili known as Maunalua and Kuli'ou'ou, who inherited these lands as a ranking descendant of her royal family.[33] Since then, these Crown Lands have been absorbed by the state of Hawai'i and private entities like the Bishop Estate.

Some research traces as far back as the whaling era during the early nineteenth century. Handy noted a "Famous Potato Planting Place":

> According to the last surviving kamaaina of Maunalua, sweet potatoes were grown in small valleys, such as Kamilonui, as well as on the coastal plain. The plain below Kamiloiki and Kealakipapa was known as Ke-Kula-o-Kamauwai. This was the famous potato-planting place from which came the potatoes traded to ships that anchored off Hahaione in whaling days. The village at this place, traces of which may still be seen, was called Wawamalu.[34]

In small valleys like Kamilonui and on the coastal plain (kula), known then as Ke-Kula-o-Kamauwai, *below* and *between* Kamiloiki and Kealakīpapa valleys, residents grew sweet potatoes, as the regional soil is less favorable for other crops. The name "Kalama Valley" is commonly found on contemporary maps, but a 1927 geological survey by the US Department of the Interior gives the name of another valley called Mauuwaii, whose existence would corroborate the expanse between Kamiloiki and Kealakīpapa as greater than just one valley (Kalama) where sweet potatoes were harvested. In any case, the borders of Ke-Kula-o-Kamauwai and Mauuwaii are unclear, and neither endonym appears in Hawaiian place-name databases. I theorize that Ke-Kula-o-Kamauwai simply means "the plain of Mauwai (Mauuwaii)," named for the valley expanse that opens to the north.

Kealakīpapa is the western lowlands and also the valley to the west between the Hawai'i Kai Golf Course and Makapu'u Head. Meaning "the road paved with stone," the origins of the ala (road or path) are unclear, according to informant Mr. J. McCombs.[35] However, a trail probably connected local residents from this area to the land and residents of Kaupō, Waimānalo, where a marine-themed amusement park called Sea Life Park exists now. Native informant Shad Kane identifies a rock wall division in Maunalua between Kona and Ko'olaupoko, asserting, "It supports a division between one geographical area and another, one of the reasons is that it supports large amounts of people living in that particular area. The reason why I say that is because, fundamentally, land divisions were based on resources, so land divisions have to do with connecting people with these resources."

Kane concludes that whatever chief oversaw the area when the moku (land division) boundary was built, there must have been large scores of people to construct it, given the level of work needed to construct the boundary wall.[36] Pukui affirms Kane's statements, referencing a chief of Wāwamalu who ordered the construction of the road by "the people who annoyed him."[37] Often, if there is a literary reference using the term "ala nui," it denotes a large, long path that circumvented the entire island, while providing access between moku. At one point in history, the ala nui and the moku boundary certainly intersected.

While Maunalua's agriculture seems to have been based, at least in part, on sweet potatoes, Sterling and Summers compile a list of at least a dozen ko'a (fishing shrines) in the area, attesting to the aquaculture and marine resources of both Maunalua Bay and Ke-Ahupua-o-Maunalua fishpond. Boundaries between the bay and pond were less defined cartographically than by oral tradition. Attempts to recover the traditional name for the lake inadvertently lead many to the endonym "Kuapā," which is actually the type of pond named for the style of the pond's construction. The traditional name for the lake, however, is "Ke-Ahupua-o-Maunalua," meaning "the shrine of young fish of Maunalua." The inclusion of Maunalua within the names of both the marine bay and the inland lake reflects the fluidity of the boundaries of both spaces, which certainly overlap (see figure 11.2). In a Hawaiian cosmology, terrestrial, demersal, and pelagic space is conceptualized holistically and flexibly, an idea to which I return later in this chapter.

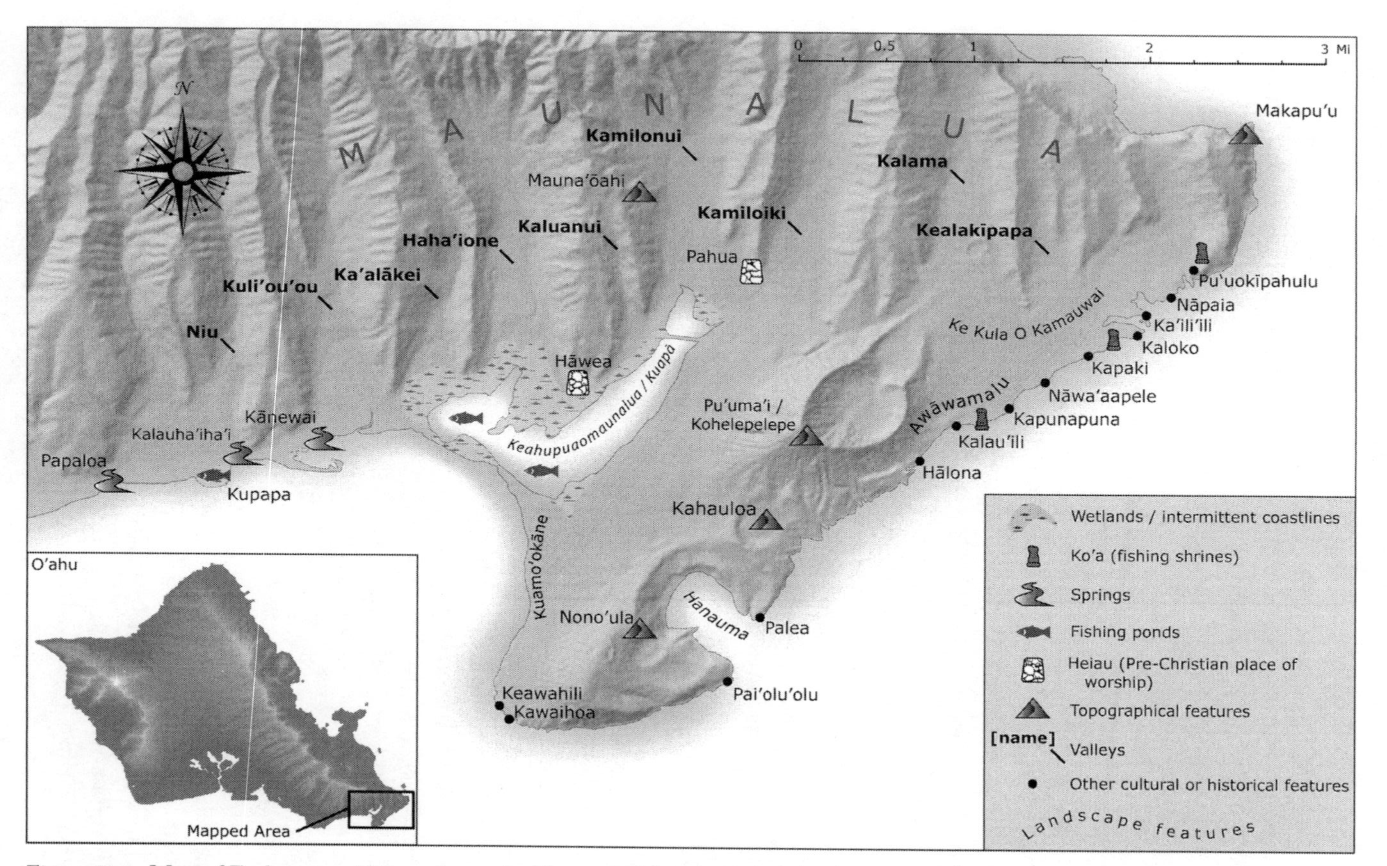

Figure 11.2. Map of Endonyms of Maunalua and a Theoretical Landscape of the Area in the 1920s. Cartography by Geoffrey Wallace. Map data from the United States Geological Survey (USGS) and other relevant sources.

Though most have fallen into disrepair in modern times, the number of fishing shrines dotting the Maunalua coastline evidences the area's abundance of marine resources. The date and manner of development of the fishpond is contested. Sterling and Summers cite one Native informant: "According to Mrs. Makea Napeahi, the pond was built by her great-grandmother, Mahoe. When the pond had been only partially completed, the menehune came and in one night finished the construction." Another account by McAllister (1930) states: "A large fishing village formerly existed in Hahaione Valley at the head of the pond (Maunalua), which, according to [Mr. Moe], was not a pond, but an arm of the sea. The people from this village fished off Maunalua in their canoes, and when the pond was built it cut off their access to the sea and the village declined."[38]

Both accounts by Native informants seem to contradict the time when the pond was constructed. Also contested is the reason the construction of a fishpond would bar access to local residents, leading to the village's decline. In either case, McAllister's account points to the out-migration from Maunalua and the temporal gaps separating the generations of local residents. The dearth of local residents and local knowledge explains the paucity of oral tradition and history that survived the passage of time and the onslaught of development in contemporary Hawaii Kai.

The previously mentioned Indigenous nomenclature of Maunalua challenges contemporary ideas of land use on landscape, or land use on *island*scape. In a Hawaiian worldview, the ahupuaʻa includes the land, the coast, the coral reefs, and beyond. Pukui defines an ahupuaʻa as a "land division usually extending from the uplands to the sea, so called because the boundary was marked by a heap (ahu) of stones surmounted by an image of a pig (puaʻa), or because a pig or other tribute was laid on the altar as tax to the chief. The landlord or owner of an ahupuaʻa might be a konohiki."[39]

For islandscapes, the uka-kai (land-sea) directionality is a continuum conceptualized in Hawaiian cosmology as an inclusive and holistic ecosystem. The constitution of ahupuaʻa land divisions in Hawaiian cosmology counters Western contemporary notions about subdivision that were conceptualized and shaped under the administration of then–US president Calvin Coolidge in the 1920s, who formed the Advisory Committee on City Planning and Zoning and published the Standard City Planning Enabling Act (SCPEA), defining "subdivision" as:

> the division of a lot, tract, or parcel of land into two or more lots, plats, sites, or other divisions of land for the purpose, whether immediate or future, of sale or of building development. For the purpose of sale or of building development: Every division of a piece of land into two or more

> lots, parcels or parts is, of course, a subdivision. The intention is to cover all subdivision of land where the immediate or ultimate purpose is that of selling the lots or building on them.[40]

The interactions between land and sea in the ahupua'a economy give breadth to the range of resources available within a single islandscape land division. Often, the inclusiveness of Hawaiian environment is encoded in spatial language such as uka-kai directionality and understanding of space. Uka marks a direction toward the coast when at sea, and toward the inland or mountain peak when on land. Oppositely, kai signifies a direction toward the sea, whether one is on the mountain peak, inland, or on the coast. Such spatial information extends into Hawaiian rhetoric through mele, traditional poetry performed orally and musically, mapping the speaker or performer onto space in relation to their surroundings. In English, translations often crudely render uka-kai as a mountain-sea boundary, that is, mountain *or* sea. As a counterexample to such binary comprehensions of space, consider the following mele, with translation by the author of this chapter:

Lele Laniloa, ua mālie	Come ashore at Laniloa, it is calm
Ke hoe a'ela e ka Moa'e	The trade wind draws its breath, fatigued
Ahu kai i nā pali, kai ko'o o lalo ē	The sea laps up toward the cliffs, rough sea below
Ua pīkai iā uka ē	Sprinkling and cleansing the upland with sea spray.[41]

This mele poetically describes the intimate interactions between the land and the sea. Such lyricism evidences the land-sea continuum as part of the Hawaiian imagination and worldview, and when extended into reality, this ideology informs Indigenous land use and stewardship as underpinned by the ahupua'a economy. Colonial logic often misconstrues islandscape directionality encoded in uka-kai as mutually exclusive when translated as a land-sea boundary. Imperial mindsets informed by continental landscapes approach this less as directionality than as an "either-or" binary. Insofar as I argue that the kai-uka Hawaiian islandscape directionality is often contemporarily misunderstood as land *or* sea, this idea advances both a cognitive and physical boundary between both spaces. I apply these ideologies in the context of land use, resource access, and food production in Maunalua. Through the Hawaiian mindset on the Hawaiian islandscape, approaches to these practices are more sustainable than those in place today. Hawaiian geographer Kapā Oliveira echoes these ideas, claiming:

The Heaven-Land-Ocean Continuum [extends] from the heavenscape through the landscape and on to the oceanscape, Hawaiian place was a multi-dimensional continuum. The various layers of place were interrelated and directly impacted one another. *Kanaka Maoli* understood that the pollution of streams affected marine life, deforestation of well-established rainforests changed the environment resulting in less rainfall, and irresponsible development near the seashore rapidly eroded the coastline.[42]

I posit that colonial ideologies devised and practiced on continental landscapes are unfit for islandscapes. Western land use practices divide the land, severing the land-sea continuum, disrupting resource availability, and repurposing the land as saleable property. Such logic has spurred devastating ecological and economic change, grossly furthered by watershed urbanization in Maunalua.

Following Hawai'i's statehood in 1959, the United States Army Corps of Engineers (USACE) extended its purview to cover the Hawaiian Islands. Engineering projects that had already been in operation on the continental United States were now being conducted in Hawai'i to address factors such as beach erosion, pollution abatement, and overall bay development. On the heels of the US River and Harbor Act of 1950, a congressional act was signed on December 31, 1970, "authorizing the construction, repair, and preservation of certain public works in rivers and harbors for navigation, flood control, and for other purposes."[43]

In the USACE's 1971 Annual Report, Statute 190 of § 106 lists the locations at which the secretary of the army authorizes the USACE to make surveys subject to all applicable provisions of the 1950 River and Harbor Act. Maunalua Bay is listed as one location where the engineers proposed a project that addressed beach erosion. Inspections were conducted in August and October 1970 in accordance with § 3, Flood Control Act of June 22, 1936, which included the "requirement that local interests maintain and operate completed flood control works in accordance with regulations prescribed by the Secretary of War."[44]

In February 1970, the USACE completed the Kuli'ou'ou Stream flood control project in Maunalua Bay ten months before Congress authorized its surveying and improvements "in the interests of pollution abatement, navigation, recreation, and overall bay development" on December 31, 1970.[45] The Kuli'ou'ou Stream project describes local cooperation as "fully complied with." That same year, flood control projects labeled "under special authorization" extended to two other streams that emptied into Maunalua Bay, Wai'alae Iki and Wailupe.

The following year, the USACE initiated the construction of a small boat harbor in Maunalua, Oʻahu. The motivations for boat harbor construction were monetary and commercial, as stated in a 2003 USACE report titled "Improving the Economic Analysis of Small Boat Harbors." This report details the importance of small boat harbors, among other factors: "The contribution of the Hawaiian commercial as well as recreation and sport fishing industry to national income is significant. The importance of the Hawaiian tourism industry (cruise ship customers, other tourists, charter boat operations (fishing, sightseeing)), and services provided to out-of-state yacht voyagers to the regional economy is also significant."[46]

These statements attest to the capitalist ambitions behind the boat harbor proposal, as well as the legislation through which it was completed. It is clear that these methods for development favor out-of-state industries without consideration of resident communities who depend on local resources to operate within their own economies. This enterprise is antithetical to the ahupuaʻa economy and has strategically facilitated other successive development projects in the area by prioritizing visitor access and tourist-oriented recreation over local resource production, distribution, and sustainability.

Ironically, the USACE seems to have acted without due consideration or research in the way of coastal engineering, an act of hubris on their behalf that seems to betray their ideas toward best practice. The following excerpt is a conclusion made in their 1973 Annual Report of the Chief of Engineers on Civil Works and Activities: "The recently completed National Shoreline Study outlines vividly the tremendous scope of the problems associated with our Nation's coastline. Less is known and understood about coastal problems than any of the other engineering problems in water resources development. Research in this area is directed toward a better understanding of coastal phenomena, with specific studies to improve current design practices."

By the close of the 1970s, the USACE completed its flood control projects by converting natural streams in Kuliʻouʻou and Waiʻalae into storm drain canals. They also imposed other measures for flood control using storm drains, channels, and canals that interfere with the area's aquifers.[47] Today, there are over seventeen urban channels that empty into Maunalua Bay.[48] Wailupe Stream is the only natural waterway unaffected by watershed development in Hawaii Kai. Such heavy-handed transformation has devastated Maunalua Bay. According to one report, "Data are lacking to quantify the apparent link between watershed development and coral reef degradation, but it is generally recognized that both physical and biological processes may contribute simultaneously to this degradation."[49]

The holistic, ongoing shift of Maunalua Bay exacerbates the biological threat, where alien species of seaweed and algae compete with native species.

Rainwater runoff from adjacent flood canals introduces invasive species. One 2016 newsletter asserts, "Land-based sources of pollution have led to sedimentation on near shore reefs and degraded water quality."[50]

The biological degradation is symbolic of Maunalua's transformation into modern-day Hawaii Kai. Today, the coastline is largely invisible from land, blocked by densely packed waterfront properties, representing a physical boundary between land and sea. This boundary is reified through colonial projects like flood control and harbor development, which afford residents the luxury of beachfront lifestyles without the threat of natural disaster. This ideology is furthered and finalized by discursive processes like the imposition of Western placenames, real estate branding, and dependence on commercial and corporate industry.

Although manifold ventures coalesced to create the Hawaii Kai Township as it exists today, such endeavors happened at different times, by different authorities, and through different processes. While the Army Corps of Engineers worked on "state land" to urbanize Hawaii Kai's watershed, other development was happening in the private sector to develop Hawaii Kai residentially. Although no direct evidence substantiates collusion between both parties, the Corps' actions certainly facilitate corporate agendas in favor of real estate development, subdivision, and urbanization.

The urban and residential development of Hawaii Kai began in the 1960s when US industrialist and shipbuilder Henry J. Kaiser moved to Hawai'i, and Bishop Estate leased 6,000 acres of private property, including Keahupuaomaunalua Lake, to development company Kaiser Aetna. The first half of Kaiser's name is evident in the exonym, Hawaii Kai—an egotistical celebration of the area's development and an insidious, deceptive exonym that appears as a Hawaiian name. When Kaiser proposed the development of Maunalua, his strategic plan labeled the space as "raw" and "undeveloped," and neglected any mention of the lagoon's history and significance.[51] The lagoon became Kaiser's focal point of spatial reconfiguration when he converted the 500-acre saltwater fishpond into an "open body of water surrounded by homes and recreational boating facilities." Political science professor Neil Milner cites the newspaper article published after Kaiser's death about the development having "left the clear impression that this land took on meaning only because of his ability to envision the land in ways that no one else was imaginative enough to see."[52]

But Kaiser's industrial ambitions backfired. His conversion of Keahupuaomaunalua into modern-day Koko Marina joined the lake to the adjacent bay to the extent that the lake became navigable waters subject to regulation by the USACE under the Commerce Clause, thereby making the lake public space. Kaiser's capitalistic avarice led him to file suit against what he perceived as government overreach into what he claimed as private property, arguing for his right to charge a fee for recreational use of the lake. Kaiser, invoking the fifth amendment, asserted that the government's "regulatory taking" mandated they compensate Kaiser in accordance with his fundamental property right to exclude. Ultimately, Congress held that "it is the public right which necessitates the exercise of regulations. Therefore, Kaiser Aetna should be required to allow public access to the pond. The court also held that, since Kuapa Pond was subject to an overriding federal navigation servitude, no compensation was required."[53]

Nonetheless, the urban development of Honolulu marched on. Milner lists the four major objectives central to the reconfiguration of Honolulu, which included Maunalua, as "the conversion of agricultural land to suburban-type housing; the conversion of urban Honolulu from high density low-rise housing to high-rise condominiums; the enormous increase in the number of Waikīkī hotels; and the evolution of public housing in the name of slum clearance."[54]

Of great import here is that the residential development of Hawaii Kai preceded its watershed urbanization by nearly a decade, prior to Kaiser's passing in 1967. Although unprovable, it is entirely relevant to question whether or not the USACE would have proposed and completed action on flood control, beach erosion, bay development, and harbor construction had Kaiser's creation of Hawaii Kai not planted residents into the area first.

Kaiser's development of Hawaii Kai was premised on a certain discourse informing housing policies, subdivision, and urbanism of the time, namely what political scientist Douglas W. Rae calls "spatial hierarchies."[55] Echoing similar ideas, Milner critically examines the impact of Kaiseresque development in that type of suburbanization in Hawaii Kai, which seems to critically reflect "the continued faith in the moral primacy of that kind of home." While this type of housing development stimulated the economy and raised real estate value, one of its goals was slum clearance; Milner argues that "none of these developments triggered significant discussions of the possibility of other lifestyles or other conceptions of home. Henry Kaiser was an exceptional real estate entrepreneur, but his job was made easier because he was selling what were already accepted cultural conventions about the modern and proper way to live."[56]

Here, it is evident that the agendas advanced by both Kaiser and the USACE worked together to transform a new kind of "livable space" that ultimately dispossessed Hawaiian and local residents of their land, livelihoods, and ways of life. While such dispossession was happening in the wake of development across the entire Hawaiian archipelago, this maneuver is harrowingly apparent here in Hawaii Kai.

By the 1960s, Kalama Valley inland of Maunalua was home to some 150 families. According to Haunani-Kay Trask—a Hawaiian activist, scholar, and poet—these families were "dependent on direct or indirect month-to-month leases from Bishop Estate."[57] On the heels of statehood in 1959, the economic, demographic, political transformation of Hawai'i impacted Hawaiians acutely.[58] Hawaiian families sought residence in rural areas or farming villages like Kalama Valley in attempts to avert the pressures of Honolulu's expanding urban core. Trask notes: "Bishop Estate had made a deal with industrialist Henry Kaiser in the 1950s to develop their entire holdings on O'ahu's east end. . . . Although the State Legislature had passed a resolution in 1959 asking the City and Bishop Estate to investigate the problem of relocation of families living in the areas scheduled for development, nothing was done."[59]

On May 11, 1971, a few dozen residents stood atop the last ramshackle home nestled in rural Kalama Valley as they refused to comply with eviction notices issued to them by Bishop Estate. The thirty-two residents were outnumbered by some seventy riot-equipped law enforcement officers who eventually arrested the residents for refusing to move from their homes.

This eviction would become a symbolic event in what photographer and Kalama Valley resident Steve Davis called "a turning point in the history of Hawai'i."[60] The implication of their resistance is echoed in the words of one of their young leaders, Linton Park, who claimed, "Hawaiian history was being made."[61] The Kalama Valley evictions are what Trask credits for representing a collective effort for preservation that "would be remembered long after as the spark that ignited the modern Hawaiian Movement, an ongoing series of land struggles throughout the decade of the seventies that was destined to change the consciousness of Hawai'i's people, especially her native people."[62]

The 1971 Kalama Valley Evictions succeeded Kaiser's development across Hawaii Kai of the 1960s while coinciding with, and prefiguring, the USACE's transformation in the 1970s. Since then, expanding subdivisions have welcomed mounting numbers of foreign transplants in American-style suburbs, while retail shops, corporate groceries, and restaurants have met the region's consumer demands. Booming real estate values climb to match the price of living in paradise. In spite of having mobilized the Hawaiian people into social action, symbolically, the Kalama Valley evictions mark the disappearance of the last vestiges of Maunalua as it existed before modern urbanization (see figure 11.3). This event represents the formal completion of Hawaii Kai, where, as of 2022, the population in the area is approximately 28,500.[63]

In spite of the drastic ecological degradation of Maunalua, there is reason to be optimistic. Several community efforts have mobilized to restore the area's environmental health. Further, many of these grassroots outreach

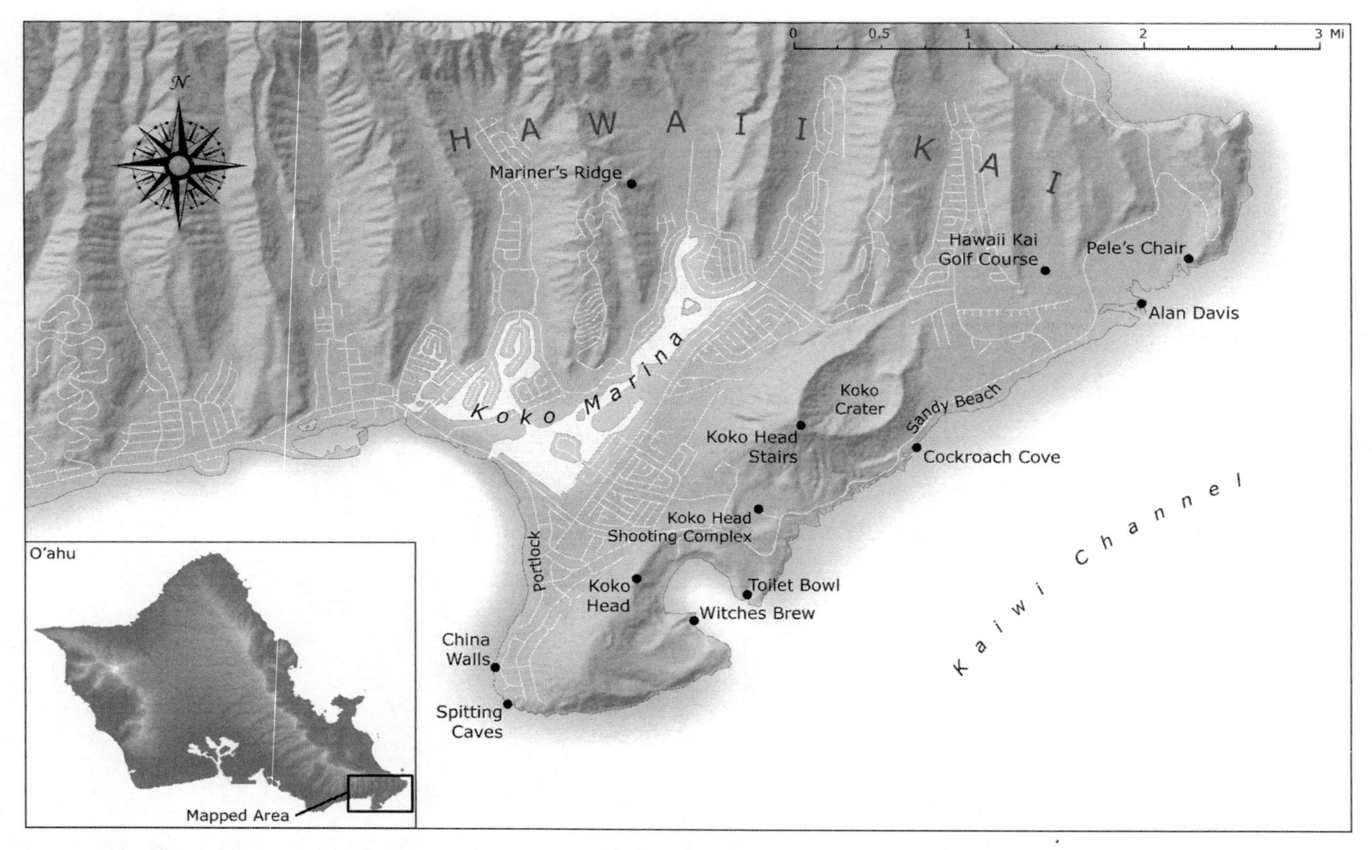

Figure 11.3. Map of Exonyms of Hawaii Kai and Current Landscape. Cartography by Geoffrey Wallace. Map data from USGS and the Honolulu Land Information System.

programs endorse traditional Hawaiian values and epistemologies regarding land stewardship. Educational outreach, habitat restoration, and volunteer opportunities have developed around objectives such as sustainability within the theoretical framework of the ahupuaʻa land-use system.

Mālama Maunalua is one such organization that hosts the monthly "Huki Project," a community-wide effort to remove invasive species in Maunalua Bay. Their website offers several educational resources and information regarding their ongoing community efforts. Strategic Plans by Mālama Maunalua list several partnering organizations that share the same mission, all of whom frame their approaches and efforts in discourse that challenges current land use in Hawaii Kai.

This discourse advances Hawaiian words in otherwise entirely English-language reports. These Hawaiian language terms include loko iʻa (fishpond for mariculture subsistence), ʻāina (sustainably cultivated land), mālama (to care for), and ahupuaʻa (terrestrial boundary contingent on a land-sea continuum). Such reports use these terms to cover and explain "local measures of ahupuaʻa health," while signifying the tenets of community and collective accountability to one's home.[64]

In contrast, the community efforts operating within such ahupuaʻa logic do not advance by way of English-language discourse that issues mere translations of their Hawaiian equivalents. It is clear that the Hawaiian terms do not translate easily into English, and should they be translated, they would not likely generate the same community response and support; here, the linguistically contrastive terms are underpinned by a Hawaiian and/or local identity rooted in Hawaiʻi's Indigenous past and sovereign political history. The Hawaiian terms are grounded in a Hawaiian cosmology, ecology, and way of life, and the success of habitat-restoration programs in Maunalua Bay is premised on these ideologies.

Further, these organizations restore endonyms in the area. Traditional place names like Maunalua, Kawaihoa, and Ke-Ahupua-o-Maunalua circulate among the local vernacular and the discourse set forth by the organizations dedicated to Maunalua's environmental and social preservation. Mālama Maunalua's vision is simple: "A Maunalua Bay where marine life is abundant, the water is clean and clear, and people take kuleana in caring for the Bay."[65]

Importantly, Mālama Maunalua reminds us all of our kuleana that we have to our homes, one that obligates our accountability to it such that we have the right to live there in a pono (fair, just, appropriate) way. Mālama Maunalua exemplifies the Hawaiian value of caring for the place named by the endonym they have chosen to restore while maintaining their commitments to taking care of that place in a way that is culturally and historically meaningful. Their work rehabilitates Maunalua Bay from beneath the layers of foreign and introduced species. As well, their work reinstates the name Maunalua after decades of sustained erasure and overwriting, while envisioning a Hawaiian sense of place of how this space used to be, and one day, how it could be again.

Notes

1. Friends of Maunalua Bay, "Special Sanctuary Management Area," http://friendsofmaunaluabay.org/#How-Will-SSMA-Affect-You (accessed March 31, 2021).

2. Eric Wolanski, Jonathan A. Martinez, and Robert H. Richmond, "Quantifying the Impact of Watershed Urbanization on a Coral Reef: Maunalua Bay, Hawaii," *Estuarine, Coastal and Shelf Science* 84, no. 2 (2009): 259–268.

3. I use land in quotation marks because of the vast amount of artificial landfill in the area that is now considered terrain.

4. The spelling of this name is intentional, as opposed to Hawai'i Kai, to reflect its vernacular pronunciation and to reflect ideologies surrounding the place name and its orthographic inconsistency.

5. Daniel Wright Clayton, *Islands of Truth: The Imperial Fashioning of Vancouver Island* (Vancouver: UBC Press, 2000), 165–235.

6. Debjani Bhattacharyya, *Empire and Ecology in the Bengal Delta: The Making of Calcutta* (New York: Cambridge University Press, 2018), 1–40.

7. Peter T. Young, "Kuapā Pond (Keahupuaomaunalua Pond)," Blog Post, September 4, 2012, https://totakeresponsibility.blogspot.com/2012/09/kuapa-pond-keahupuaomaunalua-pond.html.

8. Holly Coleman, "Ke Kula Wela La o Pahua," Research Division (Office of Hawaiian Affairs, 2014), 32.

9. Candace Fujikane, "Mapping Wonder in the Māui Mo'olelo on the Mo'o'āina," *Marvels & Tales* 30, no. 1 (2016): 47.

10. For a discussion on indigeneity in a Hawaiian context, see Kēhaulani J. Kauanui, *Hawaiian Blood: Colonialism and the Politics of Sovereignty and Indigeneity* (Durham, NC: Duke University Press, 2008), 1–24.

11. Mary Kawena Pukui and Samuel H. Elbert, *Hawaiian Dictionary: Hawaiian-English, English-Hawaiian* (Honolulu: University of Hawaii Press, 1991), 179.

12. We may also think of kuleana as an obligation that, when fulfilled, affords one rights.

13. Kahikina de Silva (Kumu hula, Hawaiian language professor) in discussion with the author, February 2014.

14. Elspeth P. Sterling and Catherine C. Summers, "Maunalua," in *Sites of Oahu* (Honolulu: Bishop Museum Press, 1978), 257–270.

15. Holly Coleman, "Ke Kula Wela La o Pahua: The Cultural and Historical Significance of Pahua Heiau, Maunalua, O'ahu" (Honolulu, Hawai'i: Office of Hawaiian Affairs, 2014), 22.

16. Mary Kawena Pukui and Dietrich Varez, in *'Ōlelo No'eau: Hawaiian Proverbs & Poetical Sayings* (Honolulu, Hawai'i: Bishop Museum Press, 2018), 199.

17. Mary Kawena Pukui, Samuel H. Elbert, and Esther T. Mookini, *Place Names of Hawaii* (Honolulu, Hawai'i: University Press of Hawaii, 1989).

18. John R. K. Clark, *Hawai'i Place Names: Shores, Beaches, and Surf Sites* (Honolulu: University of Hawai'i Press, 2002).

19. For more on this topic, see Kalani Akana, "Hei, Hawaiian String Figures: Hawaiian Memory Culture and Mnemonic Practice," *Hūlili* 8 (2012): 45–70.

20. Frances Harwood, "Myth, Memory and Oral Tradition: Cicero in the Trobriands," *American Anthropologist* 78, no. 4 (1976): 785–787.

21. Julie Cruikshank, "Getting the Words Right: Perspectives on Naming and Places in Athapaskan Oral History," *Arctic Anthropology* 27, no. 1 (1990): 55.

22. For more information on how this project started and spread to other islands, see "Ahupua'a Boundary Marker Project—Ko'olaupoko Hawaiian Civic Club," Ko'olaupoko Hawaiian Civic Club, September 14, 2018, http://www.koolaupoko-hcc.org/ahupuaa-boundary-marker-project/.

23. Coleman, "Ke Kula Wale La o Pahua," 23–33.

24. Clark, "Hawai'i Place Names," 304–305.

25. Pukui and Elbert, "Hawaiian Dictionary," 377.

26. *Arctic Anthropology* 27, no. 1 (1990): 55. R. D. K. Herman, "The Aloha State: Place Names and the Anti-Conquest of Hawai'i," *Annals of the Association of American Geographers* 89, no. 1 (March 1999): 86.

27. Interview, *Almeda Goss*, April 23, 1962.

28. J. K. Mokumaia, "Ka Aekai o Maunalua Ame Kona Mau Kuhinia," *Ka Nupepa Kuokoa*, March 4, 1921, 59th ed., sec. 9, p. 3.

29. For more detailed information on the profusion of Hawaiian national discourse, see Puakea Nogelmeier, *Mai Pa'a i Ka Leo: Historical Voice in Hawaiian Primary Materials: Looking Forward and Listening Back* (Honolulu, Hawai'i: Bishop Museum Press, 2010).

30. Hooulumahiehie, "Moolelo Hiiakaikapoli-o-Pele," *Ka Nai Aupuni*, June 28, 1906, 2nd ed., sec. 23, p. 4, https://www.papakilodatabase.com/pdnupepa/?a=d&d=KNA19060628-01.2.15&e=-------en-200--1001-byDA.rev-txt-txIN%7ctxNU%7ctxTR-hiiaka-------.

31. Lilikalā Kame'eleihiwa and Dietrich Varez, "He Mo'olelo Ka'ao o Kamapua'a: A Legendary Tradition of Kamapua'a, the Hawaiian Pig-God: An Annotated Translation of a Hawaiian Epic from Ka Leo o Ka Lahui, June 22, 1891–July 23, 1891" (Honolulu: Bishop Museum Press, 1996), 116.

32. Levi Chamberlain, "Trip Around Oahu," *Trip Around Oahu* (1826).

33. Lilikalā Kame'eleihiwa, "Ancestral Visions of 'Āina (Land That Feeds)," AVAKonohiki.org (2016), http://www.avakonohiki.org/.

34. Edward Smith Craighill Handy, *The Hawaiian Planter* (Oakland: University of California Press, 1940), 55.

35. Elspeth P. Sterling and Catherine C. Summers, "Maunalua," in *Sites of Oahu* (Honolulu: Bishop Museum Press, 1978), 259.

36. Unknown, "Exploring the History of Maunalua," maunalua.net (2014), http://www.maunalua.net/.

37. Sterling and Summers, *Sites of Oahu*, 260.

38. J. Gilbert McAllister, *Archaeology of Oahu* (New York: Kraus Reprint, 1971), 69.

39. Pukui and Elbert, *Hawaiian Dictionary*, 9.

40. *Standard City Planning Enabling Act for the Department of Commerce, for the Fiscal Year Ending June 30, 1926*.

41. Unknown, "He Moolelo No Hiiakaikapoliopele," *Ka Hoku o Ka Pakipika*, February 13, 1862, 1st ed., sec. 21, p. 4, https://www.papakilodatabase.com/pdnupepa/?a=d&d=KHP18620213-01.2.25&srpos=&dliv=none&e=-------en-20--1--txt-txIN%7ctxNU%7ctxTR-%22lele+laniloa%22-------.

42. Katrina-Ann Rose-Marie Kapāʻanaokalāokeola Nākoa Oliveira, "Ke Alanui kīkeʻekeʻe o Maui: Na Wai Hoʻi Ka ʻole o Ke Akamai, He Alanui i Maʻa i Ka Hele ʻia e Oʻu Mau mākua" (dissertation, ProQuest Dissertations Publishing, 2006), 221.

43. "An Act Authorizing the Construction, Repair, and Preservation of Certain Public Works on Rivers and Harbors for Navigation, Flood Control, and for Other Purposes" (1946).

44. William C. Westmoreland, "Chief's Annual Report," 2 Chief's Annual Report § 17 (1971), pp. 36-5-36-12, https://usace.contentdm.oclc.org/digital/collection/p16021coll6/id/486.

45. An Act authorizing the constructions, repair, and preservation of certain public works on rivers and harbors for navigation, flood control, and for other purposes, for the Fiscal Year 1970, Public Law 91-611, U.S. Statutes at Large 84 (1970): 1820.

46. Richard McDonald et al., "Improving the Economic Analysis of Small Boat Harbors" (Washington, DC, 2003), 1, https://www.iwr.usace.army.mil/portals/70/docs/iwrreports/03ps04.pdf.

47. Creighton W. Abrams, "Chief's Annual Report," 2 Chief's Annual Report § (1973), https://usace.contentdm.oclc.org/digital/collection/p16021coll6/id/486.

48. City and County of Honolulu and NOAA Fisheries, Pacific Island Regional Office (Honolulu: C&C of Honolulu, 2010).

49. Wolanski, Martinez, and Richmond, *Quantifying the Impact of Watershed Urbanization*, 266.

50. "Maunalua Bay: A Special Sanctuary Management Area," http://hawaiihumpbackwhale.noaa.gov (accessed September 2017).

51. Neal Milner, "Home, Homelessness, and Homeland in Kalama Valley: Re-Imagining a Hawaiian Nation through a Property Dispute," *Hawaiian Journal of History* 40 (2006): 152.

52. Milner, 152.

53. Janice Kelly, Kaiser Aetna v. United States, 444 U.S. 164. Accessed April 2, 2021.

54. Milner, 152.

55. Douglas W. Rae, *City: Urbanism and Its End* (New Haven, CT: Yale University Press, 2008), 254–286.

56. Milner, 153–154.

57. Haunani-Kay Trask, "The Birth of the Modern Hawaiian Movement: Kalama Valley, Oʻahu," in *Hawaiian Journal of History,* vol. 21, ed. Helen G. Chapin (Honolulu: Hawaiian Historical Society, 1987), 128.

58. For more information about this impact, see *Annotated Bibliography of Alu Like Native Hawaiian Reports, 1976–1998* (Honolulu: Research and Evaluation Unit, Alu Like, Inc., 1999).

59. Trask, 129.

60. Milner, 150.

61. Pierre Bowman, "Pig Farmer, 40 Await Eviction Action," *Honolulu Star Bulletin*, April 21, 1971, 60th ed., sec. 111, p. 2.

62. Trask, 126.

63. "Hawaii Kai, Honolulu, HI Demographics," https://www.areavibes.com/honolulu-hi/hawaii+kai/demographics/ (accessed April 8, 2022).

64. "Community Watershed Snapshot," Hawaii Conservation Alliance, July 9, 2018, https://www.hawaiiconservation.org/our-work/community-watershed-snapshot/.

65. "Restore & Conserve Malama Maunalua," Malama Maunalua (2016), https://www.malamamaunalua.org/.

Twelve

Bait and Switch

Tuna Wars, Territorial Seas, and the Eco-geography of the Eastern Tropical Pacific, 1931–1982

Kristin A. Wintersteen

> When a few lousy Ecuadorian gunboats rule the Pacific Ocean, well I don't know what this world is coming to.
>
> —*David Rico, skipper of the* A. K. Strom

On March 7, 1975, a brawl erupted aboard the *Neptune*, a San Diego–based tuna clipper that had been anchored in the harbor of Salinas, Ecuador, for more than four weeks. Crewmembers reported that thirty to forty Ecuadorian officials charged aboard without warrant, and in the ensuing confrontation the men were "beaten, kicked and jabbed with bayonets," while the officials "ransacked" the ship and made off with most of their belongings, money, food, and clothing. Ecuadorian authorities reported that the violence had broken out when drunken crewmembers returning from shore attacked Navy personnel.[1] The 950-ton *Neptune* was one of seven US tuna boats detained by authorities in early February for fishing in waters that Ecuador claimed as part of its territorial sea.[2] Denouncing the fishermen as "pirates," the Ecuadorian press noted that the latest group of seizures included several "repeat offenders."[3] At the time of the brawl, Captain John Burich and his crew were awaiting $65,000 in aid from the US government—in addition to a $281,860 fine already paid—in order to recuperate their confiscated cargo: 180 tons of yellowfin tuna (*Thunnus albacares*).[4] Such incidents were not uncommon: since the mid-1950s the United States had routinely provided aid to fishing vessels seized while operating in disputed zones off the Latin American coast.

The events of early 1975 underscored the ineffectiveness of diplomatic efforts to resolve the decades-long international conflict over control of coastal waters—a conflict in which Pacific tuna and the fleets that followed them played a prominent role. Between 1954 and 1974, Latin American nations seized at least 204 US fishing vessels, 70 percent of them tuna boats detained by Ecuador (103) or Peru (40).[5] In addition to the estimated $6,340,073 combined total cost of these indemnifications, the seven boats seized by Ecuador in February 1975 amounted to another $1,640,000.[6] The competition for Pacific tuna became particularly fierce after the mid-1950s as California fleets ventured farther from their home ports. At their peak during the 1950s and 1960s, US West Coast producers supplied an estimated 80 percent of the global tuna catch, much of it harvested south of the US border.[7] But US producers faced steep competition from cheaper imported products as Japan's industry rebounded after World War II. They also navigated increasingly thorny relationships with certain Latin American coastal nations seeking to define their role in the new international political economy by asserting control over the oceanic frontier.

Questions of sovereignty over coastal waters and the resources they contain were central to the emerging doctrine of international law in the decades following World War II, and the ongoing dispute over tuna between the United States and Pacific Latin American states reflected a broader polarization between long-distance fishing nations and those that sought to protect their fisheries from foreign encroachment. Significantly, the March 7 incident aboard the *Neptune* occurred within days of the opening of the third session of the Third United Nations Convention on the Law of the Sea (UNCLOS III), held from March 26 to May 10, 1975, in Geneva, Switzerland. The spatial definition of the territorial sea was one of the most crucial issues that delegates considered at the 1975 session. With significant known and unknown resources in the offshore realm, this concept signaled one of the most pressing foreign policy concerns of the twentieth century for coastal states, and tuna fisheries posed special challenges for those who hoped to exploit them.

While the long battle over territorial seas incited numerous diplomatic crises among international leaders, the tuna and their prey inhabited another world whose frontiers were governed instead by ecological relationships and oceanographic dynamics. Tuna swim in schools and live mostly on the open ocean in both the Atlantic and Pacific, favoring the warm waters of the tropics. Unlike coastal species, however, tuna migrate great distances to feed and spawn, traversing the maritime boundaries of multiple states as well as the high seas in their movements. Tuna therefore brought "special fishery problems" to the humans tasked with regulating them—problems that required an international cooperative approach in order to study, monitor, and protect the stocks from overfishing. As diplomatic leaders worked to establish a regime for the governance of the world oceans, the ongoing conflict between the United States and Pacific Latin American states loomed large.

One of the richest tuna fishing grounds of the mid-twentieth century, the Eastern Tropical Pacific was the subject of numerous research expeditions, especially during the 1950s and 1960s, as biologists and oceanographers sought to understand its fisheries, currents, and atmospheric variability.[8] "Eastern Tropical Pacific" broadly refers to the oceanographic region that stretches from the coast of the Americas, between Baja California (Mexico) and northern Peru, to about 150°W in the central Pacific Ocean.[9] In the southern part of this zone, off the Ecuadorian coast, converging currents around the Galápagos Islands mix with colder waters sweeping in from the Humboldt Current upwelling to the south, generating a uniquely rich marine environment.[10] Forage fish (anchovies, sardines, pilchards) and other prey species (small crustaceans) flourish in this area, often congregating in shallower coastal waters, attracting tuna, seabirds, and other predators.[11] Four of the world's seven most commercially exploited tuna species are abundant off the Latin American coast: yellowfin (*Thunnus albacares*), skipjack (*Katsuwonus pelamis*), bigeye (*Thunnus obesus*), and Pacific bluefin (*Thunnus orientalis*).[12] Along with the tuna-like Eastern Pacific bonito (*Sarda chiliensis*), these warm-water species have historically comprised an especially valuable portion of fisheries production in Ecuador and northern Peru, but are less abundant in Chilean waters.

Long-distance fishing fleets frequented the Eastern Tropical Pacific and northern Humboldt Current region to hunt the migrating tuna, sailing closer to shore to catch live bait. Before the introduction of purse seine vessels, they used the pole-and-line method, in which a "chummer" threw small fish such as sardines and anchovy into the water in order to attract the tuna, and then skilled anglers plucked the fish quickly from the water. "When they're 'biting good,' fish rain onto the deck with machine-gun rapidity," one journalist marveled.[13] This method required the storage of a large tank of up to 100 metric tons of water aboard the boat.[14] The Peruvian anchoveta (*Engraulis ringens*), which thrived in the cold upwelling region off Peru and Chile, was particularly suited for this purpose because of its ability to withstand crowding and survive for two to three months in the onboard tanks.[15] The need to fish for bait increased the likelihood of confrontations with Latin American patrols until the fleet upgraded to purse seine vessels during the 1960s.

Post–World War II advances in science, technology, and international commerce dramatically accelerated the pressure on marine ecosystems worldwide, driving a new phase of industrialization in the oceanic realm between 1950 and 1977.[16] Over the next several decades, US tuna boats would be able to detect (using sonar), harvest (with nylon nets and power winches), and preserve (in refrigerated holds) ever-greater quantities of fish to deliver to California processing plants. As early as 1946, some vessels were "equipped to stay at sea for as much as four or five months continuously, cruise as much as 9,000 miles without touching port, and bring back under refrigeration as much as 400 tons of tuna."[17] West Coast firms were poised to expand their operations along the coast of Latin America as nearby tuna populations became depleted.

Uneven access to these new technologies further deepened the divide between the United States and Latin American fishing nations. One writer noted that the "modern, steel-hulled tuna clipper, with its ability to catch yellowfin three miles offshore for consumption 5,000 miles away became a symbol of 'injustice.'"[18] Leaders in Peru, Ecuador, and Chile hoped to develop domestic industrial capacity and find new markets (both foreign and domestic) for their products, but US tariffs restricted their ability to export canned tuna to the United States.[19] The ongoing presence of foreign fleets in defiance of their claims catalyzed negative sentiment among policy makers and the public, setting a backdrop of discontent as reports of seizures filtered through the press.

This brief history of the Tuna Wars (1947–1982) is framed by the ecogeography of tuna in the Eastern Tropical Pacific, a space of competing geopolitical claims in which the range and distribution of species (both fish and humans) related to particular characteristics of the oceanic environment. The analysis focuses on the states with the highest stakes in the postwar battle for tuna, Ecuador and Peru, whose staunch defense of the 200-mile territorial sea persisted for three decades prior to the final drafting of the 1982 Law of the Sea. In a tripartite alliance with Chile, these states formally set forth the 200-mile concept in the 1952 Santiago Declaration, effectively laying claim to a combined total area of over 3,000,000 km^2 of ocean surface.[20] The declaration aimed to shift the balance of power away from the long-distance fleets of richer nations, as Ecuador, Peru, and Chile developed their own fishing industries. However, as geographer Elizabeth Havice has noted, "mobilities complicate the spatiality of sovereignty," especially in the fluid environment of the oceans, as states seek to define (or contest) the boundaries of territory and global capital.[21] Overlapping institutional and legal regimes, often at odds with one another, could not contain the dynamic and far-ranging tuna stocks or the humans who pursued them. Latin American states succeeded in shaping the emerging doctrine for ocean governance, but just as leaders reached a consensus on its terms, the industry began to shift toward new frontiers in the central and western Pacific.

A growing international consensus had long recognized that the competition for, and imminent depletion of, marine resources called for a formal regime of intergovernmental cooperation and management. In the northern oceans, whale populations had been mostly decimated before the end of the eighteenth century, but northern fleets continued to operate in the Antarctic and Southeast Pacific with few restrictions well into the twentieth century.[22] In 1931, twenty-six nations signed the first international agreement to conserve and regulate whaling.[23] The Convention for the Regulation of Whaling was an important step toward managing the exploitation of marine species whose eco-geographies do not fit easily into neatly bounded political territories. Yet the convention was also limited in scope and impact, foreshadowing the difficulties of crafting an international management regime for oceanic resources: "With constantly shifting ecosystems and market forces, participants soon recognized that it would be almost impossible to use a fixed treaty to control an industry that harvested wild animals."[24] The spatial dynamics of the whaling business presented specific challenges for the creation and enforcement of regulatory mechanisms at a global scale.

In 1935, just as the milestone whaling agreement came into effect, tuna fleets were beginning to operate on an increasingly global scale. California tuna boats ventured farther into the Pacific, often sailing to the Galápagos to fill their holds. But World War II interrupted global resource flows, allowing Latin American producers to temporarily fill the demand for canned fish and other products, such as liver oil, in US and other foreign markets.[25] It also allowed Chilean and Peruvian whaling companies to establish operations without competition from long-distance foreign fleets.[26] After the war ended, as freedom of movement returned to the high seas, rapidly increasing pressure on marine ecosystems pitted local, small-scale enterprises along the South American Pacific coast against the more capital-rich long-distance fleets of the United States, Japan, and the Soviet Union.

The postwar geopolitical restructuring of the oceanic realm revolved around national and commercial interests in gaining and maintaining access to the largely untapped resources of the sea and subsoil. On September 28, 1945, US president Harry Truman issued a set of proclamations that unilaterally laid claim to offshore resources, citing the need to protect the fisheries (Proclamation 2668), along with the subsoil and sea bed (Proclamation 2667), from "destructive exploitation."[27] The Truman Proclamations articulated a spatial definition of the coastal zone based on the physical geography of the ocean floor and the shallower seas above—the continental shelf—where the majority of marine species tend to congregate. This legal precedent soon unleashed a cascade of similar claims by other Latin American coastal states, including Mexico and Argentina. However, unlike the subsequent proclamations by Chile, Peru, and later Ecuador, these states did not specify a particular distance for their claims.[28] In 1947, Chilean president Gabriel González Videla proclaimed sovereignty over the continental shelf and its resources on or underneath the seabed up to a distance of 200 nautical miles from shore; Peruvian president José Luís Bustamante issued a similar proclamation shortly thereafter.[29] More states followed suit in subsequent years. These bold claims challenged the traditional regime of free access to underwater resources beyond a state's three-mile territorial zone.[30] The 200-mile concept, as it came to be known, became a major focal point of the Law of the Sea negotiations as participants worked to establish the rights and limits of sovereignty for coastal states.[31]

US officials faced a policy dilemma as they were forced to navigate the contradictory interests of fishing industrialists in different regions of the country. The West Coast tuna industry lobbied its government to uphold the "freedom of the seas" doctrine as their fleets laid plans for expansion, further spurred by the discovery of new fishing grounds sixty miles off the Peruvian coast that same year.[32] Having lent boats and bodies to the war effort, the industry in turn benefited from wartime investments in naval equipment and technology that enabled their fleets to more easily detect, pursue, harvest, and preserve the fish they caught.[33] At the same time, however, foreign fleets also encroached on US fishing grounds in the Northeast Pacific and Northwest Atlantic. If the United States acceded to the Latin American 200-mile claims, it weakened the basis for excluding foreign fishing off its own coasts.

Instead, the United States settled on a strategy that encouraged its tuna fleets to continue fishing in the disputed coastal zones.[34] In response, Latin Americans sent Navy patrols—often decommissioned US military vessels sold off after the war—to seize the boats they caught in violation, confiscate their cargo, and fine the companies. In 1948, US fisheries scientist Milton Lobell, then stationed in Chile, urged the State Department to negotiate a *modus vivendi* with the Latin Americans, but to no avail. By the end of the 1940s, US firms were having troubles with Mexico, Costa Rica, Panama, Colombia, and Ecuador over bait and tuna fishing in their coastal waters.[35]

Early attempts to establish a cooperative framework to address this problem took the form of bilateral and multilateral treaties. In 1949, the United States signed separate conventions with Mexico and Costa Rica, the latter of which led to the creation of the Inter-American Tropical Tuna Commission (IATTC) in 1950.[36] Its mission was to promote research and cooperation across the Eastern Tropical Pacific as well as to establish an international quota system to conserve tuna stocks. The IATTC, which was heavily influenced by California-based scientists and industry representatives in its early years, collected extensive data from fishermen throughout the region. The newly founded United Nations Food and Agriculture Organization (FAO) also worked to establish a Latin American fishery council, creating tension with leaders of the American Tunaboat Association (ATA), who suspected FAO officials of deliberately contravening its interests. Wilbert Chapman, US fisheries scientist and then–director of research for ATA, stated bluntly in a letter to FAO's director of fisheries, D. B. Finn: "It would seem that FAO is deliberately attempting to intrude itself into the field of operations of the [IATTC], and set up a competitive organization."[37] FAO leaders were insulted and distressed by these rumors and sought to correct them.[38] While all parties recognized the need to further scientific knowledge of the tuna fisheries, competing geopolitical and economic interests impeded their collaboration.

Latin American leaders were wary of both organizations, which they perceived to be dominated by foreign interests.[39] In 1952, Chile, Ecuador, and Peru jointly issued the Santiago Declaration, reaffirming sovereignty over the 200-mile maritime zone. A separate declaration pronounced the need for conservation of coastal resources.[40] They also created a new research and policy organization, the Permanent Commission of the Southeast Pacific (CPPS), to oversee the fisheries and advocate for its members' interests in the international sphere.[41] This act unified historic enemies, among whom boundary disputes had lingered since the nineteenth century, along the South American fisheries frontier.[42] It also strengthened their collective voice as they contended with the interests of the United States and long-distance fishing nations within the evolving regime for oceanic governance.

Gear improvements and flows of capital accelerated the race for fish during the 1950s and 1960s. Among its first technical assistance initiatives, the FAO sent scientists to evaluate and develop the fisheries in Chile, Peru, and Ecuador, with the express goal of promoting domestic consumption rather than "hasty ventures for industrialization of the fisheries resources for export."[43] Their reports highlighted the inadequacy of local infrastructure and scientific training. Some producers imported second-hand equipment, as fishing enterprises in the North Pacific, facing dwindling sardine and tuna stocks, transferred their boats and machinery to the South American Pacific coast.[44] During the 1950s, Van Camp Sea Foods built four canneries in Peru and one in Ecuador, along with others in Puerto Rico, Samoa, and on the African Atlantic coast.[45] The US tuna industry was expanding within and beyond the Pacific, creating a global network of canneries that increased its flexibility to respond to shifting politics and ecosystems.

As tuna fleets faced ever-greater hostility in their activities off the Latin American Pacific coast, the United States formalized its financial support of the industry by passing the Fishermen's Protective Act in August 1954.[46] The act provided for the reimbursement of fines levied by foreign governments on the basis of territorial sea claims. Although this allowed the owners of seized vessels to recuperate some losses, it did not cover the additional costs they often incurred for licenses or lost income, nor did it guarantee the prompt release of crews and boats. Ultimately, the measure was also ineffective in reducing seizures, which continued to occur at an accelerated rate in the forthcoming years.[47]

Greek-Argentine shipping magnate Aristotle Onassis soon launched a direct and very public challenge to Peru's 200-mile doctrine, and to the broader postwar project of international cooperation for the regulation of the high seas. Onassis commanded a major share of the global petroleum shipping industry with his fleet of nearly two dozen oil tankers. From 1950 to 1956, he also owned and operated the *Olympic Challenger* factory ship and its twelve "whale catchers." In November 1954, Onassis defiantly sailed the Hamburg-based whaling fleet into the waters off northern Peru.[48] Despite protestations by both Peru and Chile, whose leaders appealed unsuccessfully to Panama to stop their passage through the canal, his fleet entered the Peruvian 200-mile zone in order to hunt sperm whales on its way to the Antarctic for the Austral summer season.[49] Peruvian air and naval forces attacked and captured part of the fleet, while several vessels escaped back to Panama. An international debacle ensued over the millions in losses the seizure implied for the UK-based insurance companies backing Onassis's assets.[50] The affair also strained Peruvian relations with Panama, which protested the seizures before the United Nations and the Organization of American States.[51] Ultimately, Lloyd's of London paid a US$3 million fine to the Peruvian government on behalf of Onassis in order to release the five captured ships, which immediately departed for the Antarctic whaling season.[52] Although Onassis emerged with his fortune unscathed, the payment was an important victory for Peru in the battle for control over coastal resources. It also underscored how the seasonality of migration and hunting has shaped the temporal and geographical frame of marine resource conflicts.

When world leaders met in Rome for the 1955 International Technical Conference on the Living Resources of the Sea, tensions over coastal state rights and the territorial sea remained high. Unwilling to concede to the Latin American claims, the United States preferred to resolve individual fishery conflicts through multilateral organizations or bilateral agreements.[53] As tuna boat seizures continued with frequency, multiple meetings among representatives of the United States and the Chile-Ecuador-Peru (CEP) alliance failed to reach a formal compromise on the 200-mile question. Instead, the ATA negotiated privately with the Peruvian and Chilean governments. A 1956 Supreme Decree, drafted in cooperation with the ATA, stipulated that the Peruvian government would grant fishing licenses to foreign vessels for yellowfin and skipjack tuna, baitfish, and whales.[54] Similarly, ATA's manager Harold Cary participated directly in successful negotiations with the Chilean government in 1958.[55] Tuna migrations shifted that year with the arrival of a strong El Niño, a recurring climatic-oceanographic phenomenon caused by the periodic warming of waters in the Central and Eastern Pacific, making them more likely to be found in the abnormally warm waters off the Chilean coast.[56] Furthermore, US fishing company executives recognized that their fleets remained dependent on coastal waters for supplies: "[we must] treat a precise knife edge of policy between getting bait and fuel in a country and losing the right to fish tuna freely on the high seas."[57] This patchwork of isolated agreements allowed operations to continue in Chilean and Peruvian waters, but they did not create a permanent solution to the policy stalemate, especially with respect to Ecuador.

The conflict remained most fervent in Ecuadorian waters, even while technological and political-economic factors changed the landscape of coastal fisheries along the South American Pacific coast. US fleets began to replace the pole-and-line method with modernized purse seiners, which used a purse-like net to encircle entire schools of fish and a power winch to load them aboard, thus lessening their need to enter the shallower waters close to shore to catch bait. Peru and Chile were also entering a boom in the production of fishmeal (a high-protein commodity used in animal feeds), thus shifting their attention toward the anchoveta fishery instead of tuna processing during the 1960s.[58] On the other hand, Ecuador's crusade for the 200-mile territorial sea was the single most coherent foreign policy it pursued during its twentieth-century history.[59] Its representatives repeatedly argued for full jurisdictional control over the 200-mile zone, and Ecuadorian officials upheld this demand with frequent tuna boat seizures, with several of the incidents erupting in violence. On May 25, 1963, gunfire erupted when the Ecuadorian Navy approached a fleet thirteen miles off the coast, and all twenty-one tuna boats and their four hundred crewmembers ended up in Puerto Esmeralda, where they remained anchored for weeks awaiting diplomatic resolution.[60] Fishermen and dockworkers on the US West Coast organized in solidarity with their captive compatriots, blocking the importation of Ecuadorian goods into the port of Los Angeles.[61]

These events occurred amidst the unfolding of a national political crisis in Ecuador. On July 11, a military junta removed the left-wing president Carlos Julio Arosemena from power, and several months later the US government quietly reached an agreement with the new military regime whereby both parties would observe a twelve-mile coastal zone. This marked a historic, but temporary, reversal of Ecuador's adamant defense of the 200-mile doctrine, and a breach of the trilateral agreement with Peru and Chile. When news of the agreement became public in June 1965, it caused an uproar that contributed to the destabilization and toppling of the military junta soon thereafter.[62] In the Eastern Tropical Pacific, control of the coastal zone remained a hotly contested symbol of national sovereignty nearly two decades after Chile and Peru first pronounced their 1947 claims.

The ad hoc policy that US officials had generally pursued in dealing with specific fishing disputes prior to 1960 became more problematic as fisheries industrialization intensified throughout the world. Long-distance fleets increasingly clashed with coastal fisheries in the North Pacific, North Atlantic, and Gulf of Mexico, in addition to the Eastern Tropical Pacific, further necessitating a comprehensive international management regime. But in the United States, pressures from different sectors of its domestic fisheries complicated the ability to respond definitively to Latin American challenges in the Eastern Tropical Pacific.[63] US representatives had supported a twelve-mile contiguous fishery zone at the 1958 United Nations Convention on the Law of the Sea in Geneva (UNCLOS I), and Congress approved the twelve-mile zone in 1966.[64] These measures reflected the recognition that the international consensus was shifting away from the three-mile territorial sea and toward expanded jurisdiction, but, as one report noted, "For the United States to follow this trend is for it to choose between its coastal fisheries and its distant-water fisheries."[65] Although coastal fisheries provided 75–85 percent of total US fisheries production at the time, officials were reluctant to cede any ground in their position on the territorial sea by abandoning their support of the US tuna industry.

Departing from previous efforts to negotiate its own agreements with fishing nations, in the late 1960s, US officials sought to override the complicated framework of the existing regional organizations by proposing another, more comprehensive one for the management of Pacific fisheries. They envisioned a limited membership of states whose fleets fished in the South American Pacific—Chile, Peru, Ecuador, Japan, Canada, and the United States—in order to preserve "at least a semblance of balance" between the distant-water and coastal fishing nations.[66] Such an organization would replace the IATTC and CPPS and take into account the "special interests" of the Latin American coastal states, such as the Peruvian fishmeal industry, coordinating research and policy recommendations for the member states.[67] The evolving panoply of "more-than-territorial institutional innovations" emphasized the difficulty of developing a framework for international cooperation, particularly in the context of geopolitical inequalities and contested notions of sovereignty, to manage dynamic marine resources.[68]

Meanwhile, the State Department continued to encourage tuna fleets to fish off Latin American shores. A 1967 law revised the Fishermen's Protective Act, introducing an insurance program to compensate industrialists for the costs of boat seizures while expanding the coverage to include not only fines but also seized cargo and lost income.[69] Like its previous iteration, this law had no impact on the rate of boat seizures, especially by Ecuador and Peru. In an interview with the *Los Angeles Times*, skipper Roland Virissimo explained that he preferred capture by the Peruvians because they were "'more on the ball'" than Ecuadorians, while the latter "'don't know who is the boss. You might have to wait around in one of their ports for one week or a month until they decide what to do with you.'"[70] The familiar cat-and-mouse routine that crewmembers endured was no doubt tiresome, trapped as they were between the conflicting foreign policy interests of their home country while subject to the whims of local officials.

Not long thereafter, the Peruvian military regime of General Juan Velasco Alvarado (1968–1975) reignited tensions with the United States. Velasco reaffirmed the 200-mile doctrine. He also expropriated US-owned corporations, including the International Petroleum Company (owned by Standard Oil) and, later, five fishmeal-producing firms. In May 1969, the US State Department announced the suspension of military equipment sales to Peru in retaliation for a series of recent boat seizures.[71] Peruvian premier Ernesto Montagne then expelled the US Army, Navy, and Air Force missions and rejected the visit of Governor Nelson Rockefeller, special envoy of President Richard Nixon, on his inter-American "fact-finding tour."[72] Although the military ban lasted only a few months, the affair further strained relations with Peru.

Clashes with Ecuador also continued at sea.[73] Delegations from the United States, Chile, Ecuador, and Peru met in April 1968 (Santiago) and August 1969 (Buenos Aires) to deliberate the establishment of a new regional fishery institution and vessel registration system, but talks were inconclusive yet again. While US representatives "viewed the seizure problem as a real obstacle to greater cooperation," Latin American leaders questioned the connotation of "conservation" in the draft US proposal and were hesitant to commit to any program that might erode their rights without providing any direct economic benefit, such as the elimination of tariffs on imported tuna. Furthermore, the parties could not agree upon either the eco-geographical basis—whether spatial zones or species classifications—for a quota system or the institutional mechanism for managing its implementation in a manner that would satisfactorily distribute the balance of power.[74]

High-level diplomatic conversations were far removed from the lived experiences of antagonism among the fishing crews and sailors at sea. Solidarity among the longshoremen and fishermen of southern California ensured that the Tuna Wars also impacted commerce along the Pacific American coast more broadly. California labor unions pressured the US government to defend the tuna fishermen in the ongoing international dispute. On behalf of the Fishermen's Local 33 of Southland–San Diego, union representative John J. Royal drafted a letter to President Nixon, threatening "an all out embargo by American labor against the commodities and products coming from these countries . . . [in order to] avoid the inevitable bloodshed and probable loss of lives" that he believed the ongoing conflict would bring.[75] In 1971, five hundred picketers at the Los Angeles port of Long Beach, California, prevented the unloading of a shipment of Standard Fruit bananas from Ecuador.[76]

Ecological fluctuations further exacerbated these conflicts when the 1972 El Niño warmed coastal waters, decimating the anchoveta populations that thrived in the normally cool Humboldt Current. In Peru, amidst the economic crisis that followed the 1973 anchoveta fishery collapse, an amendment to the US Fishermen's Protective Act sparked angry reactions in Lima, where "demonstrators hurled a dead fish into the embassy compound."[77] Relations soured further when Velasco nationalized the Peruvian fishing industry and expropriated the assets of Gold Kist, Cargill, StarKist, Van Camp (Ralston Purina), Gloucester Peruvian (General Mills), and Pesqueri Meilian (International Proteins Corporation).[78] The $150 million in compensation later paid did little to ameliorate the bitter aftermath of these events.[79] The fortunes of Peruvian and Chilean fishmeal industrialists remained dim during the mid-1970s, but tuna landings were robust following the El Niño event.

Tuna boat seizures thus continued off the coast of Ecuador, where rumors swirled of a power struggle within the armed forces.[80] When the crews of the *Neptune* and six other tuna seiners found themselves detained in the port of Salinas, the antagonistic relationship between the two nations had become entrenched: "I want the United States Government to put machine guns on my boat," skipper David Rico ranted to a *New York Times* journalist, "and I want missiles—same as they're giving to those people in the Middle East—to repel these pirates."[81] One California resident echoed Rico's sentiment, complaining in a letter to the *Los Angeles Times* that "our government appears to be too chicken to deal appropriately with the pirates of Ecuador."[82] But as international deliberations over ocean governance continued to hash out the spatial and legal parameters for fishing rights and coastal state jurisdiction, Ecuador's position became increasingly isolated. At the third session of the United Nations Convention on the Law of the Sea (UNCLOS III), held in Geneva from March 17 to May 9, 1975, Ecuador once again submitted a draft proposal for a 200-mile territorial sea, but it did not generate much support, even from Chile and Peru.[83] Instead, negotiations moved toward the establishment of a 200-mile "economic zone," with the precise nature of coastal states' rights still to be determined, and special conditions for highly migratory species such as tuna.[84]

Important questions of territory, sovereignty, and cooperative management of resources nonetheless remained unresolved at the international level nearly thirty years after the 1947 Chilean and Peruvian proclamations. On January 28, 1976, the US Senate voted to establish a 200-mile fishery "conservation zone" contiguous to the territorial sea.[85] Through eleven sessions held over a nine-year period, the Third United Nations Convention on the Law of the Sea (UNCLOS III, 1973–1982) finally defined the spatial and jurisdictional characteristics of the territorial sea (12 miles), within which coastal states exercise full sovereignty over the air, sea, and subsea, and the Exclusive Economic Zone (200 miles), within which they have sovereign rights to living and non-living resources but must allow freedom of navigation.[86] In 1995, a separate agreement enacted special provisions for highly migratory species whose stocks straddle or traverse the boundaries of Exclusive Economic Zones and high seas.[87] While the conclusion of UNCLOS III marked a de facto end to the most uproarious period of gunboat diplomacy over tuna in the Eastern Tropical Pacific, it did not satisfy all demands: as of this writing, Peru has not ratified either agreement, nor has the United States ratified the 1982 Law of the Sea.[88]

The Tuna Wars of the Eastern Tropical Pacific (1947–1982) erupted in fits and starts, as fishing fleets and coastal states staked claims to offshore resources during the post–World War II period of rapid technological modernization. Both new and existing institutional and geospatial mechanisms proved inadequate in addressing this conflict over resources whose mobility and variability complicated governance frameworks. Human-designed boundaries of territory and property align poorly with the fluid ecological and biological geographies of the ocean. The long and decentered history of this conflict was shaped not only by the encounters between fishermen and sailors or policy makers and entrepreneurs with contradictory interests, but even more fundamentally, by interspecies relationships and the dynamics of their spatial distribution over time.

At the heart of these repeated clashes also lay contentious questions of national sovereignty and rights of access to, and control over, the living resources of the sea in the context of an emerging postwar and postcolonial international order. Despite their sometimes divergent diplomatic priorities, the steadfast insistence of Ecuador, Peru, and Chile on sovereignty over the

200-mile coastal zone was an important challenge to the international status quo, the "freedom of the seas" doctrine. In its final iteration, the Exclusive Economic Zone represented a compromise of the most radical demands for territorial control, but it greatly expanded the ability of coastal states to retain some autonomy over their seas and the resources within them.

During the 1980s, the geopolitics of tuna fisheries shifted toward a new center of production in the central and western Pacific, where the island nations further challenged these newly defined principles as they sought recognition for their resource and territorial claims.[89] The international legal terrain for these claims was vastly different than it had been in the 1940s and 1950s, even if the post-1982 regime for ocean governance was far from settled. With its eleven sessions and the dozens of international meetings that came before, UNCLOS III was a process that in itself held value beyond the diplomatic resolutions that it set forth. For all their ceremonious formalities, these international forums also provided a stage upon which Latin American delegates performed their discontent with the inequalities among nation-states at the global scale. In so doing, they effectively moved the battlefield from the open ocean to the conference room.

Notes

Epigraph: Quoted in Jonathan Kandell, "U.S. Tuna Men Held in Ecuador Are Bitter and in Fighting Mood," *New York Times* (February 18, 1975), 2.

1. "U.S. Crewmen Face Trial in Ecuador," *New York Times* (March 8, 1975), 6.
2. Everett R. Holles, "Tuna Fleet Asks U.S. Aid off Ecuador," *New York Times* (March 9, 1975), 20.
3. "Marina capturó al buque de bandera panameña The Dog," *El Comercio [Quito]* (February 2, 1975), 1; "Son reincidentes los dos últimos buques piratas capturadas," *El Comercio [Quito]* (February 4, 1975), 1.
4. Everett R. Holles, "Tuna Fleet Asks U.S. Aid off Ecuador," *New York Times* (February 9, 1975), 20.
5. Another fifty claims pertained to seizures by Mexico, but it is unknown how many of these pertained to tuna boats in the Pacific versus shrimp boats in the Gulf of Mexico.
6. This total cost includes an estimated $5,199,944 for Ecuador and $893,352 for Peru. A definitive accounting of all seizures is not presently available, but the numbers cited here are the most comprehensive tally I have found for the total period. They include only those claims submitted to the US Treasury as of February 1975. See Theodor Meron, "The Fishermen's Protective Act: A Case Study in Contemporary Legal Strategy of the United States," *American Journal of International Law* 69, no. 2 (1975): 308n79.
7. J. Majkowski, *Global Fishery Resources of Tuna and Tuna-like Species*, FAO Fisheries Technical Paper No. 483 (2007), 11.
8. Significantly, many of the expeditions during this period were affiliated with the Scripps Institution of Oceanography in La Jolla, California. William S. Kessler, "The

Circulation of the Eastern Tropical Pacific: A Review," *Progress in Oceanography* 69, nos. 2–4 (2006): 181.

9. The term "Eastern Tropical Pacific" is a geographical description that was favored during the period examined in this study, although it is less commonly used now. There is no single biogeographical classification system for the global marine environment. According to more recent literature, the area discussed in this study corresponds to the biogeographic realms of the "Tropical East Pacific," "Galapagos," and the northern portion of the "Warm Temperate Southeastern Pacific." See Mark D. Spalding et al., "Marine Ecoregions of the World: A Bioregionalization of Coastal and Shelf Areas," *BioScience* 57, no. 7 (2007): Box 1, https://doi.org/10.1641/b570707.

10. FishBase.org defines the Peru-Galapagos waters as the area between latitudes 2°N–17°S and longitudes of 95°W–70°W (http://www.fishbase.org/trophiceco/EcosysRef.php?ve_code=130&sp=) overlapping with the Humboldt Current marine ecosystem (4° S–54°S and 90°W–71°W) (http://www.fishbase.us/TrophicEco/EcosysRef.php?ve_code=237&sp=).

11. P. Lehodey, "The Pelagic Ecosystem of the Tropical Pacific Ocean: Dynamic Spatial Modeling and Biological Consequences of ENSO," *Progress in Oceanography* 49 (2001): 446; "The Northern Humboldt Current System: Brief History, Present Status, and a View Towards the Future," *Progress in Oceanography* 79 (2008): 95.

12. Their range varies between 45–63°N and 43–47°S, depending on the species. Two others could occur in this region but are less abundant: albacore (*Thunnus alalunga*) and Southern bluefin (*Thunnus maccoyii*). The remaining species is the Atlantic bluefin (*Thunnus thynnus*), not found in Pacific waters.

13. Lewis B. Hochstrasser, "Tuna by the Ton," *Wall Street Journal* (July 24, 1947), 1.

14. "Adventures of the Deep Sea Fishermen," *Popular Mechanics* (December 1941): 28.

15. William C. Richardson, "Fishermen of San Diego," *San Diego Historical Society Quarterly* 27, no. 4 (1981): 24.

16. David H. Cushing, *The Provident Sea* (New York: Cambridge University Press, 1988).

17. Chapman, "Tuna in the Mandated Islands," 318.

18. Wesley Marx, "The Eastern Tropical Pacific: Fishing for Cooperation," *Bulletin of the Atomic Scientists* 24, no. 9 (1968): 18–22.

19. Mary Carmel Finley, *All the Fish in the Sea* (Chicago: University of Chicago Press, 2011), 122; Kristin Wintersteen, *The Fishmeal Revolution* (Oakland: University of California Press, 2021), 50, 79.

20. The Sea Around Us project posts estimated surface area for countries' EEZs in km^2. The areas are as follows: Chile 2,000,254 km^2; Peru 906,454 km^2; and Ecuador 236,597 km^2. Sea Around Us Project, "EEZ Waters of Chile," http://www.seaaroundus.org/eez/152_87.aspx; "EEZ Waters of Peru," http://www.seaaroundus.org/eez/604.aspx; and "EEZ Waters of Ecuador," http://www.seaaroundus.org/eez/218.aspx.

21. Elizabeth Havice, "Unsettled Sovereignty and the Sea: Mobilities and More-Than-Territorial Configurations of State Power," *Annals of the American Association of Geographers* 108, no. 5 (2018): 1280–1297.

22. For a comprehensive historical treatment of the international diplomatic efforts to regulate whales, see Kurkpatrick Dorsey, *Whales and Nations* (Seattle: University of Washington Press, 2016). On northern whaling, see also Tables 16.1–16.2 in John F. Richards, "Whales and Walruses in the Northern Oceans," in *The Unending Frontier: An*

Environmental History of the Early Modern World (Berkeley: University of California Press, 2003), 613–615.

23. The convention entered into force on January 16, 1935. See "Convention for the Regulation of Whaling," League of Nations T.S. No. 880, 49 Stat. 3079, September 24, 1931, https://www.loc.gov/law/help/us-treaties/bevans/m-ust000003-0026.pdf. See also Ann L. Hollick, "The Origins of 200-Mile Offshore Zones," *American Journal of International Law* 71, no. 3 (1977): 496n11.

24. Dorsey, *Whales and Nations,* 44.

25. Arthur F. McEvoy, *The Fishermen's Problem* (New York: Cambridge University Press, 1986), 142.

26. Ann L. Hollick, "The Roots of U.S. Fisheries Policy," *Ocean Development and International Law Journal* 5, no. 1 (1978): 74; Luis Salvo González, *Historia de la industria pesquera en la región del bio* (Santiago, Chile: ASIPES, 2000).

27. The Fisheries Proclamation (2668) cited the "urgent need to protect coastal fishery resources from destructive exploitation . . ." and to protect potential future development ". . . in those areas of the high seas contiguous to the coasts of the United States wherein fishing activities have been or in the future may be developed and maintained on a substantial scale." The act also conceded the right of other coastal states to establish "conservation zones" with similar aims, "provided that corresponding recognition is given" to any US fishing interests "which may exist in such areas." See "Proclamation 2667: Policy of the United States with Respect to the Natural Resources of the Subsoil and Sea Bed of the Continental Shelf," full text at John T. Woolley and Gerhard Peters, *The American Presidency Project*, http://www.presidency.ucsb.edu/ws/?pid=12332; "Proclamation 2668: Policy of the United States with Respect to Coastal Fisheries in Certain Areas of the High Seas," full text at John T. Woolley and Gerhard Peters, *The American Presidency Project*, http://www.presidency.ucsb.edu/ws/?pid=12332.

28. Hollick distinguishes between two sets of Latin American claims; the first (Mexico, Panama, Argentina, Costa Rica) "established varying degrees of national jurisdiction or sovereignty over the continental shelf and the superadjacent waters without reference to a particular distance criterion," while the second (Chile, Peru, Ecuador) "incorporated under national sovereignty the continental shelf and offshore waters to a distance of 200 miles." A. Hollick, *U.S. Foreign Policy and the Law of the Sea* (Princeton, NJ: Princeton University Press, 1981), 68.

29. Videla cited the US, Mexican, and Argentine proclamations of the previous two years as precedent for the Chilean declaration. Goberno de Chile, Declaración de la Soberanía Marítima, 6/23/1947; Gobierno del Perú, D.S. 781, 8/1/1947.

30. The "freedom of the seas" doctrine had effectively governed the global oceanic realm since the seventeenth century. Developed by Dutch jurist Hugo Grotius amidst competition among European powers for free access to trade in the East Indies, the "territorial sea" extended out to a distance of three nautical miles from the coast, within which the state enjoyed full sovereignty and jurisdiction. For a recent, annotated translation of the original work, see Hugo Grotius, *Mare Liberum, Original Latin Text and Modern English Translation*, ed. R. Feenstra (Boston: Brill, 2009).

31. Much of this debate centered around the question of how to spatially define sovereign jurisdiction (territorial sea) vs. right of access or control of resources (economic zone) within coastal waters.

32. Letter to Ann Hollick from August Felando, American Tunaboat Association (February 22, 1977), cited in Hollick, 82n96.

33. McEvoy, *The Fishermen's Problem*, 153, 203.

34. This also continued the fleets' historic presence in these fisheries, a legal precedent that the United States could later cite as a basis for maintaining access rights as the international governance regime evolved. Meron, "The Fishermen's Protective Act," in McEvoy, 302.

35. Milton Lobell to Chapman, October 21, 1948, UW Special Collections, Chapman papers, Box 12, Folder 26; Marx, "The Eastern Tropical Pacific," 20.

36. David Loring, "The United States–Peruvian 'Fisheries' Dispute," *Stanford Law Review* 23, no. 3 (1971): 437n190.

37. Letter from W. M. Chapman to D. B. Finn, 10/14/1952, 2; RG 14FI158, Folder ATA Affair, FAO-Rome.

38. Letter from B. F. Osorio-Tafall to D. B. Finn, 3/13/1953, 2; RG 14FI158, Folder ATA Affair, FAO-Rome.

39. By 1968 neither Peru nor Chile belonged to the IATTC, while Ecuador resigned in 1961; Panama joined in 1953; Mexico in 1964, Canada in 1968, and Japan in 1970 (Loring, "The United States–Peruvian 'Fisheries' Dispute," 437n90). See also Marx, "The Eastern Tropical Pacific," 20.

40. Gobiernos de Chile, Ecuador, y Perú, Declaración conjunta relativa a los problemas de la pesquería en el Pacífico Sur, 8/18/1952, U.N. treaty no. 14757; Declaración sobre la Zona Marítima, 8/18/1952, U.N. treaty no. 14758 (1976), https://treaties.un.org/doc/Publication/UNTS/Volume%201006/v1006.pdf.

41. Gobiernos de Chile, Ecuador, y Perú, Organación de la Comisión Permanente de la Conferencia sobre Explotación y Conservación de las Riquezas Marítimas del Pacífico Sur, U.N. treaty no. 14759 (1976), https://treaties.un.org/doc/Publication/UNTS/Volume%201006/v1006.pdf.

42. The War of the Pacific (1879–1884) between Chile and the allied Peru and Bolivia claimed a large portion of the nitrate-rich Atacama Desert for Chile; Peru and Ecuador disputed their territorial boundary continuously since gaining independence from Spain, with the eruption of armed conflict occurring most in the 1941 Ecuadorian-Peruvian War.

43. Letter from B. F. Osorio-Tafall to D. B. Finn, March 13, 1953, 2; RG 14FI158, Folder ATA Affair, FAO-Rome.

44. August John Felando, interview by Robert G. Wright, San Diego Historical Society Oral History Program, September 9, 1995, http://gondolin.ucsd.edu/sio/ceo-sdhsoh/OH_felando.html.

45. "Corporations: Tuna Turnaround," *Time*, January 18, 1963, http://www.time.com/time/magazine/article/0,9171,874707,00.html. By 1948, the company had operations in at least eleven locations along the Pacific American coast (four in Latin America), between Astoria, OR, and Lima, Peru (*Pacific Fisherman's News* 4, no. 12 [June 14, 1948]), in UW Special Collections, Chapman Papers, Box 12, Folder 1.

46. Fishermen's Protective Act of 1954 (H.R. 9270, 83rd Congress).

47. Meron, "The Fishermen's Protective Act," 29192.

48. The fleet's hunting activities were also in flagrant violation of the 1946 International Convention for the Regulation of Whaling, and Onassis was later denounced by the workers who had witnessed the incident. The fleet was registered under flags-of-convenience in Panama and Honduras but managed by personnel in Hamburg and staffed by six hundred German crewmembers. See Klaus Barthelmess, "Die Gegner der 'Olympic Challenger': Wie amerikanische Geheimdienste, Norweger und Deutsche das Walfangabenteuer des Aristoteles Onassis beendeten," *Polarforschung* 79, no. 3 (2009): 155–176; "Onassis Tanker

Launched," *New York Times* (March 26, 1953), 63; "Biggest Oil Tanker Launched," *New York Times* (July 26, 1953), 12.

49. "Panama Bars Requests," *New York Times* (September 18, 1954), 3; "Limit Rejected by Panama," *New York Times* (November 17, 1954), 15.

50. At the same time, Onassis was embroiled in a separate dispute over shipping contracts with Saudi Arabian oil producers. "Peru's Navy Seizes 5 Onassis Whalers," *New York Times* (November 17, 1954), 1; "3 Onassis Ships Return," *New York Times* (December 19, 1954), 5; "British Notice to Peru," *Manchester Guardian* (November 20, 1954), 1; "Claim against Mr. Onassis," *Manchester Guardian* (November 20, 1954), 5.

51. "Onassis Whaling Upheld," *New York Times* (November 23, 1954), 9.

52. "Lloyd's Due to Pay Peru Fine on Onassis," *New York Times* (December 12, 1954), S9; "Onassis pays $3M Fine to Peru," *Washington Post and Times Herald* (December 14, 1954), 4.

53. Finley, *All the Fish in the Sea,* 138, 147.

54. Loring, "The United States–Peruvian 'Fisheries' Dispute," 443n204, citing Supreme Decree, January 05, 1956, "Regulations to Grant Fishing Permits to Foreign Ships to Fish in Jurisdictional Waters of Peru," as discussed in Hearings on Miscellaneous Fishery Legislation before the Subcomm. on Merchant Marine and Fisheries of the Senate Comm. on Commerce, 90th Cong., 1st Sess. 97–99 (1967).

55. Harold F. Cary to Luis Melo Lecaros, Letter, January 2, 1958, 2; Chapman Papers, UW Special Collections, Box 50, Folder 9; "U.S. Tuna Ships in Pact: Fishing Group to Pay Chilean Fees, Avoiding Fines," *New York Times* (January 4, 1958), 30.

56. The 1957–58 El Niño event is rated "Strong" according to its Oceanic Niño Index (ONI) value of 1.7 at its peak; see NOAA National Weather Service Climate Prediction Center, *Cold and Warm Episodes by Season (1950–2016)*, http://www.cpc.noaa.gov/products/analysis_monitoring/ensostuff/ensoyears.shtml (updated November 4, 2015).

57. American Tunaboat Association to the Resources Committee, Memo re: Chilean decree on bait and tuna fishing in territorial waters, April 2, 1959, 1; UW Special Collections, Chapman Papers, Box 50, Folder 9.

58. On the history of this industry, see Wintersteen, *The Fishmeal Revolution*.

59. Guillaume Long, "Ecuador en el mar: Materialismo, seguridad e identidad en la política exterior de un país periférico," in *Ecuador: Relaciones exteriores a la luz del bicentenario*, ed. B. Zepeda (Quito: FLACSO, 2010), 331–364.

60. "U.S. Bids Ecuador Free Tuna Boats," *New York Times* (May 30, 1963), 3; "21 U.S. Tuna Vessels in Custody: Ecuador," *Chicago Tribune* (May 30, 1963), 9; "Columbia [*sic*] Rejects U.S. Plea to Free Tuna Boats," *Washington Post* (June 1, 1963), A8; "Plea to Free Ships Spurned by Ecuador," *Chicago Tribune* (June 1, 1963), SA7.

61. "Dockers Honor Picket Line in Ecuador Row," *Los Angeles Times* (June 4, 1963), 23.

62. Long, "Ecuador en el mar"; Loring, "The United States-Peruvian 'Fisheries' Dispute"; Memo from Clarence F. Pautzke to Sec. of Interior, "Fishery Modus Vivendi with Ecuador" (January 14, 1965), Record group 22; National Archives at College Park, College Park, MD.

63. U.S. Bureau of Commercial Fisheries, Foreign Operational Briefs 67–5 (March 7, 1967), 1; Record Group 22; National Archives at College Park, College Park, MD.

64. Act to establish a contiguous fishery zone beyond the territorial sea of the United States, S. 2218, 89th Cong. 2nd Sess. (1966), https://www.govinfo.gov/content/pkg/STATUTE-80/pdf/STATUTE-80-Pg908.pdf.

65. U.S. Bureau of Commercial Fisheries, Foreign Operational Briefs 67–5 (March 7, 1967), 3; Record Group 22; National Archives at College Park, College Park, MD.

66. On the other hand, officials noted that a larger membership "offers the possibility of influencing or even overturning the extreme juridical positions of Chile, Ecuador, and Peru." U.S. Bureau of Commercial Fisheries, 5–6.

67. U.S. Bureau of Commercial Fisheries, 4.

68. Havice, "Unsettled Sovereignty," 1280.

69. Fishermen's Protective Act of 1967, H.R. Rep. No. 625, 90th Cong., 1st Sess. (1967), http://legcounsel.house.gov/Comps/Fishermen's%20Protective%20Act%20Of%20 1967.pdf.

70. "Fishing Skipper Hits at Gunboat 'Pirates,'" *Los Angeles Times* (August 27, 1967), B3.

71. "Peru Denied Arms after Seizing Boat," *Washington Post* (May 18, 1969), 18.

72. "U.S. Military Aides Expelled by Peru; Rockefeller Barred," *New York Times* (May 24, 1969), 1.

73. "U.S. Trawlers Are Released by Ecuador," *Chicago Tribune* (June 21, 1969), N2; "U.S. Tuna Boats Provocative, Ecuador Says," *Chicago Tribune* (June 22, 1969), A8.

74. David F. Belnap, "U.S.-Latin Talks Seek to Resolve 'Tuna War,'" *Los Angeles Times* (August 4, 1969), A5; William M. Terry, Notes on Santiago Talks with Chile, Ecuador, and Peru (April 17, 1968–April 19, 1968); Record Group 22; National Archives at College Park, College Park, MD.

75. Letter from John J. Royal to President Richard M. Nixon (March 3, 1970); cited in Loring, "The United States-Peruvian 'Fisheries' Dispute," 427n167.

76. Jerry Ruhlow, "Pickets Prevent Unloading of Ecuador Fruit at Long Beach," *Los Angeles Times* (March 16, 1971), 3.

77. David F. Belnap, "Latin 'Tuna War' Escalates," *Washington Post* (April 12, 1973), K7.

78. *New York Times* (May 9, 1973), 3.

79. H. J. Maidenberg, "Peruvian Govt Agrees to Pay $76 Million in Compensation," *New York Times* (February 20, 1974), 11.

80. "Tuna Boat Seizures Laid to Power Struggle in Ecuador," *Los Angeles Times* (February 14, 1975), A29.

81. Kandell, "U.S. Tuna Men Held."

82. "Tuna seizures" [Letters to the editor], *Los Angeles Times* (February 19, 1975), D4.

83. U.S. Classified Delegation Report on the Third United Nations Conference on the Law of the Sea (Geneva, Switzerland, March 17, 1975–May 9, 1975), 14; RG 43; National Archives at College Park, College Park, MD.

84. U.S. Classified Delegation Report, 44–49; RG 43; National Archives at College Park, College Park, MD.

85. The act contained an important stipulation to the reciprocity of the 200-mile zone to be observed in foreign waters, that such a zone "not be recognized whenever a nation fails to consider traditional fishing activity therein of the United States or fails to recognize that highly migratory species are to be managed by international agreements." Fishery Conservation and Management Act of 1976, H.R. 200, 94th Cong. (1976), https://www .congress.gov/94/statute/STATUTE-90/STATUTE-90-Pg331.pdf.

86. See particularly Part II on the territorial sea and Part V on the EEZ. United Nations Convention on the Law of the Sea, December 10, 1982, http://www.un.org/Depts /los/convention_agreements/texts/unclos/UNCLOS-TOC.htm.

87. The United States, Ecuador, and Chile have all ratified. United Nations, Agreement relating to the Conservation and Management of Straddling Fish Stocks and Highly Migratory Fish Stocks, September 8, 1995, https://www.un.org/ga/search/view_doc.asp?symbol=A/CONF.164/37&Lang=E.

88. Chile ratified in 1997 and Ecuador in 2012. United Nations, "Chronological Lists of Ratifications of, Accessions and Successions to the Convention and the Related Agreements," updated September 3, 2020, https://www.un.org/Depts/los/reference_files/chronological_lists_of_ratifications.htm#The%20United%20Nations%20Convention%20on%20the%20Law%20of%20the%20Sea.

89. Havice, "Unsettled Sovereignty and the Sea."

THIRTEEN

Wintering in the South
Birds, Place, and Flows

Emily O'Gorman

The wind whipped up small waves in the brackish lagoon. The warm day had turned cool, and the birds huddled into their ruffled feathers. I was standing on the edge of Fivebough Swamp, a small wetland in southern New South Wales. I could see the many birds on the water and toward the opposite shore, including what looked like a small group of Latham's snipe (*Gallinago hardwickii*). One, then another of these birds took flight. Soon the whole group was in the air, flying northward along a low mountain rage.

I was visiting Fivebough Swamp as part of a research project on the environmental history of wetlands in the Murray-Darling Basin, a large river system that covers a substantial portion of eastern Australia. Wetlands in the Murray-Darling Basin are places rich in biodiversity, cultural meanings, and contestation. This broader project aims to examine the changing and diverse uses, knowledge, and values that have shaped these places (for instance, of loggers, hunters, local Aboriginal people, governments, farmers, and ecologists) in order to develop a history of wetlands as "social-natural landscapes."[1] One aspect of this is to examine how ideas in fauna and flora protection and eradication, and wetland ecology and conservation at a variety of scales, have influenced, and been influenced by, particular wetlands sites. Another is to show the roles of plants and animals in shaping a variety of uses and values associated with wetlands by humans at these different scales. Plants and animals have also played important roles in shaping wetlands as well as possibilities for life; think, for example, of the role of mosquitoes as vectors of human and animal disease as well as the impacts of human responses to them, such as chemical treatments and drainage of wetlands. This set of approaches brings this work into conversation with multispecies studies, animal studies, and biology, and into close dialogue with work in the environmental humanities.[2]

Latham's snipe are today recognized by ornithologists as trans-equatorial migrants, although this has not always been the case. This chapter traces shifting understandings of Latham's snipe in Australia in the nineteenth and twentieth centuries and in so doing repositions environmental histories of wetlands and cultural histories of birds in Australia within wider Pacific ecologies and cultures. Historical scholarship on wetlands in Australia has largely focused on single sites or literary understandings, while cultural histories of birds in Australia have largely focused on the challenges faced by ornithologists in moving beyond British and European models of bird behavior, especially seasonal migration, and the gradual realization by ornithologists that most Australian birds were influenced more by rainfall than seasons; that is, that many birds were nomadic rather than migratory.[3] This scholarship has been important in showing the incompatibility of European understandings of animals, plants, climates, and environments, with many of those in Australia. Recent cultural histories of birds have thus focused more on species that tended not to move too far from Australia while migratory birds have been comparatively overlooked. What then of the birds that did undertake seasonal intercontinental migrations? A focus on these birds, such as Latham's snipe, can help to illuminate the diverse and wide-ranging bio-cultural networks that have connected Australia with the wider East Asian and Asia-Pacific regions and beyond and shed new light on histories of wetlands in Australia and beyond. These migrations, and how ornithologists understood them, shaped and connected distant places with a range of mixed, and important, political and ecological consequences in Australia and elsewhere.

Latham's snipe, along with many other birds around the world, have connected wetlands through their bodies as they move between them (see figure 13.1). Attentiveness to their agency can help us to think in new ways about the diverse and wide-ranging bio-cultural relationships and networks that have shaped these places and their connections, and illuminate the sometimes profound consequences of how these have been valued and understood.[4] We need to approach Pacific migrant ecologies as more than human and recognize the journeys non-humans take themselves on as well as the ones humans send them on.[5]

In 1891, a columnist in *The Australasian* newspaper revealed some exciting ornithological news about Latham's snipe to readers:

> With the first moonlight of the month [of September], sometimes in a single night, the snipe appear on a solitary marsh, but where did they wing their flight from? What was their starting point? And by what route came they? These are indeed exceedingly interesting questions that have only

Figure 13.1. A depiction of a Latham's snipe (left) with an Australian painted snipe (right). Image credit: Gracius J. Broinowski, *The Birds of Australia* (Melbourne: C. Stuart & Co., 1890–1891).

> lately been solved. It was thought that these remarkable birds came from the unknown far north-west interior . . . But, wonderful as it may seem, Henry Seebohm, in his recent work, *Birds of the Japanese Empire* [1890], tells us that: "Latham's the Australian snipe is a common visitor to the Japanese islands probably breeding in the Yezzo, and certainly doing so on the mountains of Southern Japan."[6]

This same book hypothesized the migration route of the snipe, writing that "it is probably confined to Japan for the breeding-season, but in autumn passes the Philippine Islands and the coasts of China to winter in Australia."[7] This article was likely written by Victorian ornithologist and avid egg collector A. J. Campbell. After many years of trying to find where the birds bred, Campbell could finally claim to have done so in 1898, when his hired egg collector found the birds' nests at the foot of Fujiyama (Mount Fuji) and sent some samples back to Australia.[8]

In the colonial period, many ornithologists in Australia focused on establishing the breeding places and migratory routes of birds, often between hemispheres. Historian Libby Robin has argued that this was because the migrations of the birds resonated with their own travel or migration to Australia from Britain and Europe.[9] The language of "discovery" of birds and eggs is prevalent in these ornithologists' writings and in many ways was akin to writings by colonial explorers. This was also reflected in the "discovery" and the honor of naming a new species. English ornithologist John Latham had made the first taxonomic classification of *Scolopax australis*, in 1801, giving it the common name of "New Holland snipe." He likely did this from England, using a specimen collected on one of James Cook's voyages during the early years of British colonization in Australia. Although the scientific name he had given the snipe was later dropped in favor of one made by John Gould in 1831 (*Gallinago hardwickii*), the name "Latham's snipe" stuck. The bird was also known by a range of other names in Australia in the nineteenth century, including Australian snipe and "Longbil."[10]

As historian Nancy Jacobs has noted of European ornithologists within African colonial contexts, while ornithologists sought to produce objective science, they were embedded within colonizing processes, often ignoring or appropriating Indigenous knowledge.[11] In describing the idea that Latham's snipe bred in the "unknown" inland of Australia, Campbell reinforced colonial divisions between European and Aboriginal people in Australia. Expanding on this idea, he wrote that colonists had assumed that the birds "probably . . . bred in autumn in countless companies in great marshy areas as yet only explored by the rude savage."[12] Here, racist ideologies used to support a colonial hierarchy embedded in ideas of civilizational progress are clearly at play. This view of Western scientific knowledge as superior to other kinds of knowledge was also evident in his discussions of the snipe in Japan.[13] In 1901, Campbell wrote that "[the] Japanese . . . take little interest in the natural history of their country. That is one reason why the nests and eggs remained so undiscovered, and why we know so little of the domestic matters of this feathered migrant, so full of interest to Australians."[14] People in Japan clearly knew about the birds, which were called Oh-jishigi, but not in the way Campbell wanted or needed for his scientific study.[15] Some colonial ornithologists saw themselves as adventurers, of an ilk with colonial explorers, and traveled overseas to collect specimens and observe the birds (or hired someone to do this, as with Campbell). For instance, in 1903 Robert Hall and Ernie Trebilcock undertook what Robin has called "a major ornithological expedition" to Siberia in order to observe migratory waders at their breeding grounds and collect specimens.[16]

By the turn of the twentieth century, ornithologists knew that a range of birds migrated between Australia and parts of Asia, and that they bred in the Northern Hemisphere and wintered in Australia. By 1919, Hall could publish a series a maps of birds' migratory routes based on his own and others' research and ideas. This included the route ornithologists' thought was taken by Latham's snipe.[17] International networks among ornithologists facilitated this. For instance, in 1903, Russian ornithologist Sergius Buturlin outlined his research on migratory birds between the Russian Empire and Australia in the Royal Australasian Ornithologists Union's (RAOU) journal *Emu*, stating that "so far, I know we have 48 forms in common" and giving details on his research of various collections.[18] Knowledge of bird movements was, however, far from certain. Ideas about the precise routes taken by birds like Latham's snipe were speculative and contested, and some species that did not leave Australia were thought to be migratory by both ornithologists and amateur observers. Colonists in Australia would write to newspapers with their experiences and observations of birds and nests, often in response to an article by an ornithologist.[19] This hints at a range of popular understandings of migration by colonists but also the reciprocal shaping of these and more "expert" (often middle- and upper-class) understandings by ornithologists. Gaining information about bird distribution and movements, within the narrow parameters of what was considered acceptable evidence, presented a range of challenges to ornithologists, including the time and people power it took to undertake regular observations. Some came up with inventive solutions. For example, Campbell began recording the arrival of Latham's snipe in his local area in Victoria based on the shooting of the first snipe of the season, which a local shopkeeper hung in the shop window.[20] This helped him to ascertain that Latham's snipe began arriving in Victoria, near Melbourne, in mid-August to September, while their departure was less certain but was "towards the end of the Australian autumn."[21] The establishment of an ornithological society (the ROAU) in 1901, along with a society journal, promised greater coordination for ornithological research in Australia. There remained, however, some challenges in studying migratory patterns. For example, in the 1920s and 1930s, members of the society set up a Committee on Distribution and Migration, but the committee's goal of regular observation in particular regions was at odds with constraints on some volunteers' time, and keeping volunteers involved in the project proved difficult.[22]

By the 1860s, Latham's snipe were already familiar to waterfowl hunters in southeastern Australia, who prized them for sport as well as food. Snipe required skill and patience by hunters; in fact, the mode of hunting used to kill snipe is where the word "sniper" comes from. In the early twentieth century, people began to record declines in the number of Latham's snipe in southeastern Australia, and hypothesized that it was due to a loss of habitat from wetland drainage, intensified land use, and over-hunting.[23] According to later government estimates, hunters killed up to 10,000 birds annually in Australia in this period, with many of these in Victoria and Tasmania.[24] In the late nineteenth and early twentieth centuries, concerns that birds were being over-hunted for a range of reasons—including for plumes for use in women's fashion, as agricultural pests, and for market—prompted advocacy groups to actively lobby governments for greater protection of birds.[25]

These groups often directed their efforts toward protecting native birds that were "useful" (for example, as they ate insect pests on farms) or "aesthetically pleasing."[26] Seen not to be particularly pesky, useful, or beautiful, nor truly "native" as they bred elsewhere, Latham's snipe and many other trans-equatorial migratory birds seem to have been marginal in these debates. Ornithologists, advocacy groups, and hunters also aimed to protect bird-breeding sites rather than habitat, and a range of wetlands were protected as "game reserves" in this period. But as Latham's snipe and other trans-equatorial migrants bred outside Australia, in the Northern Hemisphere, they were not included in this either. Bird protection advocates, including sport hunters, argued that birds were vulnerable and in need of protection at the time of breeding, and at other times it was, broadly speaking, expected that they could go almost anywhere. Further, while hunters viewed Latham's snipe as good sport, it was not a key game species. Historian Robert Boardman has argued that as none of the migratory birds were significant game species, Australia did not seek international agreements to protect them in this period as in North America.[27]

In the first half of the twentieth century, government research focused largely on species that farmers and fishermen viewed as "pests," like cockatoos that raided orchards and ducks that were blamed for reducing rice crops. It was the research in this period that provided scientific evidence for nomadic and opportunistic behavior in many Australian birds (already generally understood). Robin has noted that in the 1920s and 1930s this research on nomadism was influenced by international interest in "the physiology of irregular breeding" of birds, including the influence of rainfall.[28] Following an intense dry period and sand drift in the inland in the 1930s and 1940s, Australian biologists turned more directly to studying "desert" or "arid zone" birds in the 1940s and 1950s.[29] In broad terms, this extended research on nomadism and opportunism. Australia only established a government-funded national bird-banding program in the 1950s, something that the United States and Britain had done for much longer. As Robin has noted, "migration has traditionally been the major interest of banding studies," and this interest had not been significant in Australian government research.[30] In addition to migratory birds falling outside the purview of applied government science, the lack of research on migratory birds in Australia in this period might be further explained by the government's focus on "national" projects and development during and following two world wars, which had created turmoil and sensitive international relations within the Asia-Pacific region.

Throughout most of the twentieth century, migratory birds with routes into East Asia and the Pacific only seem to have gained significant popular and scientific attention as potential carriers of disease. These views became part of the contested bio-cultural terrain of immigration laws that discriminated against "non-white" migrants, including from these regions.[31] The first half of the twentieth century was a period of heightened racism in Australia, which continued in the Cold War period. A largely conservative Australian population saw links with Asia, including through birds, as undesirable. In the early 1950s, in the wake of a significant outbreak of encephalitis in the southern Australian states, medical professionals claimed that migratory birds had introduced Japanese encephalitis to Australia, which was then spread to humans by mosquitoes. Newspaper columnists interpreted the introduction of the disease within racist ideas of a clean, "white Australia" located within a diseased, "non-white" region. One stated: "Research points to migrating birds, travelling to Asia by way of the Pacific Islands north of Australia [and back], as carriers of the disease to our mainland. If this is so, it is quite possible that the disease first appeared in the islands, or in Asia rather than in white Australia."[32] Another stated that "Murray Valley encephalitis, which broke out in 1951 gave the Murray valley a bad name. But the disease is now known to be Japanese encephalitis possibly carried to Australia by migrating birds."[33] The kind of encephalitis that had caused the outbreaks in Australia was, however, later shown to be a slightly different strain, one that was local to parts of Australia.

Studying the role of migratory birds in spreading Japanese encephalitis in East and Southeast Asia was one of the key rationales behind the establishment of the Migratory Animal Pathological Survey in the 1960s. This effort was funded by the US Army and the South East Asia Treaty Organization and led by American ornithologist Elliott McClure. The study included more than fifteen countries and involved the banding of thousands of birds. Through this research McClure devised the idea of the East-Asian Flyway as a major bird migration route, which included Australia (later subsumed into the bigger East Asian-Australasian Flyway). The flyway concept had been developed by researchers in the United States in the 1920s and 1930s and had had a major influence on wetland management along migratory paths in North America. From this research, McClure argued that changes in bird habitat in Asia and Australia had altered the migration routes.[34]

It was perhaps through knowledge of this work that Australian researchers, led by Harry Frith, chief of the Wildlife Division of the Commonwealth Scientific and Industrial Research Organisation (CSIRO), began studying Latham's snipe in New South Wales in the 1970s. In 1970, they visited Japan and met with researchers at the Yamashina Institute of Ornithology to develop links between the two countries. Japanese researchers were becoming concerned over the pressures of increasing industrialization and urbanization on parts of the birds' breeding habitat, and soon after, Yoshimaro Yamashina included it as one of the endangered birds of Japan.[35] From the 1960s, bird organizations and researchers in Japan had increasingly advocated the need for international cooperation to protect birds, particularly cranes that migrated between several countries in the wider region. These efforts gathered pace in the 1970s, supported by several organizations, including the Yamashina Institute of Ornithology.[36] The strengthening of relationships with Australian researchers can be seen as an extension of these interests.

Frith and his colleagues published their findings on Latham's snipe in 1977, in which they emphasized that a loss of habitat through intensive land use, wetland drainage, and the damming and canalization of rivers, as well as hunting, had potentially altered the distribution and reduced the populations of the birds in eastern Australia. New South Wales had only recently banned hunting of the birds, and Victoria, Tasmania, and Queensland (the latter where Latham's snipe was a passage migrant) all allowed hunting during the key migration times when the birds were relatively vulnerable. Frith and his team also stressed that so little scientific research had been undertaken on the birds in Australia that making any definitive claims, even of distribution, was difficult.[37]

Indeed, widespread and growing concern over reductions in habitat and the effects of hunting for a number of species and a lack of scientific research on wildlife in Australia were the key motivations behind the establishment of a Committee on Wildlife Conservation by the Commonwealth Government in 1970. The committee included people from both major political parties, producing a report with key recommendations in 1972. This report incorporated a review of the need for the protection of migratory birds. Of the sixty-six species of birds listed as trans-equatorial migrants, the committee heard evidence that only one needed an international agreement for its protection: the Latham's snipe.[38] Frith, as chief of the Wildlife Division of CSIRO, had been one of the key people to give evidence to the committee. It is likely that he brought the birds to the committee's attention, as he had just established his research project on them. The committee members also viewed international agreements for bird protection as a politically astute move, writing that they "regard it as part of Australia's international responsibility to ensure not only that action is being taken in this country but also to demonstrate our concern internationally through the establishment of agreements with other countries."[39] In addition, they argued that species protection needed greater alignment between Australian states and territories, while different protection and hunting laws prevailed in each. Ultimately, the committee recommended that Australia "seek unilateral agreements with the Governments of Papua New Guinea, New Zealand and Japan" to protect migratory birds.[40]

Boardman has argued that Australia's involvement in the International Union for the Conservation of Nature, founded in 1948, "raised awareness of the transnational character of many issues and gave domestic conservation issues an increasingly attentive foreign constituency."[41] A range of other such agreements added to Australia's international perspective and profile in the 1970s, including the Ramsar Convention on Wetlands of International Importance in 1971 and the Bonn Convention on Migratory Animals in 1979. All of these agreements reflected a new international focus of environmental concerns that emerged in the post-war era and gathered pace in the 1970s.

In 1974, Australia entered into its first migratory bird treaty; this was with Japan, which contained the main breeding grounds of Latham's snipe, also called Japanese snipe. In forming the terms of the initial treaty, Japan required Australia to support the listing of each bird they proposed with evidence such as photographs or museum specimens and would not accept sightings as adequate proof. Here we can see the inverse of Campbell's 1901 statements that called into question Japan's knowledge of Latham's snipe, as the materials that supported Australia's knowledge of birds were instead put to the test.[42] This treaty has been used in conjunction with other international agreements, namely those with China (1986) and the Republic of Korea (2007), and extensively in Australia in wetland protection and management. These have on occasion been invoked to prevent wetland drainage and development. The treaty has been less significant in Japan, which, Boardman has argued, has placed greater political importance on supporting migratory bird treaties with other countries in Asia and prioritized regional conservation of cranes.[43] Nevertheless, in both places, the agreements were a new tool for intervening in wetlands, and opened up new political possibilities that could mobilize bird migrations to protect particular habitat. In other words, the international journeys made by the birds changed the nature of these places politically. More recently, concerns over H5N1 virus have seen another political shift. People have again become concerned that the long journeys made by birds are also possible routes for pathogens that harm humans, at the same time that immigration debates have flared.

A focus on non-human agency can draw us into different kinds of histories and into a consideration of the journeys plants and animals take themselves on, not just the flows we send them on. Through an examination of changing understandings of Latham's snipe and their movements, this chapter has sought to reposition environmental histories of wetlands and cultural histories of birds in Australia within a broader regional set of Pacific migrant ecologies. Birds like Latham's snipe can illuminate some of the many bio-cultural relationships and networks that have shaped places like wetlands and the consequences of how these relationships have been valued and understood for lives and livelihoods, environments and cultures.

Notes

1. Anna Tsing, *Friction: An Ethnography of Global Connection* (Princeton, NJ: Princeton University Press, 2005).

2. See also, Emily O'Gorman and Andrea Gaynor, "More-than-human Histories," *Environmental History* 25, no. 4 (2020): 71135; Emily O'Gorman, *Wetlands in a Dry Land: More-than-human Histories of Australia's Murray-Darling Basin* (Seattle: University of Washington Press, 2021).

3. See, for example, Rob Giblett, *Postmodern Wetlands: Culture, History, Ecology* (Edinburgh: Edinburgh University Press, 1996); Libby Robin, "Migrants and Nomads: Seasoning Zoological Knowledge in Australia," in *A Change in the Weather: Climate and Culture in Australia*, ed. Tim Sherratt, Tom Griffiths, and Libby Robin (Canberra: National Library of Australia, 2005), 42–53; Libby Robin, Robert Heinsohn, and Leo Joseph, *Boom and Bust: Bird Stories for a Dry Country* (Collingwood, Vic.: CSIRO Publishing, 2009).

4. This approach builds on other bird-centric studies focused on other parts of the world, including: Mark Cioc, *The Game of Conservation: International Treaties to Protect the World's Migratory Animals* (Athens: Ohio University Press, 2009), 58–101; Robert Wilson, *Seeking Refuge: Birds and Landscapes of the Pacific Flyway* (Seattle: University of Washington Press, 2010); Nancy Jacobs, "Africa, Europe and the Birds between Them," in *Eco-cultural Networks and the British Empire*, ed. James Beattie, Edward Melillo, and Emily O'Gorman (London, Bloomsbury, 2015), 92–120; Nancy Jacobs, *Birders of Africa: History of a Network* (New Haven, CT: Yale University Press, 2016); Kirsten Greer, "Geopolitics and the Avian Imperial Archive: The Zoogeography of Region-Making in the Nineteenth-Century British Mediterranean," *Annals of the Association of American Geographers* 103, no. 6 (2013): 1317–1331. For a discussion of the importance of considering animal mobility and migrations more generally, see Robert Wilson, "Mobile Bodies: Animal Migration in North American History," *Geoforum* 25 (2015): 465–472.

5. O'Gorman and Gaynor, "More-Than-Human Histories." In recent decades there has been a growing body of scholarship in environmental history on transnational exchanges. This has mostly examined the movements of animals and plants around the world by people. For an overview of this work, see James Beattie, Edward Melillo, and Emily O'Gorman, "Rethinking the British Empire through Eco-cultural Networks: Materialist-Cultural Environmental History, Relational Connections and Agency," *Environment and History* 20, no. 4 (2014): 561–575. See also the chapters in James Beattie, Edward Melillo, and Emily O'Gorman, eds., *Eco-cultural Networks and the British Empire: New Views on Environmental History* (London: Bloomsbury, 2015), and Joseph E. Taylor III, "Boundary Terminology," *Environmental History* 13, no. 3 (2008): 454–481.

6. *The Australasian* (September 12, 1891), 39. Attributed to "The Field Naturalist." Likely written by Campbell due to his frequent columns for the newspaper, mention of services of "Mr. Owston," and similarity to later texts by Campbell. Henry Seebohm also described the possible breeding place and migration of this snipe in an earlier work, published in 1887: Henry Seebohm, *The Geographical Distribution of the Family Charadriidae, or the Plovers, Sandpipers, Snipes, and Their Allies* (London: Henry Sotheran, 1887), 473–474.

7. Henry Seebohm, *Birds of the Japanese Empire* (London: R. H. Porter, 1890), 342.

8. A. J. Campbell, *Nests and Eggs of Australian Birds (Part III)* (Sheffield, UK: Pawson and Brailsford, 1901), 822–826. See also *The Australasian* (July 29, 1893), 32.

9. Robin, "Migrants and Nomads." Also noted by Jacobs, "Africa, Europe and the Birds between Them," 92–120.

10. Seebohm, *The Geographical Distribution of the Family Charadriidae,* 473–474; Seebohm, *Birds of the Japanese Empire* (London: R. H. Porter, 1890), 342; Campbell, *Nests and Eggs of Australian Birds (Part III)*, 822–826.

11. Nancy Jacobs, "Africa, Europe and the Birds between Them," 92–120. See also Greer, "Geopolitics and the Avian Imperial Archive."

12. *The Australasian* (July 29, 1893), 32.

13. See also, Jacobs, *Birders of Africa.*

14. Campbell, *Nests and Eggs of Australian Birds (Part III)*, 822–826.

15. M. U. Hachisuka, *A Comparative Hand List of the Birds of Japan and the British Isles* (New York: Cambridge University Press, 1925), 6.

16. Libby Robin, *Flight of the Emu* (Melbourne: Melbourne University Press, 2002), 32; R. Hall, "The Eastern Palaearctic," *Emu* 19, no. 2 (1919): 82–98.

17. Hall, "The Eastern Palaearctic," 84.

18. Sergius Buturlin, "Australian Birds in Siberia," *Emu* 11, no. 2 (1911): 95.

19. Robin, "Migrants and Nomads," 43–48. For discussions of the possible migration routes of Latham's snipe see, for example, *The Australasian* (September 12, 1891), 39; *The Australasian* (July 29, 1893), 32; *The Australasian* (December 2, 1893), 29; *The Australasian* (February 26, 1898), 36; *The Referee* (August 23, 1899), 9.

20. Campbell, *Nests and Eggs of Australian Birds (Part III)*, 822–826.

21. Ibid., 824.

22. M. Cohn, "The First Report of the Committee on the Distribution and Migration of Australian Birds," *Emu* 25, no. 2 (1925): 101–111; "The Third Report of the Migration Committee," *Emu* 30, no. 1 (1930): 22–28.

23. R. W. Legge, "Australian Snipe," *Emu* 31, no. 4 (1931): 308; H. Frith, F. Crome, and B. Brown, "Aspects of the Biology of the Japanese Snipe *Gallinago hardwickii*," *Australian Journal of Ecology* 2, no. 3 (1977): 341–368.

24. "*Gallinago hardwickii*—Latham's Snipe, Japanese Snipe," Department of Environment, Australian Government, http://www.environment.gov.au/cgi-bin/sprat/public/publicspecies.pl?taxon_id=863 (accessed January 13, 2016).

25. I discuss this more in-depth in Emily O'Gorman, "The Pelican Slaughter of 1911: Contested Values, Protection and Private Property of the Coorong, South Australia," *Geographical Research* 4, no. 3 (2016): 285–300. Opposition to the plume trade and the use of feathers in women's fashion was often highly gendered. See, for example, Carolyn Merchant, "George Bird Grinnell's Audubon Society: Bridging the Gender Divide in Conservation," *Environmental History* 15, no. 1 (2012): 3–30; Adam Rome, "'Political Hermaphrodites':

Gender and Environmental Reform in Progressive America," *Environmental History* 11, no. 3 (2006): 440–463.

26. See O'Gorman, "The Pelican Slaughter of 1911."

27. Robert Boardman, *The International Politics of Bird Conservation: Biodiversity, Regionalism and Global Governance* (Cheltenham, UK: Edward Elgar Publishing, 2006), 155.

28. Robin, "Migrants and Nomads," 46–53.

29. Ibid., 46–53.

30. Robin, *Flight of the Emu*, 160.

31. See also Peter Coates, *American Perceptions of Immigrant and Invasive Species: Strangers on the Land* (Berkeley: University of California Press, 2007).

32. *The World's News* (July 10, 1954), 14.

33. *The World's News* (October 31, 1953), 15.

34. Robin, *Flight of the Emu*, 246–247; H. Elliott McClure, *Migratory Animal Pathological Survey: Progress Report 1967* (US Army Report, 1968); Michael Lewis, "Scientists or Spies? Ecology in a Climate of Cold War Suspicion," *Economic and Political Weekly* (June 15, 2005), 2323–2332 and 2326–2330; Robert Wilson, "Directing the Flow: Migratory Waterfowl, Scale, and Mobility in Western North America," *Environmental History* 7, no. 2 (2002): 247–266.

35. Frith et al., "Aspects of the Biology of Japanese Snipe," 341.

36. Boardman, *The International Politics of Bird Conservation*, 157.

37. Frith et al., "Aspects of the Biology of Japanese Snipe."

38. *Wildlife Conservation: Report from the House of Representatives Select Committee* (Canberra: Government Printer of Australia, 1972), 55–56.

39. Ibid., 56.

40. Ibid., 56.

41. Boardman, *The International Politics of Bird Conservation*, 155.

42. "Environment—Relations with Other Countries," A1838, 703/3 PART 1, National Archives of Australia.

43. Boardman, *The International Politics of Bird Conservation*, 157, 164–165.

Fourteen

Bravo for the Pacific

Nuclear Testing, Ecosystem Ecology, and the Emergence of Direct Action Environmentalism

Frank Zelko

On Sunday, February 10, 1946, Commodore Ben H. Wyatt, a pint-sized naval officer from Williamsburg, Kentucky, and the United States' commander of the Marshall Islands, stepped onto the beach on Bikini Atoll with a Bible in his hand. A former college football player and star athlete, Wyatt became one of the Navy's first pilots in the 1920s. Legend had it that in 1936, while lost in the clouds over Germany, he landed his plane on the Nuremberg airfield during a Nazi rally. A decade later on the other side of the world, Wyatt's task was to persuade the Bikinians—all 167 of them—to leave their home and relocate to another island 125 miles away. The US military needed a place to test a mighty bomb developed by its scientists, Wyatt explained, and this device—the most powerful weapon ever created by mankind—would, paradoxically, lead to the end of all warfare. By agreeing to abandon the island, Wyatt gravely intoned, the Bikinians would be like the children of Israel, whom the Lord saved from their enemy and led into the Promised Land.[1]

The Americans were the latest in a succession of colonial powers to claim the Marshall Islands. Understandably, the Bikinians were awed by the power of the US military, which rid them of the stricter and more brutal Japanese rulers who had controlled the islands since the end of World War I. Unlike the Japanese, the US Navy fostered goodwill, providing Bikinians with food, supplies, and free medical treatment, as well as building a store, elementary school, and a medical dispensary on the island. Nevertheless, despite Wyatt's friendly tone, it was clear to the locals that "no" would not be an acceptable answer. Furthermore, as pious Christians—American and Hawaiian missionaries had converted them in the mid-nineteenth century—the Marshallese were receptive to biblical analogies. After a short deliberation,

Chief Juda Kessibuki reported their decision: "If the United States government and the scientists of the world want to use our island and atoll for furthering development, which with God's blessing will result in kindness and benefit to all mankind, my people will be pleased to go elsewhere." This, at least, was how the Navy portrayed the encounter.[2]

Prior to the US nuclear tests in the Marshall Islands, the planet had experienced only three atomic explosions: the initial Trinity test in New Mexico in July 1945 and the bombs dropped on Hiroshima and Nagasaki the following month. The United States emerged from World War II as the world's dominant military power and with a monopoly on atomic weaponry. Unsurprisingly, military strategists and scientists were keen to conduct further tests. The newly acquired islands in one of the remotest parts of the planet appeared to offer an ideal location, and Bikini Atoll was chosen as the first site. As comedian Bob Hope wryly observed: "As soon as the war ended, we located the one spot on earth that hadn't been touched by war and blew it to hell."[3]

Beyond its military implications, however, US nuclear testing in the Marshall Islands also had several unexpected environmental, political, and social consequences. In historian Paul Boyer's words, "it was Bikini, rather than Hiroshima or Nagasaki, which first brought the issue of radioactivity compellingly to the nation's consciousness."[4] For the Bikinians—and Marshallese in general—the use of their home as a nuclear testing ground had an utterly devastating impact on their way of life and long-term health.[5] Beyond that, the dozens of atomic and thermonuclear weapons that the US military detonated in the Marshall Islands between 1946 and 1958—all of them atmospheric—propelled radioactive particles into the stratosphere, where they hitched a ride on jet streams and gradually contaminated the entire planet with radioactive fallout. As evidence of this contamination accumulated—for example, in the form of strontium-90 deposits in milk and children's teeth—the peace and anti-nuclear movements began to focus increasingly on the environmental impacts of nuclear weapons testing.[6]

Ironically, the ecological worldview that undergirded this incipient environmentalism was bolstered by a seminal environmental impact assessment conducted in 1954 in the Marshall Islands. The investigators were two of the most promising young ecological scientists of the era, the brothers Eugene and Howard Odum. Their study, which was fully funded and backed by the Atomic Energy Commission (AEC), confirmed their holistic and cybernetic theory of how nature functioned. Ecosystem ecology dominated ecological thought for the next two decades, in large part due to the persuasive research and arguments of the Odum brothers. This holistic view of nature—in which humans and their technology were part of a closed circuit of natural cycles and processes—inspired popular environmental writers such as Rachel Carson. By the 1970s, it was the worldview that propelled Greenpeace's anti-nuclear protests against US and French testing in the Pacific, as well as its subsequent global environmental campaigns. For all these reasons, nuclear testing in the remotest parts of the Pacific played an important role in the history of environmentalism.

By 1945, a team of crack European and American scientists, with ample assistance from the US government and military, had successfully harnessed the power of the atom. Among many questions raised by the bomb, the issue of control was perhaps uppermost: what group or agency should be in charge of directing and coordinating atomic research and deciding what use it should be put to? As Hiroshima burned, President Truman urged Congress to pass legislation to create a new commission, to be controlled largely by the military, that would concern itself primarily with weapons production. Many in the scientific community, however, were alarmed at the prospect of extending military control of atomic power into peacetime. Senator Brien McMahon, a Democrat from Connecticut, came up with a solution: the new agency—the Atomic Energy Commission (AEC)—would be composed entirely of civilians and would concern itself with the potential non-military uses of atomic energy, as well as with weapons development. However, the generals easily circumvented McMahon's effort to ensure civilian control of atomic energy; many of the commissioners were high-ranking military officers who simply stepped down from active duty to assume their new role, and the agency's military division quickly became its dominant branch, commanding 70 percent of its budget and prioritizing weapons development for the next thirty years.[7]

Naturally, testing atomic weapons became one of the AEC's major concerns. In this sense, Bikini Atoll's misfortune was due primarily to its remoteness from other areas of human habitation and its relative proximity to Kwajalein Atoll, where the United States had already built a military air base and ship anchorage. The AEC conducted the first two tests, code-named Able (July 1, 1946) and Baker (July 23, 1946), with little concern for the safety of those involved, be they the Marshallese residents of nearby islands or US military personnel. The Navy placed numerous decommissioned and captured enemy battleships in the Bikini lagoon as part of the test, many with pigs, sheep, and goats strapped to their decks. Within hours of the first blast, military commanders sent fifteen thousand soldiers, with virtually no protective gear, into the lagoon to survey, hand-scrub, and decontaminate the ships. The government's insistence on maintaining secrecy made it difficult to obtain independent information about the effects of the tests; few journalists were permitted to witness the blasts and the subsequent cleanup, and government officials vetted their stories before they could be published.[8]

Over the next sixteen years, the AEC conducted another 108 atomic and thermonuclear bomb tests in remote regions of Oceania. The size of the blasts increased exponentially, generating a total yield of approximately 151 megatons. The tests during this period constituted approximately three-quarters of the overall yield generated by all US testing between 1945 and 1992.[9] The largest blast by far, which also took place on Bikini, was the Castle Bravo thermonuclear detonation of 1954. Despite realizing that the wind conditions were unfavorable and would likely spread radioactive fallout throughout inhabited regions, the US government went ahead with the test as scheduled. The fifteen-megaton yield was considerably larger than scientists had anticipated. Several hours after the blast, a fine white powder fell from the sky onto Rongelap, an island about ninety miles to the west of Bikini. To children of the tropics who had only heard about snow via Christmas stories, the fallout looked like snowflakes. A few minutes of happy frolicking exposed them to 175 rad of radiation and a lifetime of ill health and suffering (the maximum total body dose recommended by the International Commission on Radiological Protection is 0.5 rad per year).[10] After two months, the US government finally decided that Rongelap was too dangerous to inhabit. It relocated the locals to Ejit Island in the Majuro Atoll, where they experienced a polio epidemic and subsisted on canned food. As consolation, the former residents of Rongelap could reflect on the fact that Bravo's monumental force offered AEC scientists many special research opportunities and helped further their understanding of radio-ecological principles.[11]

By the mid-1950s, it was clear that nuclear detonations of one megaton or more propelled radioactive materials high into the stratosphere, potentially dispersing them over much of the planet.[12] Among the most troubling by-products of nuclear fission spread by the blasts was strontium 90, a radionuclide with a half-life of twenty-eight years. Strontium 90 accompanies calcium, with which it has a chemical affinity, through the food chain from soil to vegetables. It eventually accumulates in the bones of animals, where it effectively functions as a constant source of low-level internal radiation, significantly increasing the risk of bone cancer and leukemia. The body of every human now contains strontium 90, and forensic scientists can date human remains as pre- or post-Bravo, based on traces of militarized radioactive carbon in teeth.[13]

Throughout the 1950s, the AEC continued to insist that nuclear testing "created no immediate or long-range hazard to human health outside the proving ground."[14] Such reassurance, however, proved hollow and was soon undermined by the AEC's own research. In 1958, a group of AEC-contracted Columbia University scientists conducted a worldwide study of bone samples. The results flew in the face of the commission's sanguine assurance: in one year, the average level of strontium 90 in children had increased by 50 percent, and children under five had concentrations that were twenty times higher than those in adults over twenty years of age. Although the AEC insisted that such levels remained below the acceptable maximum, the media and the general public became increasingly skeptical. *Time* magazine, for example, reported that many scientists felt that the maximum had been set far too high, while the *New Republic* opined that the world had "suddenly become a small sphere too restricted in surface area for the 'safe' testing of super-bombs."[15]

That same year, Barry Commoner, a Washington University scientist and anti-nuclear campaigner who would go on to become one of the most influential environmentalists of the late twentieth century, published "The Fallout Problem" in *Science*.[16] Commoner explained what nuclear fallout was and outlined its likely long-term impact on human health and the environment. The global ecological and health consequences of Pacific nuclear testing dramatically illustrated the interconnectedness of the world's natural systems, an insight Commoner would later summarize as one of the four primary laws of ecology.[17] As the implications of fallout became increasingly clear, pacifist organizations that had protested nuclear testing since its inception began to focus increasingly on its environmental impact.[18] Furthermore, Quakers such as Albert Bigelow, inspired by Gandhi's nonviolent direct action protests against British imperialism, started planning voyages to the Marshall Islands in order to "bear witness" to nuclear testing. They adopted the slogan "No contamination without representation." None of them managed to reach their goals before being arrested, but their actions inspired future voyages such as those of Greenpeace in the 1970s.[19]

At the same time as the AEC was inspiring scientists and anti-war groups to launch campaigns against nuclear testing and the danger of fallout, it was also the chief sponsor of the branch of science that would provide environmentalists with the tools and the worldview that prompted them to challenge not only the AEC but industrialism in general. Broadly speaking, the trajectory of science since the mid-nineteenth century was increasingly mechanistic and reductive. Technological breakthroughs allowed scientists to study and manipulate organisms at the cellular level. The reductionist science of the laboratory identified diseases and promised cures; it split apart and recombined molecules into useful new materials and products. Given their efficaciousness, it is not surprising that reductionist values and assumptions became increasingly pervasive to the point of seeming self-evident. In a time of rapid industrial expansion and growing consumerism, they offered a form of science that was on the one hand practical and result-oriented, but which also promised insight into the most fundamental levels of life and matter.[20] From this perspective, atomic research was the apotheosis of modern science: it focused on the smallest known units of matter in order to maximally leverage the power of nature for military and industrial purposes.

Reductionism, unsurprisingly, spawned an oppositional trend toward a more holistic approach to science. By the mid-twentieth century, the branch of science that best represented this view of nature was ecology.[21] Arthur Tansley, an English botanist and one of the pioneers of modern ecology, argued that nature could best be understood as a series of interlocking ecological systems—a pond, a forest, the biosphere—each of which could be studied as a "whole." How could scientists understand these systems without resorting to reductionism? The key, according to the methodology developed by Eugene Odum, a young ecologist at the University of Georgia, was to examine the energy circuits and material flows that connected biotic and abiotic phenomena into a single interacting entity. And the easiest and most accurate way to measure such circuits and flows was by following radioactivity. Radioactive tracers could be used to measure the movement of materials and the flow of energy through an "ecosystem." The Odums "labeled" plants at the bottom of the food chain with radioactive isotopes. Then, at various intervals, they sampled consumers in the system for radiation. Radiation ecology, as this practice became known, enabled scientists to isolate individual food chains and determine how long it takes for energy to move through the ecosystem.[22]

Odum's holistic ecosystem ecology was, from its very inception, intimately linked with atomic research. In 1951, scientists began working on the hydrogen bomb that would be detonated on Enewetak Atoll in 1952. In order to help produce the tritium and plutonium necessary for the first full-scale thermonuclear explosion—to be code-named Ivy Mike—the AEC constructed a nuclear facility on the Savannah River in South Carolina. Odum received AEC funding to conduct an ecological survey of the region before and after the plant became operational. And he was far from being the only ecologist to benefit from AEC largesse. The commission also funded ecological research at the Oak Ridge nuclear facility in Tennessee, as well as at numerous universities and research stations throughout the country. Like environmentalism, therefore, modern ecology was very much a product of the nuclear age.[23]

In July 1954, the AEC contracted Eugene Odum to study the impact of radioactive fallout on a coral reef adjacent to Enewetak Atoll. He invited his younger brother, Howard, a recent Yale PhD who had worked under the renowned ecologist G. Evelyn Hutchinson, to accompany him, and the two spent six absorbing and fruitful weeks assaying a local reef. Altogether, the AEC conducted forty-three tests on Enewetak. The Odums began their research two years after Ivy Mike, the 10.4-megaton thermonuclear blast that had completely obliterated Elugelab Island and turned it into a giant crater. The 15-megaton Bravo test had taken place on Bikini, which was 190 miles to the east, three months prior to their arrival. And just two months before their field work began, Enewetak had been the stage for Nectar, a 1.69-megaton blast that was part of the same series as Bravo. Such was the degree of radioactivity in the area that the Odums could produce an autoradiographic image of a piece of coral by merely laying it on photographic paper.[24] Like many AEC-funded scientists, the Odums were seemingly oblivious to the moral and political implications of their work. Instead, they viewed the irradiated reef as a unique opportunity "for critical assays of the effects of radiations due to fission products on *whole populations and entire ecological systems in the field*."[25]

In addition to determining the environmental impact of nuclear testing, as per the AEC's request, the Odums planned to use Enewetak to investigate a complete ecosystem with the intent of measuring its overall metabolism—the chemical processes that maintain a living system, something that nobody had done before.[26] Using a small raft as a base, the brothers waded and dove their way around their chosen reef for hours at a time, including several night dives. "All in all," Eugene wrote several years later, "there is no better way to become impressed with the functional operation of a community than to put on a face mask and explore a coral reef."[27] "Functional" was a key term in the Odums' lexicon. Most ecologists up to that point had taken for granted that investigators would painstakingly develop a thorough familiarity with the majority of species in the ecosystem they were studying in order to be able to describe its structure. The Odums, however, could identify very few species on the reef. Moreover, they firmly believed that such detailed knowledge was not necessary in order to trace energy flows and measure a system's metabolism. In other words, an ecologist could understand the way an ecosystem—especially a radioactive one—functioned without necessarily having intimate familiarity with all of its components. Furthermore, the reef research validated their theory that natural selection favored ecological stability or, in layperson's terms, the balance of nature. "It seems clear," they concluded, "that the vast coral reef community is highly productive and not far from a steady state balance of growth and decay."[28] This belief in ecosystem stability, orderly succession, and mutualism was shared by Evelyn Hutchinson and other prominent mid-century ecologists, and the Odums' Enewetak research appeared to provide the empirical data to confirm it.[29]

The Odums' landmark ecological study, "Trophic Structure and Productivity of a Windward Coral Reef Community on Eniwetok [*sic*] Atoll," won the 1956 Mercer Award from the Ecological Society of America, inspiring numerous similar studies of mutualism and influencing ecology for the next several decades.[30] Aldo Leopold, the seminal environmental writer of the first half of the twentieth century, had urged scientists to study the "fountain of energy flowing through a circuit of soils, plants, and animals."[31] The Odums' study fit well with Leopold's injunction. Although they may not have been fully aware of the metaphysical implications of their theory, philosophers like Baird Callicott subsequently noted how ecosystem research, and the theories it supported, undermined more atomistic scientific worldviews in favor of one that was relational, holistic, and based on ebbs and flows rather than individual organisms.[32] From a more practical perspective, Eugene Odum was convinced that his brand of functional, holistic ecology, by broadly measuring the metabolism of whole ecosystems, was well suited to the kind of environmental impact studies that would be increasingly necessary as the United States embarked upon a large-scale nuclear energy program.[33]

The Odums' Enewetak research constituted an important addition to the second edition of Eugene's textbook, *Fundamentals of Ecology*, published in 1959. From the publication of the first edition in 1953 until well into the 1970s, *Fundamentals* was by far the most popular and influential textbook in university ecology courses, selling over 112,000 copies by 1970.[34] By then, according to *Time* and *Newsweek*, *ecosystem* had become a household word.[35] In her 1962 classic, *Silent Spring*, Rachel Carson employed the concept to describe how chemicals moved along food chains, explicitly comparing pesticide fallout to nuclear fallout and arguing that numerous products of the chemical industry were irreparably disturbing the balance of nature.[36] In 1963, Barry Commoner and a group of other scientists changed the name of the Committee for Nuclear Information, which they had formed in 1958, to the Committee for Environmental Information.[37] The change signaled the committee's emerging interest in a host of broader environmental issues, in the process demonstrating how ecosystem ecology had amplified anxieties over nuclear fallout into broader concerns about human impact on the global environment.

Despite the popularity of *Fundamentals*, some scientists found Eugene Odum's metaphors and concepts deeply problematic. Most evolutionary biologists, for example, were convinced that individual fitness was the key to understanding how life functioned and evolved. They were thus deeply suspicious of the group adaptation theories embedded in Odum's ecosystem concept, as well as the notion that the elements of nature "cooperated" in an effort to achieve a balanced state. Nevertheless, Odum's metaphors resonated with broader cultural trends. His insistence that even a spacecraft constituted an "ecosystem," a self-contained "life support system" in which everything needed for survival was contained in a single vessel, was a powerful image for a public that was fascinated with the space program and beginning to see the first photos of the earth taken from outer space.[38] Furthermore, Odum was quite happy to see ecosystem ecology conflated with environmentalism; in fact, he actively promoted this conflation in numerous lectures and publications throughout the United States and the world, and his ecosystem evangelism resonated with students in particular. As his biographer Betty Jean Craige noted, "the left-leaning students who believed that ecology would enable them to 'save the earth' liked Odum's environmentalist message, populist political posture, vision of nature as inherently orderly, and desire for a peaceful and harmonious society in which humans would cooperate with one another rather than compete."[39]

While the Marshall Islands bore the greatest brunt of weapons testing in the remote Pacific, it was not the only region to host multiple atomic and thermonuclear blasts. With the signing of the Partial Nuclear Test Ban Treaty in 1963, which banned atmospheric testing among its signatories, the United States shifted its underground tests to the Aleutian Islands in the far north Pacific.[40] Meanwhile France, which was not a signatory, decided to continue atmospheric tests. Between 1960 and 1966, the French exploded seventeen bombs in Algeria. However, the Algerians' anti-colonial struggle forced France to look further afield for test sites, and it eventually settled on an area of French Polynesia not far from Tahiti.[41] Both US and French testing policies inspired numerous protest campaigns throughout the Pacific and the rest of the world. The most important and durable movement to emerge from these protests, at least as far as the history of environmentalism is concerned, was Greenpeace.

The founders of Greenpeace were products of both the anti-nuclear movement and the holistic worldview that spilled over from ecosystem ecology into environmentalism. The AEC and its French equivalent, the Centre d'expérimentation du Pacifique (CEP), were their chief antagonists, environmental vandals preparing for an unwinnable war and poisoning the planet from remote Pacific outposts where no civilians could witness their crimes. The founding of Greenpeace, which in its early years existed in a fluid state between a social movement and a non-governmental organization, is a complicated story, but the short version goes like this: In the late 1960s, numerous Americans found themselves living in Canada due, in one way or another, to various disagreements with their government's foreign policy. In addition to young draft evaders, there were older immigrants from the WWII generation who wanted to ensure that their sons would not get drafted into the US military once they came of age. Others left because they found US preparations for nuclear war to be unconscionable. Quite a few were Quakers. In Vancouver, a fertile center of the Canadian counterculture, these older Americans came into contact with numerous hippies and radical activists who shared their misgivings about issues such as nuclear warfare and the malign influence of the US military-industrial complex. Many were also concerned about issues such as pollution, while some of the Americans were Sierra Club members.[42]

This disparate array of anti-war activists, environmentalists, and the politically disaffected members of the counterculture were galvanized by US nuclear testing on Amchitka Island, a small grassy island in the faraway Aleutians. Apart from their general opposition to nuclear weapons and their concerns about fallout, many feared that the tests—conducted in a geologically unstable area—could set off earthquakes and a tsunami that would, in the words of journalist and Greenpeace founder Bob Hunter, "slam the lips of the Pacific Rim like a series of karate chops."[43] Between 1969 and 1971, the tests inspired much opposition and numerous protests. On October 2, 1969, for example, thousands of protesters descended on the United States–Canadian border, disrupting the smooth flow of people and goods for the day. It was at one such protest on the British Columbia–Washington State border that the nucleus of the Greenpeace coalition was formed. It was here that two older American activists—Irving Stowe from Rhode Island and Jim Bohlen from Pennsylvania—met up with various student radicals and other young protest groups and decided to form an organization that would try to stop the next major nuclear test, scheduled for late 1971. They gave themselves the rather vivid, if somewhat clumsy moniker the Don't Make a Wave Committee (DMWC) and began meeting regularly at Stowe's house in Vancouver. After many fruitless discussions, Bohlen, recalling the Quaker efforts of the late 1950s, came up with a plan: they would charter a boat and sail it into the nuclear test zone, thereby bearing witness to the ecological crime and putting political pressure on both the US and Canadian governments.

Appealing to popular ideas of ecology and to broadly held notions of peace, security, and human rights, the DMWC used evocative slogans and pithy catchphrases that could be picked up by the media and that would resonate with the masses. For example, they characterized the AEC as "ecological vandals" and argued, "Amchitka may be the link in the chain of events which will bring human history to an end."[44] Bohlen spoke of the US defense umbrella as a "death canopy for Canada," while Stowe charged that the AEC was creating a "pocket of poison" on Amchitka that was "filled with the most lethal and terrible kinds of polluting radiation on the planet."[45] The AEC, Stowe proclaimed, demonstrated "that power pollutes and nuclear power pollutes absolutely."[46] Patrick Moore, a young ecology graduate student at the University of British Columbia, argued that if the US government wished to "indulge itself" and test a device it claimed was safe, "why not explode it in the geographic center of the United States in central Kansas?"[47]

The *Greenpeace*, as the activists called their boat, set sail from Vancouver on September 15, 1971, bound for Amchitka. The twelve crewmen spent six weeks on the storm-tossed waters of the far north Pacific, stopping at numerous Canadian and Alaskan villages along the way. The AEC kept postponing the blast, most likely assuming that the protestors would not be foolhardy enough to sail their old halibut seiner through the churning October sea. With much down time, the core group of Greenpeace founders spent many hours in animated conversation about what they hoped to achieve and how their campaign might evolve into a larger movement. Their environmental discussions clearly show a metaphysical debt to the holistic ecology of Eugene Odum, particularly the moral and political inflection given to it by the likes of Rachel Carson and Barry Commoner, as well as the spiritual dimension characterized by countercultural writers like Gary Snyder.[48] While hiking on the Aleutian island of Akutan, for example, ecology graduate student Patrick Moore began to gently dig into the island's moss and soil with his bare hands. Others kneeled down and joined him, marveling at the miniature ecosystem that existed below the surface. Moore began to give a spontaneous lecture on the interconnectedness of life, how all species were, at base, interdependent. The reductionist view of nature that characterized modern science, Moore argued, had served to obscure this holism, which to the men kneeling in Moore's little circle was never more apparent than at that moment. A wide grin appeared on Moore's face as he found the perfect hippie metaphor to describe this holistic ecosystem. It means, he exclaimed jubilantly, "that a flower is your brother!"[49]

In the end, the *Greenpeace*, stymied by the weather and the US Navy, never made it to Amchitka. Nevertheless, the campaign generated the embryonic stirrings of a broad international trans-political alliance. Despite their failure to reach their destination and the flakiness that characterized some aspects of the campaign, it was nonetheless a substantial achievement. Unlike similar voyages in the past, such as the Quaker anti-nuclear protests of the 1950s, the *Greenpeace* managed to attract considerable media attention. Furthermore, as well as employing the direct action tactics of its predecessors, the campaign, which was almost two years in the making, was instrumental in uniting two of the major social movements of the twentieth century—environmentalism and the peace movement.

From 1972 to 1974, Greenpeace continued its Pacific anti-nuclear campaigns, this time protesting against French testing on Mururoa Atoll near Tahiti.[50] In the process, core Greenpeacers from Vancouver traveled to New Zealand to organize a protest ship that would sail to Mururoa. Patrick Moore and Jim Bohlen went to New York to lobby the United Nations, while other activists flew to Paris and London to help organize protests there, before ending up at the UN Conference on the Human Environment in Stockholm, one of the groundbreaking events in the history of international environmentalism. In the process, Greenpeace activists established contact with anti-nuclear activists throughout North America, Australasia, and Western Europe, who embraced Greenpeace's nonviolent direct action approach and their holistic ecological worldview.[51]

Greenpeace continued campaigning against French nuclear testing until the CEP detonated its 193rd and final bomb in 1996. Throughout almost a quarter of a century of protest, the French military expressed its irritation with Greenpeace by ramming its boats and, most notoriously, bombing the *Rainbow Warrior* in Auckland, New Zealand, in 1985 and killing one of the activists onboard.[52] In 1975, Greenpeace broadened its environmental scope and began protesting Soviet and Japanese whaling, first in the Pacific and subsequently in the North Atlantic and the Antarctic. That Greenpeace had the capacity for such actions—or that it could even contemplate them in the first place—was due to the fact that it had cut its teeth protesting nuclear testing in the remote Pacific. The same could be said about its subsequent campaigns against offshore nuclear and chemical waste dumping and oil exploration. Ultimately, this influential approach to environmental activism has its origins in the Cold War and the AEC's decision to test atomic bombs in the Marshall Islands. The fear of nuclear warfare and radioactive fallout prompted opposition from peace groups and scientists like Barry commoner, simultaneously forcing activists to consider how they could organize effective protests in remote parts of the Pacific Ocean. In addition, the commission's support for ecology as an instrument for studying the environmental impact of radiation inadvertently promoted a holistic ecological worldview that would animate those who opposed the AEC and the kind of future it represented. In this sense, modern environmentalism is the AEC's bastard child.

Thirty-one years after radioactive "snow" fell onto Rongelap, 95 percent of the population alive between 1948 and 1954 had contracted thyroid cancer, and a high proportion of their children suffered from genetic defects. In 1957, three years after its evacuation, the US government determined that Rongelap Atoll could be safely re-inhabited, so long as people stayed away from the northernmost islands and imported their food. Over the next three decades, the Rongelapese became convinced that their high rates of illness, premature deaths, and birth defects were due to continuous exposure to their island's contaminated soil.[53] In response, the Parliament (Nitijela) of the Marshall Islands passed a unanimous resolution asking the US government to relocate the Rongelapese. Despite extensive evidence to the contrary, the United States continued to insist that the island was safe and refused to offer assistance. The Rongelapese turned to Greenpeace for help. In May 1985, two months before the French blew her up, the *Rainbow Warrior* transported the Rongelapese and all their belongings to Mejatto, a small island on the western side of Kwajalein Atoll.[54] Though not exactly a happy ending, the relocation at least gave the people of Rongelap a measure of relief, as well as constituting a poignant reminder of the intertwined histories of Pacific nuclear testing, environmentalism, and the ecological health of the planet.

Notes

1. Jonathan M. Weisgall, *Operation Crossroads: The Atomic Tests at Bikini Atoll* (Annapolis, MD: Naval Institute Press, 1994), 106. A film of Wyatt's speech, along with other interesting footage from Operation Crossroads, can be seen in Robert Stone's sobering 1988 documentary, *Radio Bikini*. As if being told to abandon their island wasn't painful enough, the Bikinians had to endure several takes of the speech as the director continually pushed Wyatt to improve his delivery.

2. David Hanlon, "Patterns of Colonial Rule in Micronesia," in *Tides of History: The Pacific Islands in the Twentieth Century*, ed. K. R. Howe, Robert C. Kiste, and Brij V. Lal (Honolulu: University of Hawai'i Press, 1994). Juda's speech is quoted in Weisgall, *Operation Crossroads*, 107. Geographer Sasha Davis is skeptical of the Navy's version of Bikinian acquiescence. See *The Empire's Edge: Militarization, Resistance, and Transcending Hegemony in the Pacific* (Athens: University of Georgia Press, 2015), 62–65.

3. Bob Hope quote: Mike Moore introduction to "The Able-Baker-Where's-Charlie Follies," *Bulletin of the Atomic Scientists* 50, no. 3 (May/June 1994): 26.

4. Paul Boyer, *By the Bomb's Early Light: American Thought and Culture at the Dawn of the Atomic Age* (Chapel Hill: University of North Carolina Press, 1985), 90.

5. Robert C. Kiste, *The Bikinians: A Study of Forced Migration* (Menlo Park, CA: Cummings Publishing, 1974); Holly M. Barker, *Bravo for the Marshallese: Regaining Control in a Post-Nuclear, Post-Colonial World* (Belmont, CA: Wadsworth, Cengage Learning, 2013); Jane Dibblin, *Day of Two Suns: Nuclear Testing and the Pacific Islanders* (London: Virago, 1988).

6. Michael Egan, *Barry Commoner and the Science of Survival: The Remaking of American Environmentalism* (Cambridge, MA: MIT Press, 2007), 71–75.

7. Dan O'Neill, *The Firecracker Boys* (New York: St. Martin's Press, 1994), 14–15. The classic work on the history of the atom bomb is Richard Rhodes, *The Making of the Atom Bomb* (New York: Simon & Schuster, 1986).

8. Baker, *Bravo*, 21–22.

9. Mark D. Merlin and Ricardo M. Gonzalez, "Environmental Impacts of Nuclear Testing in Remote Oceania, 1946–1996," in *Environmental Histories of the Cold War*, ed. J. R. McNeill and Corinna R. Unger (New York: Cambridge University Press, 2010), 170–171.

10. Ibid., 193. Bravo was the second-largest nuclear explosion in history thus far. The largest, estimated at a whopping 50 megatons, was the so-called "Tsar Bomba" detonated by the Soviet Union in Novaya Zemlya in 1961. See Vitaly I. Khalturin, Tatyana G. Rautian, Paul G. Richards, and William S. Leith, "A Review of Nuclear Testing by the Soviet Union at Novaya Zemlya, 1955–1990," *Science and Global Security* 13 (2005): 1–42.

11. US actions at the time, along with subsequent declassified documents, strongly suggest that AEC scientists were keen to see what impact nuclear fallout would have on humans. If not a deliberate policy of exposure, the AEC is at the very least guilty of a form

of calculated neglect in which the population of Rongelap in particular was exposed to heavy doses of radiation, which did, indeed, provide useful data for scientists. See Barker, *Bravo.*

12. Elizabeth DeLoughrey argues that the AEC promoted the notion that nuclear testing sites were like isolated laboratories in which bombs could be exploded without doing any harm beyond the immediate blast site. "The Myth of Isolates: Ecosystem Ecologies in the Nuclear Pacific," *Cultural Geographies* 20, no. 2 (2012): 167–184.

13. Eileen Welsome, *The Plutonium Files: America's Secret Medical Experiments in the Cold War* (New York: Dial Press, 1999), 30; K. Spalding et al., "Forensics: Age Written in Teeth by Nuclear Tests," *Nature* 437, no. 7057 (2005): 333–334.

14. The quote is from a 1953 AEC report, quoted in Egan, *Barry Commoner and the Science of Survival*, 51.

15. Quoted in Thomas Jundt, *Greening the Red, White, and Blue: The Bomb, Big Business, and Consumer Resistance in Postwar America* (New York: Oxford University Press, 2014), 94–95.

16. Commoner, "The Fallout Problem," *Science* 127 (May 2, 1958): 1023–1026.

17. Barry Commoner, *The Closing Circle: Nature, Man, and Technology* (New York: Knopf, 1971). Rule one was: "Everything is connected to everything else." This was hardly a new insight, of course, but Commoner was nonetheless an important figure in conveying it to the broader public. See Egan, *Barry Commoner*, 126–127.

18. Thomas Jundt, *Greening the Red, White, and Blue*: *The Bomb, Big Business, and Consumer Resistance in Postwar America* (New York: Oxford University Press, 2014), 94–98.

19. Albert Bigelow, *The Voyage of the Golden Rule: An Experiment with Truth* (Garden City, NY: Doubleday, 1959); Earle L. Reynolds, *The Forbidden Voyage* (New York: D. McKay Co., 1961). For more on the history of early anti-nuclear activism, see Lawrence S. Wittner, *Resisting the Bomb: A History of the World Nuclear Disarmament Movement, 1954–1970* (Stanford, CA: Stanford University Press, 1997); Robert Divine, *Blowing in the Wind: The Nuclear Test Ban Debate, 1954–1960* (New York: Oxford University Press, 1978); and Allan M. Winkler, *Life under a Cloud: American Anxiety about the Atom* (Urbana: University of Illinois Press, 1993).

20. Charles Rosenberg, "Holism in Twentieth-Century Medicine," in *Greater Than the Parts: Holism in Biomedicine, 1920–1950*, ed. Christopher Lawrence and George Weisz (New York: Oxford University Press, 1998), 336.

21. Anne Harrington, *Reenchanted Science: Holism in German Culture from Wilhelm II to Hitler* (Princeton, NJ: Princeton University Press, 1996); Mitchell Ash, *Gestalt Psychology in German Culture, 1890–1967: Holism and the Quest for Objectivity* (Cambridge: Cambridge University Press, 1995); Frank Golley, *A History of the Ecosystem Concept in Ecology: More Than the Sum of the Parts* (New Haven, CT: Yale University Press, 1993).

22. Joel B. Hagen, *An Entangled Bank: The Origins of Ecosystem Ecology* (New Brunswick, NJ: Rutgers University Press, 1992), 114; Chunglin Kwa, "Radiation Ecology, Systems Ecology, and the Management of the Environment," in *Science and Nature: Essays in the History of the Environmental Sciences*, ed. Michael Shortland (Stanford-in-the-Vale, UK: British Society for the History of Science, 1993), 213–249.

23. Hagen, *Entangled Bank*, 108–110. For more on the linkages between ecosystem ecology and the AEC, see Stephen Bocking, *Ecologists and Environmental Politics: A History of Contemporary Ecology* (New Haven, CT: Yale University Press, 1997), especially chapters 4 and 5.

24. Hagen, *Entangled Bank*, 102.

25. Howard T. Odum and Eugene P. Odum, "Trophic Structure and Productivity of a Windward Coral Reef Community on Eniwetok Atoll," *Ecological Monographs* 25 (1955): 291. Emphasis in original.

26. Donald Worster, *Nature's Economy: A History of Ecological Ideas,* 2nd ed. (New York: Cambridge University Press, 1994), 363–365; Sharon Kingsland, *The Evolution of American Ecology, 1890–2000* (Baltimore: Johns Hopkins University Press, 2005), 194–195; Hagen, *Entangled Bank*, 103–107.

27. Eugene Odum and Howard T. Odum, *Fundamentals of Ecology*, 2nd ed. (Philadelphia: W. B. Saunders, 1959), 358.

28. Odum & Odum, "Trophic Structure," 318.

29. Hagen, *Entangled Bank*, 106.

30. Ibid.

31. Aldo Leopold, *A Sand County Almanac, and Sketches Here and There* (New York: Oxford University Press, 1987), 202. The book was originally published in 1949.

32. J. Baird Callicott, "The Metaphysical Implications of Ecology," in *In Defense of the Land Ethic* (Albany: State University of New York Press, 1989), 109.

33. Kingsland, *Evolution of American Ecology*, 194.

34. Craige, *Eugene Odum*, 79; Joel B. Hagen, "Teaching Ecology during the Environmental Age, 1965–1980," *Environmental History* 13 (2008): 675–694.

35. "Ecology: The New Jeremiahs," *Time* 94, no. 7 (August 15, 1969): 38–39; "Dawn for the Age of Ecology," *Newsweek* 75, no. 4 (January 26, 1970): 35–36.

36. Rachel Carson, *Silent Spring* (New York: Houghton Mifflin, 1962). For more on how Odum and other mid-twentieth-century ecologists influenced Carson, see William Dritschilo, "Rachel Carson and Mid-Twentieth Century Ecology," *Bulletin of the Ecological Society of America* 87, no. 4 (October 2006): 357–367.

37. Worster, *Nature's Economy*, 354.

38. Worster, *Nature's Economy*, 366–367; Hagen, "Teaching Ecology," 705–706. The impact of *Fundamentals of Ecology* was not limited to North America; it was translated into twelve other languages. See Craige, *Eugene Odum*, xii.

39. Craige, *Eugene Odum*, 123.

40. Dean W. Kohlhoff, *Amchitka and the Bomb: Nuclear Testing in Alaska* (Seattle: University of Washington Press, 2002).

41. Jean-Marc Regnault, "France's Search for Nuclear Test Sites, 1957–1963," *Journal of Military History* 67, no. 4 (2003): 1223–1248.

42. For the longer, more complicated version, see Frank Zelko, *Make It a Green Peace! The Rise of Countercultural Environmentalism* (New York: Oxford University Press, 2013).

43. Hunter article in the *Vancouver Sun* (September 24, 1969).

44. Bohlen testimony at the AEC hearings in Alaska, *Congressional Record—Senate*, "Planned Nuclear Bomb Tests in Alaska This Year," V.117 (92–1), June 4, 1971, 18091.

45. *Vancouver Sun* (February 5, 1970).

46. *Georgia Straight* (November 11–18, 1971), 12–13.

47. *Wall Street Journal* (June 24, 1971), 1.

48. For Snyder's contribution to environmentalism, see Timothy Gray, *Gary Snyder and the Pacific Rim: Creating Counter-Cultural Community* (Iowa City: University of Iowa Press, 2006), especially ch. 4.

49. Frank Zelko, "A Flower Is Your Brother! Holism, Nature, and the (Non-Ironic) Enchantment of Modernity," *Intellectual History Review* 23, no. 4 (2013): 531–532. Eugene Odum, in fact, was briefly in Vancouver at the time that the Greenpeace coalition was taking

shape. His son, Bill, was a postdoctoral fellow in ecology at the University of British Columbia in 1970, and Eugene visited him that spring. I did not come upon any evidence to suggest that either Bill or Eugene had any connections with the people involved in the anti-Amchitka campaign, although one imagines that Bill would have at least heard about it through others in the UBC ecology program. Sadly, he passed away in 1991, eleven years before his father. Craige, *Eugene Odum*, 103, 134.

50. Miriam Khan discusses the impact of French nuclear colonialism on Tahiti. See *Tahiti beyond the Postcard: Power, Place, and Everyday Life* (Seattle: University of Washington Press, 2011).

51. For more detail, see Zelko, *Make It a Green Peace.* Several early Greenpeace activists published first-hand accounts of their participation in various anti-nuclear campaigns of this era. Among the most useful are Robert Hunter, *Warriors of the Rainbow: A Chronicle of the Greenpeace Movement* (New York: Holt, Rinehart and Winston, 1979); Jim Bohlen, *Making Waves: The Origins and Future of Greenpeace* (Montreal: Black Rose Books, 2001); and David McTaggart, *Outrage! The Ordeal of Greenpeace III* (Vancouver: J. J. Douglas, 1973).

52. Greenpeace, *Testimonies: Witnesses of French Nuclear Testing in the South Pacific* (Auckland: Greenpeace International, 1990); Sunday Times Insight Team, *Rainbow Warrior: The French Attempt to Sink Greenpeace* (London: Hutchinson, 1986).

53. Subsequent tests proved the Rongelapese correct. The natural leaching of radioactive cesium 137 from soil is inhibited by the mica-rich dust that for millions of years has blown from the arid regions of East Asia and settled on Pacific islands. The dust bonds effectively with cesium 137, particularly in drier areas such as Rongelap, and prevents it from breaking down over time. See Merlin and Gonzalez, "Environmental Impacts," 184.

54. Barker, *Bravo*, 64–66; Greenpeace, *Report on the Marshall Islands by Henk Haazen and Bunny McDiarmid* (Auckland: Greenpeace New Zealand, 1986).

FIFTEEN

A Pacific Anthropocene

Ruth A. Morgan

We are living in "the middle of a storm of our own making," observed the Indigenous Australian author Tony Birch in the wake of the September 2019 global climate strike, where he had watched a hundred thousand protestors take to the streets of Melbourne.[1] His assessment echoed the provocative assessment that Nobel Laureate chemist Paul Crutzen and ecologist Eugene Stoermer had made twenty years earlier. They argued that since the late eighteenth century, the enormous expansion in the use of fossil fuels had transformed the planet such that the earth's biophysical systems are no longer independent of humans. Collectively, humans have become a geophysical force causing planetary change: we are living in the Anthropocene.[2]

Whether this so-called "Age of Man" represents a new geological epoch remains a topic of debate. So too is the timing of its onset, with the invention of the steam engine, the Columbian exchange, and the Neolithic Revolution among the possible turning points in the planet's trajectory.[3] While geologists debate these planetary thresholds, scholars in the humanities and social sciences have reflected on the usefulness, historical accuracy, and implications of the term "Anthropocene."[4] As environmental historian Sverker Soerlin has observed, the concept has become "more a metaphor and a historical, symbolical, and now a political concept that speaks to the underlying environmental and climate impacts of human societies."[5] These conversations have produced alternative conceptualizations to clarify the particular socioeconomic structures that produced this moment, most notably the Capitalocene, the Plantationocene, and the Chthulucene.[6]

Although the resulting litany of alternative -ocenes is both generative and speculative, the Anthropocene itself has a "silver lining," as geographer Laura Pulido argues, because it "forces us to reckon with history."[7] For Pulido, the uneven racial geography of the Anthropocene demands closer historical analysis, while feminist anthropologist Anna Tsing questions the single universalizing narrative or timeline that the Anthropocene implies.[8] These critiques align with those that highlight the dominance of male "northern voices" in planetary science circles.[9] In this "hegemonic Anthropocene narrative," historian Stefania Barca argues, "the forces of production (science and industrial technology) are maintained as the only possible tool for understanding the errors and for repairing them. The system itself is not under question; its gender, class, spatial and racial inequalities are either invisible or irrelevant: no paradigm shift is necessary."[10]

That the Anthropocene is the product of historical processes has attracted environmental historians to its analysis. It is a concept that speaks to the very project of the field, that is, to show that all human history is environmental. Their early engagements with the concept (and its critiques) focused on the intellectual history of the idea; questions of disciplinary expertise and authority regarding its definition; and representations of the Anthropocene in material culture and historical narratives; and have since turned to moral questions of historical responsibility and environmental justice. In this chapter, I open with a reflection on the contributions of settler Australian environmental historians in the articulation of the Anthropocene; I then situate this historiography in terms of the southwest Pacific and its peoples. Placing the Anthropocene in this way traces the connecting threads between the concept and its (uneven) materiality, recognizing the historical processes that have produced this moment and the material differences in its manifestation around the globe. Focusing on the southwest Pacific, as this chapter suggests, takes a regional approach to placing the Anthropocene that brings such material differences into sharp relief.

This attempt to situate the Anthropocene in a particular place complements the call of historian Gabrielle Hecht to treat the concept and its critiques as scalar projects.[11] Following geographers' efforts to "ground" the Anthropocene, Hecht seeks in her study of an African Anthropocene a "means of holding *the planet* and *a place on the planet* on the same analytic plane."[12] The purpose of doing so is twofold: first, it allows for the acknowledgment of the "unequal weight of human communities possessing disparate earth-changing powers."[13] Second, it makes the planetary scale of the Anthropocene "mentally manageable," which may foster a sense of agency at more local scales.[14] Situating what environmental historian Andrea Gaynor calls "radical remembering" in place encourages not only temporal thinking, but also collective thinking of shared places, of community with people and other creatures, as an antidote to the prevalence of atomistic, anthropogenic self-interest.[15]

For environmental historians at least, this task of engaging with human and non-human scales of time and space is a familiar one. The very method of historical writing provides highways for what Hecht describes as "interscalar vehicles." As environmental historian Tom Griffiths argues, "Environmental history frequently makes more sense on a regional or global scale than it does on a national one. It uniquely bridges planetary and deeply local perspectives, staking a claim for historians that are bound intimately to place and also embrace the natural world, histories that are deeply attentive to human biological parochialism."[16]

He attempts this himself in an autobiographical sketch that details his own "coming of age in the Great Acceleration."[17] Between the growing suburbia of 1950s Melbourne, Australia, and the space race, Griffiths finds common ground in the very source of these advances—their mobilization "by the same unsustainable energy systems." In short, he demonstrates how historians can render the planetary personal, and the personal planetary. Writing across these scales might help to illuminate the conundrum of our environmental crisis: although we are collectively the cause of anthropogenic climate change, we experience climate change unevenly and our political agency complicates collective action.[18]

Who writes history at these different scales (and for whom) was a key question of expertise for the Australian historian of science Libby Robin. Not even a decade after the concept of the Anthropocene had emerged, Robin observed that scientists of global change had become just as much interested in the study of global history as historians. Environmental pasts, she observed, "are integral to discussions about the environmental future of the planet."[19] Although their studies differed in terms of scale and audience, both scientists and historians have been drawn to the study of the rapid rate of global change that the Anthropocene represents, particularly since the end of World War II.[20] As her North American colleague John R. McNeill had observed in the year 2000 (coinciding with the publication of Crutzen and Stoermer's Anthropocene hypothesis), the ecological changes humans had wrought over the twentieth century were unprecedented to the extent that they represented "something new under the sun."[21]

These shared interests in the nature and causes of planetary change have fostered cross-disciplinary collaboration that contribute to, and challenge, northern conversations. Robin herself, for example, collaborated with earth system scientist Will Steffen, her American-Australian colleague at the Australian National University in Canberra, where he also served as a senior science adviser to the Commonwealth government.[22] Together, they were also joined by the environmental historian Tom Griffiths and archaeologist Mike Smith, to contribute Australian perspectives to the international Integrated History and Future of People on Earth (IHOPE) project. Their work in this capacity reflects a broader interest in showing how Australia's deep time challenges "global" human and ecological histories.[23] Meanwhile, the Australian historian David Christian offers an alternate perspective through his vision of Big History, whereby human history unfolds in the context of the history of the universe.[24]

That Australian environmental historians should be so engaged with the elucidation of the Anthropocene comes as no surprise to Australian historian of science Alison Bashford. She argues that local imperatives have long demanded that historians come to terms with deep time, "prehistory" and modern history, or natural, Indigenous, and non-Indigenous Australia, which the Anthropocene idea has helped to reinvigorate.[25] Citing the IHOPE project, Bashford argues that the human history of the island continent of Australia "confounds" the (Northern) idea of "'civilization' as a historical marker."[26] Bashford also notes the relationship between Australia's colonial history and the history of the Anthropocene: the British colonization of Australia from 1788 is "tantalisingly close," she observes, to James Watt's improvements to the steam engine—the date Crutzen and Stoermer originally proposed for the advent of the Anthropocene. Attending to the Anthropocene's imperial history, as economic historian Andreas Malm has argued, shifts the responsibility for planetary change from the species to the architects and agents of Victorian Britain's fossil economy.[27] For its part, settler Australia's own "coalopolis" of Newcastle exported coal across the Pacific to California, Peru, and Chile from the mid-nineteenth century.[28] As a consequence of these shared paths of settler colonialism and coal extraction, Bashford wonders, "Is modern Australia the Anthropocene's twin?"[29]

If not the "Anthropocene's twin," Australia is certainly family. According to the latest State of the Environment Report (2016), the nation's per capita carbon emissions are the second highest in the OECD and among the highest in the world.[30] Among the biggest sources of these greenhouse gas emissions are the nation's energy, transport, and agricultural sectors. Meanwhile, the nearby low-lying islands of the Pacific have become as much symbols of global warming and rising sea levels as are polar bears and melting glaciers. These islands, as the Lifou Declaration of 2015 states, "are among the most severely affected in the world [by anthropogenic climate change]. However, Pacific Island countries and territories' emissions account for merely 0.03 per cent of global greenhouse gas emissions."[31]

As the Pacific Climate Warriors warn, however, the peoples of the Pacific are not victims—"We are not drowning, we are fighting."[32] So too are the Torres Strait 8, the eight Traditional Owners from across the Torres Strait who, in 2019, lodged a complaint against the Australian Government with the UN's Human Rights Committee, highlighting the threat of climate change to their culture and their ability to live on their home islands, which lie in Australia's northeast waters.[33] The Australian government has attempted to block this complaint, dismissing the issue as a problem for the future, rather than the present. The divergent ways in which the Anthropocene manifests in the Global South and Global North are starkly evident in this southwest corner of the Pacific.

As the concept of the Anthropocene proliferated beyond scientific and technological circles, questions as to its historical and ongoing relationship to settler colonialism have intensified. These critiques have come to interrogate not only the Eurocentric concept of the Anthropocene but also the ways in which universalizing narratives as to its extent and origin were falling far short of acknowledging the violence, injustice, and dispossession associated with the colonization of the New World, which produced the so-called "Orbis Spike" of 1610.[34] The legacies of these events endure. For many of the world's Indigenous peoples, as Potawatomi scholar Kyle Powys Whyte puts it, anthropogenic climate change represents "colonial déjà vu."[35] Furthermore, where narratives of global environmental change are told (and by whom) are "mingled" with "the land and stories *of this place*," as Métis anthropologist Zoe Todd argues.[36] She asks,

> What does it mean to have a reciprocal discourse on catastrophic end times and apocalyptic environmental change in a place, where, over the last five hundred years, Indigenous peoples faced (and face) the end of worlds with the violent incursion of colonial ideologies and actions? What does it mean to hold, in simultaneous tension, stories of the Anthropocene in the past, present, and future?[37]

Todd has elsewhere argued, with Heather Davis, for the input of Indigenous knowledges in disrupting the universalizing impulses inherent to the Anthropocene concept.[38] As a settler environmental historian myself, writing this chapter on the unceded lands of the Kulin Nation, the Ngunnawal people, and the Dharawal people, I am sensitive to the sovereign knowledge and authority of First Nations peoples, which Tony Birch (and many others) argue is vital to climate justice.[39] I rather follow Todd and Davis' approach, that is, "to refuse to write from an un-embodied or universal position."[40] Such an outlook infuses the Australian-edited collection, *Living with the Anthropocene*, published in the wake of the 2019/20 Savage Summer and the commemoration of 250 years since the landing of Britain's Captain James Cook at Kamay, Sydney's Botany Bay, with Tahitian navigator Tupaia, in 1770. By attempting to place the Anthropocene in this chapter, I echo the editors' goal for their collection, to "shift the lens of the Anthropocene from the global and systemic . . . to the realm of everyday experience. It is also to shorten our gaze, focusing less on how our time will be remembered in the fossil record . . . and more on how the Anthropocene is."[41]

Although the Anthropocene was initially conceived as the product of Western industrialization in the late eighteenth century, the post–World War II era has emerged not only as a second phase to the Anthropocene but now as an alternate starting point.[42] Termed the "Great Acceleration," the end of World War II marks the dawn of the atomic age and the moment when the cumulative impact of human activity underwent exponential growth at a planetary level.[43] From population growth to fertilizer consumption, from deforestation to land domestication, graphs produced by the International Geosphere-Biosphere Program in 2015 clearly show what McNeill calls the "screeching acceleration" of both socio-economic and earth system trends in the post-war era.[44] For McNeill, these exponential trends of the twentieth century are the product of humankind "play[ing] dice with the planet, without knowing all the rules of the game." Echoing the warnings of scientists Roger Revelle and Hans Suess in the 1950s, and Wally Broecker in the 1980s,[45] McNeill argues:

> The human race without intending anything of the sort, has undertaken a gigantic uncontrolled experiment on the earth. In time, this will appear as the most important aspect of twentieth-century history, more so than World War II, the communist enterprise, the rise of mass literacy, the spread of democracy, or the growing emancipation of women.[46]

A casino for this game of roulette, a laboratory for this "gigantic uncontrolled experiment," has been the Pacific—if not since Captain Cook, then certainly since the dawn of the Atomic Age.

Although the significance of the end of World War II to Pacific peoples cannot be overstated, the planetary impact of the nuclear bombs detonated in 1945 has come under question. Proponents of the stratigraphic significance of the Anthropocene suggest that the detonation of atomic devices in 1945 in New Mexico had only local impacts, whereas thermonuclear weapons tests between 1952 and 1980 have left a clear global imprint.[47] Yet the human, cultural, and geopolitical impacts of the atomic bombs unleashed in 1945, particularly over Hiroshima and Nagasaki, place the Pacific at the epicenter of the Great Acceleration. The Pacific's atomic age might represent a stratigraphical rupture, but the violence enacted upon colonized or powerless peoples, as well as their lands and waters, was a continuation of the past.[48] Over the following five decades, the United States, the United Kingdom, and France subjected an area stretching across the Indo-Pacific—from the Monte Bello Islands in the west, to Maralinga in the Australian central desert, and the Marshall Islands in the east—to over three hundred nuclear tests, with France persisting into the 1990s, well after the Partial Nuclear Test Ban Treaty of 1963.[49]

Despite this developing nuclear regime in the region, such was the perceived isolation and exoticism of the southwest Pacific that in response to widespread anxiety about the possibility of nuclear war, the area became imagined in literary classics such as William Golding's *Lord of the Flies* (1954) and Nevil Shute's *On the Beach* (1957). These authors reprised a Western cultural imaginary of the Pacific as a place to escape modern life.[50] In *Lord of the Flies*, a remote, unspecified Pacific island is a refuge from the outbreak of nuclear war between the United Kingdom and the Soviet Union.[51] In *On the Beach*, the survivors of nuclear war in the Northern Hemisphere—saved only by the isolation of southeastern Australia—grimly await their fate as clouds of radiation drift south across the equator.[52]

The same year that Nevil Shute published *On the Beach*, the Soviet Union launched its satellite Sputnik into space. Sputnik 1, the first artificial Earth satellite, was launched on October 4, 1957, and an anxious United States promptly followed suit with Explorer 1 months later. The Space Race had begun.[53] These breakthroughs have since overshadowed the very endeavor that made them possible, the International Geophysical Year (IGY) of 1957–1958.[54] The IGY heralded an unprecedented degree of international cooperation, bringing together some 40,000 scientists and technicians from nearly 70 nations, working at some 4,000 observation stations across the globe.[55]

During the IGY, the Pacific Ocean, covering a third of the earth's surface, came under closer scrutiny than ever before. Despite arguments by Svante Arrhenius at the turn of the twentieth century, and Guy Stewart Callendar in the 1930s, about the impact of atmospheric carbon dioxide on the global climate, in the 1950s most scientists believed that the world's oceans acted as an enormous sink that would trap the excess carbon dioxide arising from human activities.[56] The development of radiocarbon dating after World War II allowed oceanographer Roger Revelle and chemist Hans Suess to put this assumption to the test. Revelle had been closely involved in measuring the impacts of nuclear weapons testing at the Bikini Atoll in the Marshall Islands in 1946 and had since become head of the Scripps Institution of Oceanography in California, which was flourishing under its post-war patronage of the University of California and the US Navy.[57] Together,

Revelle and Suess found that the oceans would not retain all the extra carbon dioxide after all. In their famous 1957 paper, they warned, "Human beings are now carrying out a large-scale geophysical experiment of a kind that could not have happened in the past nor be reproduced in the future."[58]

But there were few measurements of the concentration of carbon dioxide in the atmosphere to confirm their finding. Revelle and Suess sought to establish a baseline of carbon dioxide values around the world to see if it would rise in the coming decades. They hired geochemist Charles David Keeling to set up instruments to undertake such measurements at the weather observatory on the volcano Mauna Loa in Hawai'i, which had been built in 1956. The observatory's unique location offered Keeling the chance to take a "snapshot" of global carbon dioxide.[59] Thanks to the availability of IGY funds, Keeling was soon able to report that carbon dioxide levels were rising, as Callendar had long argued.[60] It was only after the 1972 UN Stockholm Conference on the Human Environment that scientific and political concerns about the changing composition of the atmosphere led to the establishment of an Australian air pollution station in order to provide a "pole-to-pole line through the Pacific," joining the US-monitored stations at Barrow (Alaska), Mauna Loa (Hawai'i), American Samoa, and the South Pole.[61]

Despite the depictions of Shute and Golding, the Southern Hemisphere remained largely absent from scientific considerations of the planetary impacts of human activity. Although the IGY had been a boon for Antarctic exploration, the Pacific remained "embarrassingly unknown" south of the equator nearly a decade later.[62] To stimulate research in the region, the Scientific Committee on Oceanic Research convened a symposium on the "Scientific Exploration of the South Pacific" at the Scripps Research Institute in California in June 1968.[63] Delegates learned that "except for the coastal regions off Australia and South America, our knowledge of the surface currents in the Pacific was actually not much better than Alexander Findlay's was over 100 years ago."[64] Similarly, "the areas between the equator and 35 degrees south . . . still can be considered, biologically, as *terra incognita*."[65]

For one delegate at least, the reason for this neglect was clear. Egyptian oceanographer Sayed El-Sayed, a professor at Texas A&M, argued:

> The vast expanse of the South Pacific presents a formidable problem to any investigator or to any one nation. . . . [I]t is highly unlikely that an all-out attack on the biological oceanography of the region will be seriously considered in the near future. This, together with the sparse human population scattered in the islands of the South Pacific, would not make it politically expedient to suggest an ambitious undertaking similar to that of the International Indian Ocean Expedition.[66]

Modeled on the IGY, the International Indian Ocean Expedition (1962–1965) brought together expertise from developed as well as developing countries to benefit both marine science and the heavily populated nations of the region.[67] Although such scientific and development agendas were also relevant to Oceania, its geopolitical significance hindered further investment in the region.

Similar sentiments found their expression in the following decade as a series of catastrophic climate events drew scientific and political attention to the possibility of global climate change. The Indian Monsoon failed, droughts plagued the Sahel and Ukraine, while in the Pacific, the 1972 El Niño decimated Peruvian fisheries.[68] These calamities coincided with the publication of the Club of Rome's Malthusian *Limits to Growth* (1972), which intensified global concerns that the world was rapidly depleting its limited resources.[69] Many scientists, meanwhile, were debating the causes and consequences of what appeared to be a global cooling trend. Some climatologists suggested that the world's climate was progressing toward another glacial phase. With an ever-growing world population and most arable land already under cultivation, the scientific consensus was that any change in the climate, whether warming or cooling, would severely affect the world's food supply.[70]

After the World Food Conference in November 1974, climatologists and meteorologists from Australia, the United Kingdom, the United States, and New Zealand met in Melbourne, Australia, to examine the state of knowledge about the causes and mechanisms of climatic change and variability in the Southern Hemisphere. As the editors of the conference proceedings observed, "Much has been written about Northern Hemisphere climate, sometimes as if it were the global story, but relatively little is available which focusses on the south."[71] Australian climate scientist A. Barrie Pittock, an editor of the proceedings, noted, "The geographical, historical and cultural reasons for this bias are understandable, but the consequences in terms of a true global understanding of climate are serious. This is of global rather than merely regional concern."

He continued, highlighting the geopolitical, economic, and historical influences on meteorological research:

> The reasons for Northern Hemisphere bias in the literature are not merely northern parochialism and the remoteness of the Southern Hemisphere from the major centres of modern scientific culture, but also more fundamental limitations determined until the advent of meteorological satellites by a much smaller and more recent ground-based network of meteorological stations, and a much smaller area and latitudinal range over which to establish a land-based palaeoclimatic record in the south.[72]

Although reviewers noted that the proceedings fell short in their purported coverage of the whole Southern Hemisphere, its material on Australia and Antarctica in particular offered a "long-awaited" corrective "to the more familiar North Atlantic sector."[73] Its international reception highlighted the growing contribution of Australasian-based scientists and their findings to the study of atmospheric change and its influence on the climate.

While research advanced on the implications of rising carbon dioxide levels, the lack of scientific knowledge about the interactions of the ocean and atmosphere in the Pacific proved to be catastrophic in the early 1980s. The strongest episode yet experienced during the twentieth century, the 1982–1983 El Niño was responsible for record droughts, fires, floods, and hurricanes in South America, the United States, South Asia, Southeast Asia, West Africa, and Australia.[74] The severity of this particular event prompted renewed studies of this equatorial Pacific phenomenon, prompting the United States and Japan, with France, Taiwan, and South Korea, to develop a network of ocean buoys across the tropical Pacific to measure atmospheric and oceanographic data and to help predict El Niño events.[75]

By the early 1980s, the Southern Hemisphere was no longer deemed safe from atmospheric problems in the north. In the event of nuclear war, scientists predicted that the spread of smoke across the Northern Hemisphere would cause temperatures to drop and produce global cooling effects, precipitating a "nuclear winter." This phenomenon, another of Paul Crutzen's projects, had global implications.[76] According to the authors of a report published by the Scientific Committee on Problems of the Environment,

> [T]he indirect effects on populations of a large-scale nuclear war, particularly the climatic effects caused by smoke, could be potentially more consequential globally than the direct effects, and *the risks of unprecedented consequences are great for non-combatant and combatant countries alike*.[77]

The South Pacific was no longer a refuge from superpower rivalries; Shute's dystopian vision had become a reality. Based on local research, the Australian Department of Foreign Affairs warned that "even if the war was confined to the northern hemisphere, even if Australia was not hit by a single nuclear weapon, we would still suffer a nuclear winter effect in the southern hemisphere."[78] The recent McClelland Royal Commission into British Nuclear Tests in Australia, meanwhile, had recently found that the Maralinga test sites of the 1950s were still, decades later, contaminated with low levels of plutonium.[79]

By the end of the 1980s, the fear of nuclear winter had been surpassed by fears of the greenhouse effect. While National Aeronautics and Space Administration (NASA) scientist James Hansen presented his testimony to the US Senate Committee on Energy and Natural Resources in June 1988, the Greenhouse 87 conference in Melbourne focused on how the Australian region could "plan" for what the changed climate might be like in the year 2030.[80] Delegates from Australia and New Zealand agreed that "research undertaken in the northern hemisphere to produce regional assessments will have only limited application here" in the South Pacific. More local collaborations were necessary to improve climate modeling in the region so that a more detailed picture of local climate impacts might be prepared.[81] Scientists' growing awareness of the diverse manifestations of anthropogenic climate change combined with their improving technical capacity to produce regional climate models that could inform policy making and preparedness.

The anxieties these environmental problems produced found creative expression in George Turner's 1987 science fiction novel *The Sea and Summer*, which was published as *Drowning Towers* in the United States. Turner, a winner of the prestigious Australian literature prize the Miles Franklin Award, envisioned a Melbourne drowned as a result of rising sea levels in the middle of the twenty-first century. The population of this Melbourne would be cleaved into haves and have-nots, the Sweet and the Swill. This Melbourne, the reader learns, is the product of the "Greenhouse Culture"—where the forces of population growth, industrialization, and capitalism have gone unchecked. Its watery fate is human caused and, the characters suggest, avoidable.[82]

Although the prospect of rising sea levels was a distant one for many Australians, their island neighbors in the South Pacific were becoming increasingly concerned about their long-term futures.[83] The Australian Government established a sea level monitoring system across the Pacific Islands in 1991 to measure the variability of sea levels in the region. Their very real anxieties about rising sea levels reflected ongoing concerns in the South Pacific about the region's future in the wake of decolonization and the emerging agenda of sustainable development.[84] In terms of its climate diplomacy, however, Australia's subsequent efforts at the 1997 Third Conference of the Parties to the United Nations Framework Convention on Climate Change in Kyoto, Japan, fell far short of the hopes of their Pacific neighbors. As Cook Islands prime minister Sir Geoffrey Henry observed during the meeting, "Australia's insistence on protecting its coal and energy intensive industries was self-serving."[85]

Over a decade later, the fears of rising sea levels in the Pacific had intensified. At the 2009 Copenhagen climate change conference, the lead negotiator for the island nation of Tuvalu, Australian Ian Fry, frustrated at the meeting's lack of progress, tearfully urged his fellow delegates to sign a legally binding treaty to limit the rise in global temperature to 1.5°C: "The fate of my country (Tuvalu) rests in your hands."[86] In 2014, the Fijian village of Vunidogoloa relocated inland to reduce its susceptibility to flooding and sea level rise. Environmental campaigners declared the village to be the "first" community to relocate due to climate change.[87] That same year, then Australian prime minister Tony Abbott had declared coal as "good for humanity" at the opening of another mine.[88] Meanwhile, Ioane Teitiota of the South Pacific island nation of Kiribati had sought refugee status in New Zealand, arguing that he feared for his future because of the consequences of rising sea levels for his home.[89] Despite an appeal, New Zealand's courts rejected his claim to refugee status, and in September 2015, the world's "first climate change refugee" was deported.[90]

That Australia and New Zealand are invested in the climate futures of their island neighbors is not simply a product of proximity. Rather, these relationships derive from over a century of imperial governance and extraction, which continue in the form of foreign aid and refugee policy. From the turn of the twentieth century, the extractive enterprise of colonial mining companies transformed the Pacific islands of Nauru, Christmas Island, and Banaba (in Kiribati) into what anthropologist Kạterina Teaiwa calls a "sea of phosphate."[91] This phosphate imperialism forged deep lithospheric connections between the ancient soils of Australia's farmlands and the phosphate-rich sediments of these Pacific islands.[92] Australian and New Zealand soils were enriched for the profit of industrial agriculture, with devastating consequences for the peoples and ecologies of this archipelago. Banabans were exiled to Fiji from 1945 as their home became a phosphate quarry; less than a decade later, their island's resources depleted, Nauruan leaders considered a plan to resettle their people in Australia.[93] Might they have been among the first refugees of the Great Acceleration?

In his 2013 book *The Reef: A Passionate History*, historian Iain McCalman shares the stories of twelve individuals for whom the Great Barrier Reef has been a source of "terror," "nurture," or "wonder." Among the stories of wonder is the story of J. E. N. "Charlie" Veron. A coral expert and former chief scientist of the Australian Institute of Marine Science, Veron has researched the world's largest coral reef ecosystem since the 1970s, and discovered and described over 20 percent of known coral species.[94] At the Royal Society in London in July 2009, he sounded an alarm for the World Heritage Area in a provocative lecture titled "Is the Great Barrier Reef on Death Row?"[95] To an audience including Sir David Attenborough, he described how he had come to see corals as "canaries of climate change," as organic thermometers of a warming world.[96] In addition to their role as "custodians of geological history," corals are subject to mass bleaching and erosion arising from abnormally high sea temperatures, elevated levels of carbon dioxide, and ocean acidification.[97] Grimly concluding his lecture, Veron likened the experience of watching the steady decline of the Great Barrier Reef to "seeing a house on fire in slow motion . . . there's a fire to end all fires, and you're watching it in slow motion, and you have been for years."[98]

Reflecting on his early career in the 1970s, Veron recalls a time when it was commonplace to believe that "the oceans [were] limitless and the marine world indestructible."[99] Such a mindset had sustained a faith in the oceans to act as a planetary sponge for carbon dioxide. Five decades after Revelle and Suess proclaimed that humankind was conducting a "large scale geophysical experiment," the world's oceans have reached a third of their capacity to absorb greenhouse gases.[100] But what if the clock could be wound back? What if ocean-atmospheric processes could be harnessed to mitigate anthropogenic climate change? Californian entrepreneur Russ George tried to do just that. In 2012, he dumped about one hundred tons of iron sulfate into the Pacific Ocean as part of a "rogue" geoengineering scheme off the west coast of Canada. The iron spawned an artificial plankton bloom that spread across 10,000 square kilometers.

According to media reports, his intention was for the plankton to absorb carbon dioxide and then sink to the ocean bed, and in time, cool the planet. George also hoped that this geoengineering technique known as ocean fertilization might generate carbon credits. The dump took place in the Haida eddies about two hundred nautical miles west of the islands of Haida Gwaii, where he had convinced the council of the Haida village of Old Massett to help fund his scheme on the grounds that it would stimulate local salmon fisheries.[101] A year later, in the face of allegations that the project had contravened the United Nations Convention on Biological Diversity and the London Convention on the Prevention of Marine Pollution by Dumping of Wastes and Other Matter, the Haida Nation severed ties with George.[102] Unfazed, George continues "working to pioneer new methods to save the world," while the Pacific has become a laboratory for new projects for planetary change.[103]

This wider project to make the planetary personal, and the personal planetary, was echoed at the 2017 United Nations Climate Change Conference in Bonn, Germany, just fifty kilometers from one of Europe's biggest sources of greenhouse gas emissions, the Hambach open-cast lignite coal mine. Presiding over the meeting was the Pacific island nation of Fiji; so great were the demands of hosting such a conference that the island nation was unable to hold the event at home. The display and performance of Fijian culture and history symbolized just what was at stake at these negotiations—the past, present, and the future. Fijian prime minister Frank Bainimarama reminded the attendees that we are "all in the same canoe," the Pacific is planetary.[104]

Aside from the moral and ethical reasons to consider the challenges facing the peoples of the Pacific, centering their experiences of climate change helps to overcome what geographer Mike Hulme describes as a new form of climate determinism, what he calls "climate reductionism." Hulme locates such reductionism in the claims of scientists, analysts, and commentators who have elevated and isolated climate as the primary determinant of the past, present, and future. In these narratives of a "climate-shaped destiny" that derive from the hegemony of the natural sciences, he argues, the complexities of human and non-human interactions are lost, contingency overlooked, and human agency ignored.[105] Although the impact of rising sea levels on the future of the Pacific Islands and the Torres Strait are indeed the product of anthropogenic climate change, Hulme's critique of climate reductionism encourages us to look further—that their condition is so much more than these calculations might suggest. Focusing on these human experiences of a warming world makes tangible what literary scholar Rob Nixon describes as the "slow violence" of anthropogenic climate change, capitalism, colonialism, and extractivism, which separately and together "threaten in slow motion."[106]

The detonation of the atomic bombs over Hiroshima and Nagasaki in August 1945 transformed the Pacific into a planetary space. Already underway were the formative processes of the Pacific Anthropocene in the extractive enterprises of colonialism, first to fuel industrialization, and later, to sustain settler agriculture, with its devastating legacies for Indigenous and First Nations peoples. The unfolding Great Acceleration in the Pacific demands from historians and scientists working in the region closer attention to the ways their Anthropocene has unfolded in ways peculiar to this particular place. After all, it is here that human history has unfolded in deep time, over some 65,000 years or more, during which humans survived the massive temperature and sea level changes of the last ice age.[107] By placing the Anthropocene in the Pacific, settler environmental histories narrate across multiple scales the measurements, representations, and diverse drivers, manifestations, and experiences of planetary change. Listening more carefully to the region's Indigenous and First Nations peoples, and learning more of their Anthropocene, may further encourage closer epistemological and ontological reflection on the field's Eurocentric methods and approaches.[108]

Since writing earlier drafts of this chapter, Australia has weathered the long Savage Summer of 2019/20. Bushfires swept across the continent, making the air a focus for anxiety and control even before COVID-19 gripped the world. Winds took the bushfire smoke across the Pacific—New Zealand's glaciers turned brown from the smoke, ash, and dust that would likely accelerate the season's glacier melt. Still, further east, the smoke reached South America and continued to circumnavigate the globe over the Southern Ocean. For weeks the cities of Brisbane, Melbourne, Sydney, and Canberra were shrouded in smoke. Those that could stayed indoors, bought masks, and even packed their bags for the island state of Tasmania, where they hoped to protect themselves and their children. Air quality index readings above 200 are considered hazardous to health; on some days, these cities and their suburbs had the worst air quality in the world.[109]

Six months after the smoke cleared, Euahlayi researcher Bhiamie Eckford-Williamson gave evidence at the Australian Royal Commission into National Natural Disaster Arrangements, which the federal government had established in early 2020 to assess the nation's disaster preparedness in the wake of the recent bushfires. There, he shared the findings of a co-authored working paper that studied the impacts of the bushfires on Aboriginal peoples.[110] His research found that the Indigenous population in areas hit by the bushfires was double the Indigenous population in the affected states as a whole, meaning that Aboriginal people had been "disproportionately affected" by the inferno. Further still, he reported, one in ten children affected by the bushfires is Indigenous. Of the twenty-two Aboriginal communities living in rural fire-affected areas, twenty are in New South Wales, where they often live on the mission lands to which they had once been forcibly removed under assimilationist government policies.[111] As the intensity and frequency of such extreme bushfire events are likely to increase in a rapidly warming world, a reckoning with colonial pasts is long overdue. By placing the Anthropocene, environmental historians can contribute to more just and equitable futures.

Notes

1. Tony Birch, "Having Gone, I Will Come Back," in *Living with the Anthropocene: Love, Loss and Hope in the Face of Environmental Crisis*, ed. Cameron Muir, Kirsten Wehner, and Jenny Newell (Sydney: New South Publishing, 2020), 27.

2. Paul Crutzen and Eugene Stoermer, "The Anthropocene," *Global Change Newsletter* 41 (2000): 1317.

3. For a historical overview of these turning points, see Nancy J. Jacobs, Danielle Johnstone, and Christopher Kelly, "The Anthropocene from Below," in *World Histories from Below: Disruption and Dissent, 1750 to the Present*, ed. Antoinette Burton and Tony Ballantyne (London: Bloomsbury, 2016), 197–230.

4. For critiques of the Anthropocene, see Dipesh Chakrabarty, "The Climate of History: Four Theses," *Critical Inquiry* 35, no. 2 (2009): 197–222; Andreas Malm and Alf

Hornborg, "The Geology of Mankind? A Critique of the Anthropocene Narrative," *Anthropocene Review* 1 (2014): 62–69.

5. Sverker Sörlin, "The Anthropocene: What Is It?" *IAS Letter* (Summer 2014): 13.

6. Donna Haraway, "Anthropocene, Capitalocene, Plantationocene, Chthulucene: Making Kin," *Environmental Humanities* 5, no. 1 (2015): 159–165; Jason W. Moore, "The Capitalocene, Part I: On the Nature and Origins of Our Ecological Crisis," *Journal of Peasant Studies* 44 (2017): 594–630.

7. Laura Pulido, "Racism and the Anthropocene," in *Future Remains: A Cabinet of Curiosities for the Anthropocene,* ed. Gregg Mitman, Marco Armiero, and Robert Emmett (Chicago: University of Chicago Press, 2018), 124.

8. Anna Lowenhaupt Tsing, "A Feminist Approach to the Anthropocene: Earth Stalked by Man," Helen Pond McIntyre '48 Lecture, New York, Barnard College, December 18, 2015, www.youtube.com/watch?v=ps8J6a7g_BA (accessed February 26, 2021).

9. See Sherilyn McGregor and Nicole Seymour, "Introduction," *RCC Perspectives: Men and Nature—Hegemonic Masculinities and Environmental Change* 4 (2017): 9.

10. Stefania Barca, *Forces of Reproduction: Notes for a Counter-Hegemonic Anthropocene* (New York: Cambridge University Press, 2020).

11. Gabrielle Hecht, "Interscalar Vehicles for an African Anthropocene: On Waste, Temporality, and Violence," *Cultural Anthropology* 33, no. 1 (2018): 111.

12. For example, Mark Whitehead, *Environmental Transformations: A Geography of the Anthropocene* (New York: Routledge, 2014); Bruce Braun, Mat Coleman, Mary Thomas, and Kathryn Yusoff, "Grounding the Anthropocene: Sites, Subjects, Struggles in the Bakken Oil Fields: A Report from an Antipode Foundation International Workshop," *Antipode* (November 3, 2015), https://wp.me/p16RPC-1gw (accessed February 26, 2021).

13. Rob Nixon, "The Anthropocene: The Promise and Pitfalls of an Epochal Idea," in Mitman, Armiero, and Emmett, *Future Remains*, 1.

14. Lawrence Buell, *The Future of Environmental Criticism: Environmental Crisis and Literary Imagination* (Oxford: Blackwell, 2005), 68. Anthropologist Mark J. Hudson has also examined the place of Asia in the Anthropocene; see "Placing Asia in the Anthropocene: Histories, Vulnerabilities, Responses," *Journal of Asian Studies* 73, no. 4 (2014): 941–962.

15. Katie Holmes, Andrea Gaynor, and Ruth A. Morgan, "Doing Environmental History in Urgent Times," *History Australia* 17, no. 2 (2020): 234.

16. Tom Griffiths, "How Many Trees Make a Forest? Cultural Debates about Vegetation Change in Australia," *Australian Journal of Botany* 50 (2002): 377–378.

17. Tom Griffiths, "Coming of Age in the Great Acceleration," *Australian Book Review* no. 366 (2014): 8, 10–11.

18. Dipesh Chakrabarty, "The Climate of History: Four Theses," *Critical Inquiry* 35, no. 2 (2009): 197–222.

19. Libby Robin, "History for Global Anxiety," *RCC Perspectives: The Future of Environmental History—Needs and Opportunities* (2011): 42.

20. Libby Robin and Will Steffen, "History for the Anthropocene," *History Compass* 5, no. 5 (2007): 1694–1719.

21. J. R. McNeill, *Something New under the Sun: An Environmental History of the Twentieth Century World* (New York: W. W. Norton, 2000). See also, J. R. McNeill and Peter Engelke, *The Great Acceleration: An Environmental History of the Anthropocene since 1945* (Cambridge, MA: Harvard University Press, 2016).

22. Robin and Steffen, "History for the Anthropocene," 1694–1719.

23. Libby Robin, "Histories for Changing Times: Entering the Anthropocene?" *Australian Historical Studies* 44, no. 2 (2013): 329–340.

24. David Christian, *Maps of Time: An Introduction to Big History* (Berkeley: University of California Press, 2011).

25. Alison Bashford, "The Anthropocene Is Modern History: Reflections on Climate and Australian Deep Time," *Australian Historical Studies* 44, no. 2 (2013): 341–349; Chris Otter, Alison Bashford, John L. Brooke, Frederik Jonsson, and Jason M. Kelly, "Roundtable: The Anthropocene in British History," *Journal of British Studies* 57, no. 3 (2018): 568–596; Ann McGrath and Mary-Anne Jebb, eds., *Long History, Deep Time: Deepening Histories of Place* (Canberra: ANU Press, 2015); Billy Griffiths, *Deep Time Dreaming: Uncovering Ancient Australia* (Melbourne: Black Inc., 2018).

26. Bashford, "The Anthropocene Is Modern History," 342.

27. See, for example, Andreas Malm, "Who Lit This Fire? Approaching the History of the Fossil Economy," *Critical Historical Studies* 3, no. 2 (2016): 215–248.

28. On Newcastle's role in the Pacific coal trade, see Michael Clark, "'Bound out for Callao!' The Pacific Coal Trade, 1876–1897," *Great Circle* 28, no. 2 (2006): 26–45; L. E. Fredman, "Coals from Newcastle: Aspects of the Trade with California," *Australian Journal of Politics and History* 29, no. 3 (1983): 440–447.

29. Bashford, "The Anthropocene Is Modern History," 347.

30. Melita Keywood, Kathryn Emmerson, and Mark Hibberd, "Climate: Australia's Emissions in Context," *Australia: State of the Environment 2016*, Australian Government Department of the Environment and Energy, Canberra, https://doi.org/10.4226/94/58b65c70bc372.

31. *Lifou Declaration: Paris 2015: Save Oceania!*, 3rd Oceania 21 Summit, New Caledonia, April 30, 2015, http://www.sprep.org/attachments/VirLib/New_Caledonia/Lifou_Declaration_2015.pdf (accessed February 26, 2021).

32. "The Pacific Warrior Journey," *The Pacific Climate Warriors*, https://world.350.org/pacificwarriors/the-pacific-warrior-journey/ (accessed February 26, 2021). For a wider view of Pacific resistance in the context of climate change, see Katerina Teaiwa, "Our Rising Sea of Islands: Pan-Pacific Regionalism in the Age of Climate Change," *Pacific Studies* 41, no. 1–2 (2018): 26–54.

33. Marian Faa, "Torres Strait Islander Complaint against Climate Change Inaction Wins Backing of UN Legal Experts," *ABC Pacific Beat*, December 11, 2020, https://www.abc.net.au/news/2020-12-11/torres-strait-islander-complaint-against-climate-change-inaction/12972926 (accessed February 26, 2021).

34. Simon L. Lewis and Mark A. Maslin, "Defining the Anthropocene," *Nature* 519 (2015): 171–180.

35. Kyle Powys Whyte, "Is It Colonial Déjà Vu? Indigenous Peoples and Climate Injustice," in *Humanities for the Environment*, 88–105.

36. Zoe Todd, "Relationships," in *Anthropocene Unseen: A Lexicon*, ed. Cymene Howe and Anand Pandian (Brooklyn, NY: Punctum Books, 2020), 381–384 (italics in original).

37. Todd, "Relationships," 382.

38. Heather Davis and Zoe Todd, "On the Importance of a Date, or Decolonizing the Anthropocene," *ACME: An International Journal for Critical Geographies* 16, no. 4 (2017): 761–780.

39. Tony Birch, "'We've Seen the End of the World and We Don't Accept It': Protection of Indigenous Country and Climate Justice," in *Towards a Just and Ecologically Sus-*

tainable Peace, ed. Joseph Camilleri and Deborah Guess (Singapore: Palgrave Macmillan, 2020), 251–273.

40. Davis and Todd, 764; See also Zoe Todd, "An Indigenous Feminist's Take on the Ontological Turn: 'Ontology' Is Just Another Word for Colonialism," *Journal of Historical Sociology* 29, no. 1 (2016): 4–22.

41. Cameron Muir, Jennifer Newell, and Kirsten Wehner, "A Storm of Our Own Making," in *Living with the Anthropocene*, 7.

42. Jan Zalasiewicz, Mark Williams, and Colin N. Waters, "Can an Anthropocene Series Be Defined and Recognised?" *Geological Society, London, Special Publications* 395 (2014): 39–53.

43. Will Steffen et al., "The Trajectory of the Anthropocene: The Great Acceleration," *Anthropocene Review* (2015): 1–18; Jan Zalasiewicz et al., "When Did the Anthropocene Begin? A Mid-Twentieth Century Boundary Level Is Stratigraphically Optimal," *Quaternary International* 383 (2015): 195–203.

44. McNeill, *Something New under the Sun*, 4; Steffen et al., "The Trajectory of the Anthropocene," 4, 6–7.

45. Roger Revelle and Hans E. Suess, "Carbon Dioxide Exchange between Atmosphere and Ocean and the Question of an Increase of Atmospheric CO_2 during the Past Decades," *Tellus* 9 (1957): 18–27; Wallace S. Broecker, "Unpleasant Surprises in the Greenhouse?" *Nature* 328 (1987): 123–126.

46. McNeill, *Something New under the Sun*, 3–4.

47. Waters et al., "The Anthropocene Is Functionally and Stratigraphically Distinct from the Holocene," https://doi.org/10.1126/science.aad2622.

48. Tony Birch, "'The Lifting of the Sky': Life Outside the Anthropocene," in *Humanities for the Environment: Integrating Knowledge, Forging New Constellations of Practice*, ed. Joni Adamson and Michael Davis (New York: Routledge, 2016), 196–209; Heather Goodall, "Damage and Dispossession: Indigenous People and Nuclear Weapons on Bikini Atoll and the Pitjantjatjara Lands, 1946 to 1988," in *The Routledge Companion to Indigenous Global History*, ed. Lynette Russell and Ann McGrath (London: Routledge, forthcoming), with many thanks to Heather for generously sharing her unpublished work.

49. Jean-Marc Regnault, "The Nuclear Issue in the South Pacific: Labor Parties, Trade Union Movements, and Pacific Island Churches in International Relations," *Contemporary Pacific* 17, no. 2 (2005): 339–357.

50. David Seed, *Under the Shadow: The Atomic Bomb and Cold War Narratives* (Kent, OH: Kent State University Press, 2012), 45.

51. William Golding, *Lord of the Flies* (London: Faber and Faber, 1954).

52. Nevil Shute, *On the Beach* (Melbourne: Heinemann, 1957).

53. For an Australian reflection on the launch of Sputnik, see Griffiths, "Coming of Age in the Great Acceleration," 8, 10–11.

54. Rip Bulkeley, "The Sputniks and the IGY," in *Reconsidering Sputnik: Forty Years since the Soviet Satellite*, ed. R. D. Launius, J. M. Logsdon, and R. W. Smith (London: Routledge, 2002), 125–160; Roger D. Launius, James Rodger Fleming, and David H. DeVorkin, "Rise of Global Scientific Inquiry in the International Polar and Geophysical Years," in *Globalizing Polar Science: Reconsidering the International Polar and Geophysical Years*, ed. R. D. Launius, J. R. Fleming, and D. H. DeVorkin (New York: Palgrave Macmillan, 2010), 1–12.

55. Herbert Friedman, "The Legacy of the IGY," *Eos* 64, no. 32 (1983): 497–499.

56. Spencer R. Weart, *The Discovery of Global Warming* (Cambridge, MA: Harvard University Press, 2008), 26–27.

57. Jacob Darwin Hamblin, "Seeing the Oceans in the Shadow of Bergen Values," *Isis* 105, no. 2 (2014): 352–363.

58. Weart, *The Discovery of Global Warming*, 27–28; Revelle and Suess, "Carbon Dioxide Exchange between Atmosphere and Ocean and the Question of an Increase of Atmospheric CO_2 during the Past Decades," 18–27.

59. Charles D. Keeling, "Rewards and Penalties of Monitoring the Earth," *Annual Review of Energy and the Environment* 23 (1998): 2–82. See also, James Rodger Fleming, *Inventing Atmospheric Science: Bjerknes, Rossby, Wexler and the Foundations of Modern Meteorology* (Cambridge, MA: MIT Press, 2016).

60. Weart, *The Discovery of Global Warming*, 35.

61. D. J. Pack, Letter to A. J. Dyer, July 19, 1974, A8520, WE3/1 Part 1, National Archives of Australia, Canberra; Ruth A. Morgan, "Southern Skies: Australian Atmospheric Research and Global Climate Change," *Disaster Prevention and Management* 30, no. 1 (2021): 47–63.

62. Warren S. Wooster, ed., *Scientific Exploration of the South Pacific: Proceedings of a Symposium Held during the Ninth General Meeting of the Scientific Committee on Oceanic Research, June 18–20, 1968* (Washington, DC: National Academy of Sciences, 1970), 1.

63. Wooster, ed., *Scientific Exploration of the South Pacific.*

64. Bruce A. Warren, "General Circulation of the South Pacific," in *Scientific Exploration of the South Pacific*, 33.

65. Sayed Z. El-Sayed, "Phytoplankton Production of the South Pacific and the Pacific Sector of the Antarctic," in *Scientific Exploration of the South Pacific*, 208.

66. Ibid.

67. Jacob Darwin Hamblin, *Oceanographers and the Cold War: Disciples of Marine Science* (Seattle: University of Washington Press, 2005).

68. William J. Burroughs, *Does the Weather Really Matter? The Social Implications of Climate Change* (Cambridge: Cambridge University Press, 1997), 88.

69. Donella H. Meadows, Dennis L. Meadows, Jørgen Randers, and William W. Behrens, *The Limits to Growth: A Report for the Club of Rome's Project on the Predicament of Mankind* (New York: Universe Books, 1972).

70. William W. Kellogg, "Mankind's Impact on Climate: The Evolution of an Awareness," *Climatic Change* 10 (1987): 121–122.

71. A. Barrie Pittock, Lawrence A. Frakes, Dick Jenssen, Jim A. Peterson, and John W. Zillman, eds., "Preface," in *Climatic Change and Variability* (New York: Cambridge University Press, 1980), np.

72. A. Barrie Pittock, "An Overview," in *Climatic Change and Variability*, ed. Pittock, Frakes, Jenssen, Peterson, and Zillman, 4.

73. Jim Salinger, "Climatic Change and Variability," *New Zealand Geographer* (October 1979); F. Alayne Street, "Book Reviews," *Progress in Physical Geography: Earth and Environment* 3 (1979): 158.

74. Michael H. Glantz, *Currents of Change: Impact of El Niño and La Niña on Climate and Society* (New York: Cambridge University Press, 2001), 84–101.

75. Gregory T. Cushman, "Choosing between Centers of Action: Instrument Buoys, El Niño, and Scientific Internationalism in the Pacific, 1957–1982," in *The Machine in Neptune's Garden: Historical Perspectives on Technology and the Marine Environment*, ed. H. M. Rozwadowski and D. K. van Keuren (Sagamore Beach, MA: Science History Publications, 2004), 162–164; Michael J. McPhaden, "El Niño and Ocean Observations," in *Physical*

Oceanography: Developments since 1950, ed. M. Jochum and R. Murtugudde (New York: Springer, 2006), 79–99.

76. Paul J. Crutzen and John W. Birks, "The Atmosphere after a Nuclear War: Twilight at Noon," *Ambio* 11, no. 114 (1982): 114–125; Matthias Dorries, "The Politics of Atmospheric Sciences: 'Nuclear Winter' and Global Climate Change," *Osiris* 26, no. 1 (2011): 198–223.

77. A. Barrie Pittock et al., *SCOPE 28: Environmental Consequences of Nuclear War, Vol. 1—Physical and Atmospheric Effects* (Chichester, UK: Wiley & Sons, 1986), xxvi. (Emphasis in original.)

78. Cited in Lawrence Badash, *A Nuclear Winter's Tale: Science and Politics in the 1980s* (Cambridge, MA: MIT Press, 2009), 262.

79. James R. McClelland (Chair), *Royal Commission into British Nuclear Tests in Australia (1984–1985)* (Canberra: AGPS, 1985).

80. James Hansen, "Statement of Dr James Hansen, Director, NASA Goddard Institute for Space Studies," and "The Greenhouse Effect: Impacts on Current Global Temperature and Regional Heat Waves," Testimony before the Committee on Energy and Natural Resources, Washington, DC, US Senate, June 23, 1988.

81. C. Bee, W. Furler, and N. Quinn, "Greenhouse: International and National Policy Approaches," in *Greenhouse: Planning for Climate Change*, ed. Graeme I. Pearman (East Melbourne, Vic: CSIRO Publications, 1988), 728.

82. George Turner, *The Sea and Summer* (London: Orion Publishing Group, 2013 [1987]). Elsewhere, I have compared the scientific and literary imaginings of CSIRO and Turner, "Imagining a Greenhouse Future: Scientific and Literary Depictions of Climate Change in 1980s Australia," *Australian Humanities Review* 57 (2014): 4360.

83. Jean-Marc Verstraete, "Low Frequency Sea Level Variability in the Western Tropical Pacific, 1992–1998," in *Sea Level Changes and Their Effects*, ed. J. Noye and M. Grzechnik (Singapore: World Scientific Publishing, 2001), 125–214.

84. Geoffrey Bertram, "'Sustainable Development' in Pacific Micro-economies," *World Development* 14, no. 7 (1986): 809–822; Florian Gubon, "Steps Taken by South Pacific Island States to Preserve and Protect Ocean Resources," in *Freedom for the Seas in the 21st Century: Ocean Governance and Environmental Harmony*, ed. J. M. van Dyke, D. Zaelke, and G. Hewison (Washington, DC: Island Press, 1993), 121–130.

85. Cited in Xiaojiang Yu and Roslyn Taplin, "The Australian Position at the Kyoto Conference," in *Climate Change in the South Pacific: Impacts and Responses in Australia, New Zealand, and Small Island States*, ed. Alexander Gillespie and William C. G. Burns (Dordrecht: Springer, 2000), 113.

86. Ben Block, "Interview with Tuvalu Climate Negotiator Ian Fry," *Worldwatch Institute*, 2013, http://forestindustries.eu/de/content/interview-tuvalu-climate-negotiator-ian-fry (accessed February 26, 2021).

87. Michael Green, "Contested Territory," *Nature Climate Change* 6 (2016): 817–820.

88. "Coal 'Good for Humanity,' Prime Minister Tony Abbott Says at $3.9b Queensland Mine Opening," *ABC News*, October 13, 2014, https://www.abc.net.au/news/2014-10-13/coal-is-good-for-humanity-pm-tony-abbott-says/5810244 (accessed February 26, 2021).

89. Bernard Lagan, "Australia Urged to Prepare for Influx of People Displaced by Climate Change," *Guardian*, April 16, 2013, http://www.theguardian.com/environment/2013/apr/16/australia-climate-change-refugee-status (accessed February 26, 2021).

90. Tim McDonald, "The Man Who Would Be the World's First Climate Change Refugee," *BBC News*, November 5, 2015, http://www.bbc.com/news/world-asia-34674374 (accessed February 26, 2021). Although New Zealand rejected climate change as grounds

for refugee status for a Tuvalu family in 2014, the country's Immigration and Protection Tribunal allowed the family to remain in New Zealand due to "exceptional circumstances of a humanitarian nature." See Kelly Buchanan, *New Zealand: 'Climate Change Refugee' Case Overview* (Washington, DC: Law Library of Congress, 2015).

91. Katerina Teaiwa, "Ruining Pacific Islands: Australia's Phosphate Imperialism," *Australian Historical Studies* 46, no. 3 (2015): 374–391.

92. See also Zachary Caple and Gregory T. Cushman, "The Phosphorus Apparatus," *Technosphere Magazine* (2016), https://technosphere-magazine.hkw.de/p/1-The-Phosphorus-Apparatus-czfdPRXcpUj4nxj8aQQ1GZ.

93. Katerina Teaiwa, *Consuming Ocean Island: Stories of People and Phosphate from Banaba* (Bloomington: Indiana University Press, 2015); Gil Marvel Tabucanon and Brian Opeskin, "The Resettlement of Nauruans in Austraila," *Journal of Pacific History* 46, no. 3 (2011): 337–356.

94. Iain McCalman, *The Reef: A Passionate History* (Melbourne: Penguin, 2013), 238.

95. McCalman, *The Reef*, 249–274; J. E. N. Veron, *A Reef in Time: The Great Barrier Reef from Beginning to End* (Cambridge, MA: Belknap Press, 2008).

96. Veron cited in McCalman, *The Reef*, 258.

97. J. E. N. Veron et al., "The Coral Reef Crisis: The Critical Importance of <350ppm CO_2," *Marine Pollution Bulletin* 58 (2009): 1428–1436.

98. Veron cited in McCalman, *The Reef*, 261.

99. Veron cited in McCalman, *The Reef*, 257.

100. IPCC, "Summary for Policymakers," in *Climate Change 2014: Synthesis Report. Contribution of Working Groups I, II, and III to the Fifth Assessment Report of the Intergovernmental Panel on Climate Change*, ed. IPCC (Geneva: IPCC, 2015), 4.

101. Bruce Falconer, "Can Anyone Stop the Man Who Will Try Just about Anything to Put an End to Climate Change?" *Pacific Standard* (January 16, 2018), http://www.psmag.com/books-and-culture/battlefield-earth-can-anyone-stop-man-will-try-just-anything-fix-climate-78957 (accessed February 26, 2021).

102. David Biello, "Pacific Ocean Hacker Speaks Out," *Scientific American*, October 24, 2012, http://www.scientificamerican.com/article/questions-and-answers-with-rogue-geoengineer-carbon-entrepreneur-russ-george/ (accessed February 26, 2021); Martin Lukacs, "World's Biggest Geoengineering Experiment 'Violates' UN Rules," *Guardian*, October 15, 2012, http://www.theguardian.com/environment/2012/oct/15/pacific-iron-fertilisation-geoengineering (accessed 26 February 2021). For a study of local responses to the scheme, see Kate E. Gannon and Mike Hulme, "Geoengineering at the 'Edge of the World': Exploring Perceptions of Ocean Fertilisation through the Haida Salmon Restoration Corporation," *Geo* (2018): doi.10.1002/geo2.54.

103. Russ George, "Living Outside the Box," *Russ George*, 2021, http://russgeorge.net (accessed February 26, 2021).

104. "Incoming COP23 President Launches Partnership Days in Fiji," *UN Climate Change Conference: COP23*, October 16, 2017, https://cop23.com.fj/incoming-cop23-president-launches-partnership-days-fiji/ (accessed February 26, 2021).

105. Mike Hulme, "Reducing the Future to Climate: A Story of Climate Determinism and Reductionism," *Osiris* 26, no. 1 (2011): 245–266.

106. Rob Nixon, *Slow Violence and the Environmentalism of the Poor* (Cambridge, MA: Harvard University Press, 2011).

107. Tom Griffiths, "Environmental History, Australian Style," *Australian Historical Studies* 46, no. 2 (2015): 173; Patrick Nunn, *The Edge of Memory: Ancient Stories, Oral Tradition*

and the Post-Glacial World (London: Bloomsbury, 2018); Ruth A. Morgan, "The Continent without a Cryo-History? Deep Time and Water Scarcity in Arid Settler Australia," *Journal of Northern Studies* 13, no. 2 (2019): 43–70.

108. See, for example, in the Canadian context, Lianne Leddy, "Intersections of Indigenous and Environmental History," *Canadian Historical Review* 98, no. 1 (2017): 83–95; Lianne Leddy, "Historical Sources and the Beothuk: Questioning Settler Interpretations," in *Tracing Ochre: Changing Perspectives on the Beothuk*, ed. Fiona Polack (Toronto: University of Toronto Press, 2018), 199–219.

109. Climate Council, "Air Quality: What You Need to Know," *Climate Council*, January 14, 2020, https://www.climatecouncil.org.au/resources/air-quality/ (accessed February 26, 2021).

110. Bhiamie Williamson, Francis Markham, and Jessica K. Weir, *Aboriginal Peoples and the Response to the 2019–2020 Bushfires*, CAEPR Working Paper no. 134 (Canberra: ANU Centre for Aboriginal Economic Policy Research, 2020).

111. Bhiamie Williamson, Francis Markham, and Jessica K. Weir, "1 in 10 Children Affected by the Bushfires Is Indigenous: We've Been Ignoring Them for Too Long," *The Conversation (Australia)*, April 2, 2020, https://theconversation.com/1-in-10-children-affected-by-bushfires-is-indigenous-weve-been-ignoring-them-for-too-long-135212 (accessed March 1, 2021).

About the Contributors

James Beattie is an award-winning environmental and world historian whose work focuses on the Asia-Pacific region, mostly over the last two hundred years. His many books and articles explore cross-cultural exchanges occasioned by British imperialism, especially the nexus between environment, gardens, health, and art. His current book projects are an environmental history of Chinese migration in the Pacific; a history of Chinese art collecting; and a historical archaeology of a "Chinatown." Beattie is also founding editor of *International Review of Environmental History*, and he co-edits the book series Palgrave Studies in World Environmental History and Routledge Research on Gardens in History. He is associate professor, Victoria University of Wellington, and holds various research associateships, including Senior Research Associate, Department of History, Faculty of Humanities, University of Johannesburg and Visiting Professor, Department of History, Sun Yat-sen University.

William Matt Cavert is a faculty member at the University of Hawai'i–West O'ahu. His prior publications have appeared in the *Journal of Pacific History* and the *Contemporary Pacific*, as well as the *Teaching Oceania* series from the Center for Pacific Island Studies. He is a member of the Biocultural Initiative of the Pacific at the University of Hawai'i at Mānoa. His research is primarily centered on the French colonial presence in the Pacific and its intersections with health, medicine, science, and the environment.

Hannah Cutting-Jones received her PhD in history from the University of Auckland in 2018. She studies food history with a focus on the impact of colonization and religion on foodways in the Pacific and meat production in settler-colonial societies. Cutting-Jones is currently teaching in the Department of History and the Clark Honors College at the University of Oregon in Eugene, Oregon, where she has developed and taught several courses, including the History of Vegetarianism, the Philosophy of Food, and a senior seminar, Food and Empire. She published an article in the *Journal of Pacific History* in 2020 and is working on a book manuscript based on her thesis, *Feasts of Change: Food and History in the Cook Islands*, and a new project on the social and cultural history of protein.

HOLGER DROESSLER is an assistant professor of history at Worcester Polytechnic Institute in Massachusetts, United States. He is a historian of nineteenth- and twentieth-century US history, with a special focus on imperialism, capitalism, and the Pacific Ocean. In his book *Coconut Colonialism: Workers and the Globalization of Samoa* (Harvard University Press, 2021), he argues that the globalization of Samoa at the turn of the twentieth century was driven by a diverse group of working people on and off the islands. He has published on a variety of topics, including labor history, Pacific history, environmental history, and global hip-hop.

KATSUYA HIRANO teaches history at the University of California, Los Angeles. He is the author of *The Politics of Dialogic Imagination: Power and Popular Culture in Early Modern Japan* (University of Chicago Press, 2014). He has published numerous articles and book chapters on the cultural and intellectual history of early modern and modern Japan, the Fukushima nuclear disaster, settler colonialism, and critical theory, including "Thanatopolitics in the Making of Japan's Hokkaido: Settler Colonialism and Primitive Accumulation" (*Critical Historical Studies*).

RYAN TUCKER JONES is Ann Swindells Associate Professor at the University of Oregon. He is the author of *Empire of Extinction: Russians and the North Pacific's Strange Beasts of the Sea, 1741–1867* (Oxford University Press, 2014) and *Red Leviathan: The Secret History of Soviet Whaling* (University of Chicago Press, 2022).

BENJAMIN MADLEY is associate professor of history at the University of California, Los Angeles. Educated at Yale and Oxford, he writes about Native American history as well as colonialism in Africa, Australia, and Europe, often applying a transnational and comparative approach. His work has appeared in journals ranging from *The American Historical Review* to *The Western Historical Quarterly* as well as several edited volumes. In 2016, Yale University Press published his prize-winning first book, *An American Genocide: The United States and the California Indian Catastrophe, 1846–1873.*

JOHN R. MCNEILL is University Professor at Georgetown University. His latest book, written together with his former student, Peter Engelke, is *The Great Acceleration: An Environmental History of the Anthropocene since 1945*

(Harvard University Press, 2016). His previous one, *Mosquito Empires: Ecology and War in the Greater Caribbean, 1620–1914* (Cambridge University Press, 2010) won the Beveridge Prize of the American Historical Association for the best book on the history of the Western Hemisphere.

EDWARD DALLAM MELILLO is the William R. Kenan, Jr. Professor of History and Environmental Studies at Amherst College. His books include *The Butterfly Effect: Insects and the Making of the Modern World* (Penguin Random House, 2020) and *Strangers on Familiar Soil: Rediscovering the Chile-California Connection* (Yale University Press, 2015), which won the Western History Association's 2016 Caughey Prize for the most distinguished book on the American West. He is the co-editor (with James Beattie and Emily O'Gorman) of *Eco-cultural Networks in the British Empire: New Views on Environmental History* (Bloomsbury Press, 2015). In 2017, Melillo received a Mellon New Directions Fellowship to spend a year studying ʻōlelo Hawaiʻi on Oʻahu. He received his PhD and his MPhil from Yale University and his BA from Swarthmore College.

RUTH A. MORGAN is an environmental historian and historian of science at the Australian National University, where she is the director of the Centre for Environmental History. She has published widely on the climate and water histories of Australia and the British Empire, including her award-winning book *Running Out? Water in Western Australia* (University of Western Australia Press, 2015). Her current project, on environmental exchanges between British India and the Australian colonies, has been generously supported by the Australian Research Council and the Alexander von Humboldt Foundation. She is co-author of the forthcoming book *Cities in a Sunburnt Country: Water and the Making of Urban Australia* (Cambridge University Press), and a lead author in Working Group II of the Intergovernmental Panel on Climate Change's Assessment Report 6.

EMILY O'GORMAN situates her research within environmental history, more-than-human geography, and the interdisciplinary environmental humanities, and is primarily concerned with contested knowledges within broader cultural framings of authority, expertise, and landscapes. She holds a PhD from the Australian National University and is currently a senior lecturer at Macquarie University in Sydney, Australia. Her research has been supported by nationally competitive research grants as well as a Carson Writing Fellowship at the Rachel Carson Center in Munich, Germany, from 2014 to 2015. She is the author of *Flood Country: An Environmental History of the Murray-Darling Basin* (CSIRO Publishing, 2012) and *Wetlands in a Dry Land: More-Than-Human*

Histories of Australia's Murray-Darling Basin (University of Washington Press, 2021), and co-editor (with James Beattie and Matthew Henry) of *Climate, Science, and Colonization: Histories from Australia and New Zealand* (Palgrave Macmillan, 2014) and (with James Beattie and Edward Melillo) *Eco-cultural Networks and the British Empire: New Views on Environmental History* (Bloomsbury, 2015). She co-leads the Environmental Humanities research group at Macquarie University, was a founding associate editor of the journal *Environmental Humanities* (2012–2014) and a founding co-editor of the Living Lexicon in that journal (2014–2020).

Gregory Samantha Rosenthal (they/them or she/her) is associate professor of history and coordinator of the Public History Concentration at Roanoke College in Salem, Virginia. She is the author of two books, *Beyond Hawai'i: Native Labor in the Pacific World* (University of California Press, 2018), and *Living Queer History: Remembrance and Belonging in a Southern City* (University of North Carolina Press, 2021). Their work has received awards and recognition from the National Council on Public History, the Committee on LGBT History, the American Society for Environmental History, and the Working Class Studies Association.

N. Ha'alilio Solomon is an instructor in the Kawaihuelani Center for Hawaiian Language at the University of Hawai'i at Mānoa, where he is currently a doctoral student in linguistics. Ha'alilio advocates for the reclamation of 'ōlelo Hawai'i, the Hawaiian language, through education, outreach, and translation. His dissertation focuses on linguistic ideologies, pedagogy, and revitalization.

Lissa Wadewitz is professor of history and environmental studies at Linfield University in McMinnville, Oregon. She received her PhD in US history from the University of California, Los Angeles. Her first book, *The Nature of Borders: Salmon, Boundaries, and Bandits on the Salish Sea* (University of Washington Press, 2012) won book prizes from the American Historical Association, the Western History Association, and the North American Society for Oceanic History. Her current research is focused on the intersections of race, sexuality, labor, and the environment in the nineteenth-century US Pacific whaling fleet. She teaches classes on environmental history, Native American history, and the history of the US West.

Kristin A. Wintersteen is a scholar of modern Latin America, environmental history, and global food studies. Her book *The Fishmeal Revolution: The Industrialization of the Humboldt Current Ecosystem*, was published by the

University of California Press (2021). She earned her PhD in history from Duke University.

Frank Zelko is associate professor of history at the University of Hawai'i at Mānoa. His research focuses on the global history of environmentalism and the nexus between environment and health. He is the author of *Make It a Green Peace: The Rise of Countercultural Environmentalism* (Oxford University Press, 2013) and the recipient of the 2019 Alice Hamilton Prize from the American Society for Environmental History.

Index

Page numbers in boldface type refer to illustrations.

 **Perspectives on the Global Past**

Anand A. Yang and Kieko Matteson
SERIES EDITORS

Interactions: Transregional Perspectives on World History
Edited by Jerry H. Bentley, Renate Bridenthal, and Anand A. Yang

Contact and Exchange in the Ancient World
Edited by Victor H. Mair

Seascapes: Maritime Histories, Littoral Cultures, and Transoceanic Exchanges
Edited by Jerry H. Bentley, Renate Bridenthal, and Kären Wigen

Anthropology's Global Histories: The Ethnographic Frontier in German New Guinea, 1870–1935
Rainer F. Buschmann

Creating the New Man: From Enlightenment Ideals to Socialist Realities
Yinghong Cheng

Glamour in the Pacific: Cultural Internationalism and Race Politics in the Women's Pan-Pacific
Fiona Paisley

The Qing Opening to the Ocean: Chinese Maritime Policies, 1684–1757
Gang Zhao

Navigating the Spanish Lake: The Pacific in the Iberian World, 1521–1898
Rainer F. Buschmann, Edward R. Slack Jr., and James B. Tueller

Nomads as Agents of Cultural Change: The Mongols and Their Eurasian Predecessors
Edited by Reuven Amitai and Michal Biran

Sea Rovers, Silver, and Samurai: Maritime East Asia in Global History, 1550–1700
Edited by Tonio Andrade and Xing Hang

Exile in Colonial Asia: Kings, Convicts, Commemoration
Edited by Ronit Ricci

Burnt by the Sun: The Koreans of the Russian Far East
Jon K. Chang

Shipped but Not Sold: Material Culture and the Social Protocols of Trade during Yemen's Age of Coffee
Nancy Um

Encounters Old and New in World History: Essays Inspired by Jerry H. Bentley
Edited by Alan Karras and Laura J. Mitchell

At the Edge of the Nation: The Southern Kurils and the Search for Russia's National Identity
Paul B. Richardson

Liminality of the Japanese Empire: Border Crossings from Okinawa to Colonial Taiwan
Hiroko Matsuda

Sudden Appearances: The Mongol Turn in Commerce, Belief, and Art
Roxann Prazniak

A Power in the World: The Hawaiian Kingdom in Oceania
Lorenz Gonschor

Transcending Patterns: Silk Road Cultural and Artistic Interactions through Central Asian Textile Images
Mariachiara Gasparini

Land of Plants in Motion: Japanese Botany and the World
Thomas R. H. Havens

Spreading Protestant Modernity: Global Perspectives on the Social Work of the YMCA and YWCA, 1870–1970
Edited by Harald Fischer-Tiné, Stefan Huebner, and Ian Tyrrell

Imperial Islands: Art, Architecture, and Visual Experience in the US Insular Empire after 1898
Edited by Joseph R. Hartman

Migrant Ecologies: Environmental Histories of the Pacific World
Edited by James Beattie, Ryan Tucker Jones, and Edward Dallam Melillo